W0018614

Sustainable Fisheries Management

Sustainable Fisheries Management

Contributors

Hans-Joachim Rätz et al.

AURIS
Reference

www.aurisreference.com

Sustainable Fisheries Management

Contributors: Hans-Joachim Rätz et al.

Published by Auris Reference Limited

www.aurisreference.com

United Kingdom

Copyright 2016
Printed in 2017 for Sale in the Indian Subcontinent

The information in this book has been obtained from highly regarded resources. The copyrights for individual articles remain with the authors, as indicated. All chapters are distributed under the terms of the Creative Commons Attribution License, which permit unrestricted use, distribution, and reproduction in any medium, provided the original author and source are credited.

Notice

Contributors, whose names have been given on the book cover, are not associated with the Publisher. The editors and the Publisher have attempted to trace the copyright holders of all material reproduced in this publication and apologise to copyright holders if permission has not been obtained. If any copyright holder has not been acknowledged, please write to us so we may rectify.

Reasonable efforts have been made to publish reliable data. The views articulated in the chapters are those of the individual contributors, and not necessarily those of the editors or the Publisher. Editors and/or the Publisher are not responsible for the accuracy of the information in the published chapters or consequences from their use. The Publisher accepts no responsibility for any damage or grievance to individual(s) or property arising out of the use of any material(s), instruction(s), methods or thoughts in the book.

Sustainable Fisheries Management

ISBN: 978-1-78154-977-3

British Library Cataloguing in Publication Data
A CIP record for this book is available from the British Library

Printed in the United Kingdom

Exclusively distributed by CBS Publishers & Distributors Pvt. Ltd.

Sales & Distribution Rights only for India, Pakistan, Bangladesh, Sri Lanka, Nepal and Bhutan.This book is not to be sold outside these territories.

Contents

List of Abbreviations

AVHRR	Advanced Very High Resolution Radiometer
ARMA	Autoregressive moving average
BC	Biocapacity
CCMs	Carbon concentrating mechanisms
CPUE	Catch per unit effort
CERES	Clouds and the Earth's Radiant Energy System
CPR	Common pool resources
CMS	Connectivity Modeling SystemDay/Night Band DNB, 1, 2, 12
DMSP	Defense Meteorological Support Program
DFO	Department of Fisheries and Oceans
DIC	Dissolved inorganic carbon
EF	Ecological footprint
ELOHA	Ecological limits of hydrologic alteration
EBM	Ecosystem-based management
EN	Endangered
EEZs	Exclusive Economic Zones
FP	Fish print
FAO	Food and Agriculture Organization
FWI	Freshwater inflow
FPCA	Fuzzy principal component analysis
GAMS	General Algebraic Modeling System
GME	Generalized maximum entropy
GINA	Geographic Information Network of Alaska's
GIS	Geographic information system
GPS	Global Positioning Satellite
IHA	Hydrologic Alteration
IFM	Index of flow modification
ITQ	Individually transferable quotas
ICES	International Council for the Exploration of the Sea
IUCN	International Union for Conservation of Nature
JPSS	Joint Polar Satellite System
LEK	Local Ecological Knowledge
MER	Marine Extractive Reserve
MPAs	Marine protected areas
MSC	Marine Stewardship Council
MEY	Maximum economic yield
MSY	Maximum sustainable yield
MOFA	Ministry of Food and Agriculture
NGDC	National Geophysical Data Center
NPP	National Polar-orbiting Partnership
NSR	Northern Sea Route
OVM	Ontogenetic vertical migration

OLS	Operational Linescan System
OFT	Optimal foraging theory
OMPS	Ozone Mapping and Profiler Suite
PLD	Pelagic larval duration
PMZ	Provisional Measures Zone
QCS	Queen Charlotte Sound
ROK	Republic of Korea
SBS	Shifting baseline syndrome
TACs	Total allowable catches
USGS	United States Geological Survey
UGC	Upper Gulf of California
VOS	Voluntary Observing Ship

List of Contributors

Hans-Joachim Rätz
European Commission, Joint Research Centre, Institute for Protection and Security of the Citizen, Ispra (VA), Italy
Johann Heinrich von Thünen-Institut, Federal Research Institute for Rural Areas, Forestry and Fisheries, Institute for Sea Fisheries, Hamburg, Germany

Javier Sánchez-Hernández
Department of Zoology and Physical Anthropology, Faculty of Biology, University of Santiago de Compostela, Spain
Station of Hydrobiology "Encoro do Con", Castroagudín s/n, Vilagarcía de Arousa, Pontevedra, Spain

María J. Servia
Department of Animal Biology, Vegetal Biology and Ecology, Faculty of Science, University of A Coruña, Spain

Vieira-Lanero
Station of Hydrobiology "Encoro do Con", Castroagudín s/n, Vilagarcía de Arousa, Pontevedra, Spain

Fernando Cobo
Department of Zoology and Physical Anthropology, Faculty of Biology, University of Santiago de
Compostela, Spain
Station of Hydrobiology "Encoro do Con", Castroagudín s/n, Vilagarcía de Arousa, Pontevedra, Spain

Peter C. Sakaris
Department of Biology and Chemistry, Southern Polytechnic State University, Marietta, GA, USA

William C. Straka III
Cooperative Institute for Meteorological Satellite Studies, University of Wisconsin-Madison, Madison, WI 53706, USA

Curtis J. Seama
Cooperative Institute for Research in the Atmosphere, Colorado State University, Fort Collins, CO 80523, USA

Kimberly Baugh
Cooperative Institute for Research in Environmental Sciences, University of Colorado-Boulder, Boulder, CO 80309, USA

Kathleen Cole
National Weather Service, Alaska Sea Ice Program, Anchorage, AK 99502, USA

Eric Stevens
Geographic Information Network of Alaska, Fairbanks, AK 99775, USA

Steven D. Miller
Cooperative Institute for Research in the Atmosphere, Colorado State University, Fort Collins, CO 80523,
USA

Myrna Leticia Bravo-Olivas
Biological Sciences Department, Centro Universitario de la Costa, Universidad de Guadalajara, Puerto Vallarta, Jalisco 48280, Mexico

Rosa María Chávez-Dagostino
Biological Sciences Department, Centro Universitario de la Costa, Universidad de Guadalajara, Puerto Vallarta, Jalisco 48280, Mexico

Carlos Antonio López-Fletes
Master's Program in Human Ecology, Vrije Universiteit Brussel, Brussel 1050, and Belgium

Elaine Espino-Barr
Regional Center of Fisheries Research, National Fisheries Institute, Manzanillo, Colima 28200, Mexico

Teunis Jansen
DTU AQUA - National Institute of Aquatic Resources, Technical University of Denmark, Charlottenlund, Denmark

Andrew Campbell
Fisheries Ecosystems Advisory Services, Marine Institute, Galway, Ireland

Ciarán Kelly
Fisheries Ecosystems Advisory Services, Marine Institute, Galway, Ireland

Hjálmar Hátún
Faroe Marine Research Institute, To´rshavn, Faroe Islands

Mark R. Payne
DTU AQUA - National Institute of Aquatic Resources, Technical University of Denmark, Charlottenlund, Denmark

ETHZ Swiss Federal Institute of Technology, Zurich, Switzerland

Mariana G. Bender
Departamento de Ecologia e Zoologia, Universidade Federal de Santa Catarina, Floriano´polis, SC, Brazil

Gustavo R. Machado
Departamento de Biologia Marinha, Universidade Federal Fluminense, Niteroi, RJ, Brazil

Paulo Jose´ de Azevedo Silva
Fundac¸a~o Instituto de Pesca, Arraial do Cabo, RJ, Brazil

Sergio R. Floeter
Departamento de Ecologia e Zoologia, Universidade Federal de Santa Catarina, Floriano´polis, SC, Brazil

Cassiano Monteiro-Netto
Departamento de Biologia Marinha, Universidade Federal Fluminense, Niteroi, RJ, Brazil

Osmar J.Luiz
Department of Biological Sciences, Macquarie University, Sydney, NSW, Australia

Sari M. Oksanen
Department of Biology, University of Eastern Finland, Joensuu, Finland

Markus P. Ahola
Natural Resources Institute Finland, Turku, Finland

Jyrki Oikarinen
Perämeren Kalatalousyhteisöjen Liitto ry, Oulu, Finland

Mervi Kunnasranta
Department of Biology, University of Eastern Finland, Joensuu, Finland

Andrew S. Kough
Rosenstiel School of Marine and Atmospheric Sciences, University of Miami, Miami, Florida, United States of America

Claire B. Paris
Rosenstiel School of Marine and Atmospheric Sciences, University of Miami, Miami, Florida, United States of America

Mark J. Butler IV
Department of Biological Sciences, Old Dominion University, Norfolk, Virginia, United States of America

James M. Tolan
Texas Parks and Wildlife Department, Coastal Fisheries Division, Natural Resource Center 2501, Unit 5846, Corpus Christi, TX, USA

Gerardo Rodríguez-Quiroz
Centro Interdisciplinario de Investigaciones para el Desarrollo Integral Regional, Unidad Sinaloa

Eugenio Alberto Aragón-Noriega
Centro de Investigaciones Biológicas del Noroeste, Unidad Sonora

Miguel A. Cisneros-Mata
Centro de Investigaciones Biológicas del Noroeste, Unidad La Paz, México

Alfredo Ortega Rubio[4]
Centro de Investigaciones Biológicas del Noroeste, Unidad La Paz, México

Wisdom Akpalu
United Nations University—World Institute for Development Economics Research (UNU-WIDER),
University of Ghana, Legon-Accra, Ghana

Isaac Dasmani
Economics Department , University of Cape Coast, University Post Office, Cape Coast, Ghana
Center for Environmental Economics Research & Consultancy (CEERAC), Accra, Ghana

Ametefee K. Normanyo
Ho Polytechnic, Ho, Ghana
Center for Environmental Economics Research & Consultancy (CEERAC), Accra, Ghana

Rowan Haigh
Pacific Biological Station, Fisheries and Oceans Canada, 3190 Hammond Bay Road, Nanaimo, British Columbia, V9T 6N7, Canada

Debby Ianson
Institute of Ocean Sciences, Fisheries and Oceans Canada, 9860 West Saanich Road, Sidney, British Columbia, V8L 4B2, Canada

Carrie A. Holt
Pacific Biological Station, Fisheries and Oceans Canada, 3190 Hammond Bay Road, Nanaimo, British Columbia, V9T 6N7, Canada

Holly E. Neate
Pacific Biological Station, Fisheries and Oceans Canada, 3190 Hammond Bay Road, Nanaimo, British Columbia, V9T 6N7, Canada
Department of Biology, University of Victoria, Station CSC, Victoria, British Columbia, V8W 2Y2, Canada

Andrew M. Edwards
Pacific Biological Station, Fisheries and Oceans Canada, 3190 Hammond Bay Road, Nanaimo, British Columbia, V9T 6N7, Canada
Department of Biology, University of Victoria, Station CSC, Victoria, British Columbia, V8W 2Y2, Canada

Rudi Voss
Department of Economics, University of Kiel, Kiel, Germany

Martin F. Quaas
Department of Economics, University of Kiel, Kiel, Germany
Kiel Institute for the World Economy, Kiel, Germany

Jo¨ rn O. Schmidt
Department of Economics, University of Kiel, Kiel, Germany

Olli Tahvonen
Department of Forest Sciences, University of Helsinki, Helsinki, Finland

Martin Lindegren
Scripps Institution of Oceanography, University of California San Diego, San Diego, California, United
States of America

Christian Mo¨ llmann
Institute for Hydrobiology and Fisheries Science, Center for Earth System Research and Sustainability (CEN), University of Hamburg, Hamburg, Germany

Preface

The text Sustainable Fisheries Management describes the key features of fisheries management, and provides important guidance on how we can make the transition towards sustainable fisheries. A conventional idea of a sustainable fishery is that it is one that is harvested at a sustainable rate, where the fish population does not decline over time because of fishing practices. Sustainability in fisheries combines theoretical disciplines, such as the population dynamics of fisheries, with practical strategies. First chapter focuses on obligation of sustainable fisheries management. Second chapter supports the hypothesis differences in the feeding habits and habitat utilization of different age classes of trout could reduce competition for food, by allowing food resource partitioning. Third chapter provides a general overview of the impacts of hydrologic alteration and presents several management approaches, which have been developed to address it. Fourth chapter examines the day/night band's (DNB's) usage to observe anthropogenic light emission sources, with particular emphasis on the activity of maritime vessels. The aim of fifth chapter is to evaluate the fisheries sustainability in the Jalisco coast through the fishing footprint, or fishprint (FP), based on the primary productivity required (PPR) and the appropriated surface by the activity (biocapacity). In sixth chapter, we demonstrate correlation between temperature and mackerel migration/distribution as proxied by mackerel catch data from both scientific bottom trawl surveys and commercial fisheries. Seventh chapter discusses how local ecological knowledge and scientific data reveal overexploitation by multigear artisanal fisheries in the Southwestern Atlantic. A novel tool mitigated by-catch mortality of baltic seals in coastal fyke net fishery has been outlined in eighth chapter. Ninth chapter focuses on larval connectivity and the international management of fisheries. The goal of tenth chapter is to expand the focus of interest of the resource-based approach beyond the limited number of fisheries target species and to include juvenile stages of fisheries species to examine the functional role of FWI in shaping the total nekton assemblage structure in estuaries. In eleventh chapter, we identify and analyze the most important artisanal fisheries of the Upper Gulf of California, which are in continuous interaction with the vaquita. Twelfth chapter investigates the effect of climate variation on biophysical parameters and yields. The effects of ocean acidification on temperate coastal marine ecosystems and fisheries in the northeast pacific have been investigated in last chapter.

Chapter 1

THE OBLIGATION OF SUSTAINABLE FISHERIES MANAGEMENT: REVIEW OF ENDURED FAILURES AND CHALLENGES IN EXPLOITATION OF THE LIVING SEA

Hans-Joachim Rätz[1, 2]

[1]European Commission, Joint Research Centre, Institute for Protection and Security of the Citizen, Ispra (VA), Italy

[2]Johann Heinrich von Thünen-Institut, Federal Research Institute for Rural Areas, Forestry and Fisheries, Institute for Sea Fisheries, Hamburg, Germany

INTRODUCTION

Fishing is defined as a specific use of our living environment, the extensive or intensive activity to hunt or collect aquatic species for a huge variety of motivations related to leisure, nutrition or profit. More generally interpreted, the term can also be applied as a metaphor for any passive and active advantage taking from our surroundings, from "fishing" for monetary values to compliments. So we (creatures) and many of our actions are concerned. The world's confined biosphere is composed of living and non-living constituents, which are interlinked by a complex web of relations at different levels and with different intensities. We subdivide the various constituents and their effects into ecosystems. Their common and main feature appears to be a dynamic change at all scales in time and space, which provokes the vital evolution through everlasting mutation and adaptation (Pickett et al., 2007). We need to realize that the non-equilibrium is a major feature of ecology and widely accepted as paradigm (Lévêque, 2003). Humanity, as highly developed constituent, intensively exploits living and non-living resources with high impacts (footprints), and thus highly depends on stability and resilience when optimizing its exploitation strategies over short or medium term, i.e. periods of generations or beyond with increasing ethical concerns. The initiation of the concept of sustainability in the 20th century reflects the increasing global awareness of the threat posed by the human-induced effects and thus can be interpreted as a logical consequence by defining limits to achieve a stable and optimized use of any sort of common or private goods. While such conceptual thinking is not new and appears easily comprehendible, the reality largely differs regarding both common property (Hardin, 1968) as well as private property (common experiences, I guess). Despite sufficient knowledge leading to various definitions and requests of precautionary approaches, principles and time frames towards sustainability, human management normally fails and results in crisis management to minimize damages at all levels, from personal to international dimensions. The desperate try to constrain climate changes and mitigate their consequences are an impressive example. Fisheries do not represent an exemption (Cochrane et al., 2009), and different arguments are used driven by multifaceted objectives of various interest groups, such as non-governmental organizations, stakeholders, politicians and their international frameworks, and even scientists.

OVERFISHING AS ECOLOGICAL FOOTPRINT: THE FACTS AND DEFINITIONS

Historically, hunting of whales was among the first human activities which proved that marine resources are limited. Commercial bowhead (Balaena mysticetus) whaling began in the 1840s, and within two decades caught over 60 percent of the bowheads (Braham, 1984). It's noteworthy that the populations appear still not fully recovered as the International Union for Conservation of Nature (IUCN) assigns their status still as 'lower risk' to 'critically endangered'. More than 100 years ago, Garstang (1900) demonstrated that increased fishing could reduce fish abundance, which is seen as the basis for Graham's (1943) fishing law (Hart & Reynolds, 2002). Schaefer (1954) formulated the first general production model to be applied to fisheries data for the quantification of the surplus, which is still interpreted as the sustainable yield from a given living resource. Further milestones with increased understanding and precision were the growth modeling (v. Bertalanffy, 1938) and the development of age structured dynamic pool models (Gulland, 1965; Pope, 1972) to estimate the past and future stock production (Beverton & Holt, 1957; Ricker, 1975).

The drastic short term changes in the ecosystems and their components is reflected in the relative high amount of energy many marine species invest for reproduction, i.e. the amount of eggs and prolonged reproductive seasons in tropical, boreal and polar regions. The species are classified based on the number and quality of offsprings (MacArthur & Wilson, 1967) into the so-called r-strategists (high number of eggs and short lived) versus the kstrategists (low number of eggs and long lived). Higher variability is expected and can be seen in the fisheries targeting the r-strategists of the pelagic habitats. Taking this into account, the recent scientific challenge is characterized by the move from the individual stocks to the ecosystem approach (Jennings, 2001) to fisheries management, which shall provide a wider understanding of the human impact through exploitation of living marine resources. Hilborn (2010) defines the most important elements of ecosystem based fisheries management as keeping fishing mortality rates low enough to prevent ecosystem-wide overfishing, reducing or eliminating by-catch and avoiding habitat-destroying fishing methods. The recent state of the marine ecosystems has been continuously assessed by many authors and institutions.

The largely biased general public opinion is that the sea is empty, the food webs are fished down to small species and there is a general loss in biodiversity impairing the oceans' capacity to provide food, maintain water quality, and recover from perturbations (Pauly et al., 1998; Pauly & Palomares, 2005; Worm et al., 2006). But the facts prove that the oceans are surprisingly resistant, despite the destructive and ongoing illegal, unreported and unregulated fishing practices (IUU fishing), which are officially condemned and combated as a serious global problem regarding habitat destruction and fish stock depletion. The destructive and incidental catch of sharks, seabirds, turtles and marine mammals has to be avoided by selective devices (FAO, 2008). However, the latest assessment of the Food and Agriculture Organization of the United Nations (FAO, 2010 a) concludes that global production of marine capture fisheries reached a peak of 86.3 million tons in 1996 and then declined slightly to 79.5 million tons in 2008, with high annual fluctuations and changes in contributions of the major species.

While there are severe concerns regarding the human impacts through capture fisheries on marine ecosystems, we may realize that 63% of assessed fish stocks worldwide still require rebuilding to optimize the productivity, and even lower exploitation rates are needed to reverse the collapse of vulnerable species (Worm et al., 2009). In summary, overfishing is an ecological footprint of our recent society but we are far from the apocalypse of collapsed world's fisheries (Hilborn, 2011). After having assessed the world's fishing resource situation, we now need to define sustainability and then review the development of recent management reference points consistent with sustainability. In accordance with the definition by Costanza & Patten (1995) sustainability is generally interpreted as the capacity of ecological, economic or social systems to endure under stress, e.g. exploitation. However, it appears clearer when sustainability is compared with resilience (Ludwig et al., 1997).

Sustainability encompasses resilience but also requires a predefined goal in addition. Sustainable goals or reference points are commonly set at high or optimized levels. However, in order for sustainability to be a useful criterion for guiding changes, its characterization should be literal, systemoriented, quantitative, predictive, stochastic and diagnostic (Hansen, 1996). The international requirement for marine protection is stipulated in the Treaty on the Convention on the Law of the Sea (UN, 1982), where all States enjoy the traditional freedoms of navigation, overflight, scientific research and fishing on the high seas and they are obliged to adopt, or cooperate with other States in adopting, measures to manage and conserve living resources. Coastal States are granted sovereign rights in a 200-nautical mile exclusive economic zone (EEZ) with respect to natural resources and certain economic activities, and exercise jurisdiction over marine science research and environmental protection.

During the Earth Summit held in Rio de Janerio, Brazil, 3 to 14 June 1992, the Agenda 21, the Rio Declaration on Environment and Development, and the Statement of principles for the Sustainable Management of Forests were adopted by more than 178 governments. The "Rio Principles" represent the international guidelines calling specifically for the reduction of unsustainable patterns of production and consumption and capacity building for sustainable development (UN, 1992). In 1995, the United Nations agreed upon the implementation of the provisions of the convention on the Law of the Sea of 10 December 1982 relating to the conservation and management of straddling fish stocks and highly migratory fish stocks.

The implementation of limit reference points was requested, which are intended to constrain harvesting within safe biological limits within which the stocks can produce maximum sustainable yield (MSY), while target reference points are intended to meet management objectives (UN, 1995). Assigning the maximum yield a long term perspective immediately turns the underlying intention towards maximum conservation, as only well protected stocks can produce high yields over a long time. However, any stock size status indicator shall gain less weight in comparison with the exploitation indicator in the decision making progress as the actual stock size underlies and is considered the outcome of many ecological effects in addition to the human impacts through fishing. In the same year the idea of such reference levels for the fisheries management was more widely applied in the Code of Conduct for Responsible Fisheries by the FAO (1995). The sustainability goal for fisheries management was re-confirmed during the Sustainability Summit in Johannesburg (UN, 2002), interpreted as the core publication. The MSY of all exploited stocks has now to be implemented by the specific date of the year 2015, a clear ecological target. Undoubtedly such ratified political design, which is based on the principle of short term losses in the view of long term gains, requires major and continued efforts towards transparent information and protection against the unsustainable solution of short term gains versus long term losses.

The internationally agreed fishing mortality FMSY that produces MSY is defined as

$$F_{MSY} = r/2 , \tag{1}$$

where r is intrinsic rate of population growth in the logistic population growth model (Prager, 1994), in which the change in stock biomass over time (dB_t/dt) is a quadratic function of biomass (B) and K is defined as the carrying capacity.

$$dB_t/dt = rB_t - (r/K)B_t^2 , \tag{2}$$

After all these considerations we are in the position to defend the conclusion that the world fisheries do unsustainably exploit many of the living marine resources, and have a long history and prominent examples to do so with disastrous socio-economic consequences. In particular, the hardly or non-reversible damages caused by fishery effects in the deep sea or hard substrate habitats (coral reefs) have to be avoided by all means (FAO, 2008).

Discarding, throwing back into the sea the whole or selected parts of the unwanted catch, appears an unacceptable performance, recognizing the ethical concerns regarding the waste if biological resources through discards in the magnitude of 7 million tons in the world's fisheries (FAO, 2010 a). In addition, unknown and thus unaccounted discarding implies increased uncertainty in assessments of exploited stocks, scientific advice and fisheries management. However, a discard ban or landing obligation is already implemented in national fisheries regimes but unsurprisingly appears difficult to control after all. While the best practice is to avoid discarding by not catching the potentially unwanted fish, a discard ban might incentivize improved technical selection through appropriate gear specifications or closing of sensitive areas.

Probably the most impressive relation between humans and fish is the story of Atantic cod (Gadus morhua) fisheries which spans a thousand years and four continents (Kurlansky, 1997). Before the 1970s, the annual capture production exceeded 3 million tons and rapidly fell below 1 million tons at the beginning of the millennium 30 years later. We could continue with many examples, e.g. collapsed and recovered herring (Clupea harengus) fisheries or the recent annual bluefin tuna (Thunnus thynnus) battles heavily debated in the international press.

Europe ranges among the poor regions when it comes to the status of its common fishery resources as the great majority of the European fish stocks (88%) remain overexploited with regard to high long term yields (EU, 2009). The deep-rooted problem of overcapacity and imprecise policy objectives and will are identified as the main structural failings, in particular at the operative level of individual fishermen and their fishing strategies. The joint exploitation of fish and shell fish stocks in European marine waters underlie the Common Fisheries Policy (CFP; EU, 2002), which apparently lacks a specific definition of the sustainability. The European Parliament acts as co-legislator under the Lisbon Treaty (EU, 2007a), with the exception of measures on fixing prices, levies, aid and quantitative limitations and on the fixing and allocation of fishing opportunities, that will remain as in the EC Treaty, where they have to be adopted by the European Council on a proposal from the European Commission. However, the CFP is due to a reform by 2012, after a standard 10 years interval. The management goals of the reformed CFP requires consistency with the European Marine Strategy Framework Directive (MSFD; EU, 2008a) and its focus on good environmental status of all exploited fish and shellfish stocks in all European regions including the EEZs and territorial waters by 2020 (Rätz et al., 2010). The complementary decision by the European Commission (EU, 2010) identified the so-called fishing mortality F to generate MSY as primary indicator of sustainability. F is defined as the famous coefficient of the annual rate of dead or removed fish caused by fishing as a function of the annual rate of dead fish caused by natural reasons (see following section). Such stock specific level of exploitation needs to be identified and set considering all ecological effects, as the sustainable production differs not only among species but also among stocks. Rätz & Lloret (2003) demonstrated that the cod stocks in the warm regions of the Northeast Atlantic are more productive in terms of growth and recruitment and can sustain higher fishing rates as compared with the stocks inhabiting the colder Northwest Atlantic, which appear more vulnerable through lower productivity.

LESS IS MORE: WE KNOW ENOUGH TO MOVE TOWARDS IT!

The major challenge of the modern fisheries management is not any longer to define sustainable exploitation levels and to best approach them but to correct the errors made in the past decades, mainly related to the reduction of fishing power accounting for increased efficiency and technological creep, which can reach 2% per year (Rijnsdorp et al., 2006). Whatever kind of regulations are chosen, ranging from the suite of technical measures (e.g. closures, gear configurations) and direct fishing restrictions (total allowable catches TAC or total allowable fishing effort TAE), they shall effectively control the fisheries induced mortality and shall consider of potential future effects based on experience.

By fishing less the expected yields from an overexploited resource will be increasing as can be seen from the classical yield curve of the prominent and continuously overfished North Sea cod stock (including Skagerrak and Eastern Channel, Fig. 1). It must be acknowledged that there exists no experience in stock dynamics during sustainable exploitation at the maximum sustainable term yield or below it, even after 50 years of data. Therefore, the estimation of the maximum sustainable yield of 700,000 tons per year at a stock weight of all spawning fish around 800,000 tons is a result of an extrapolation and thus appears quite hypothetical. These high values can be considered overestimated if under such favorable conditions the ecological processes gain a more dominant role, i.e. intra-specific (cannibalism) and inter-specific consumption (predation by other species).

In the cases of heavy overfishing and depleted stocks it may be advisable to search agreement among interested parties by designing multi-annual plans considering an adaptive stepwise mitigation process rather than the short term solution with drastic consequences for all involved parties. However, the mitigation process should be significant enough to ensure a transparent monitoring in order to justify the measures taken. This can be achieved even if the scientific advice regarding the final goal is imprecise but clearly quantifies the problem and the direction to solve it (Patterson et al., 2001).

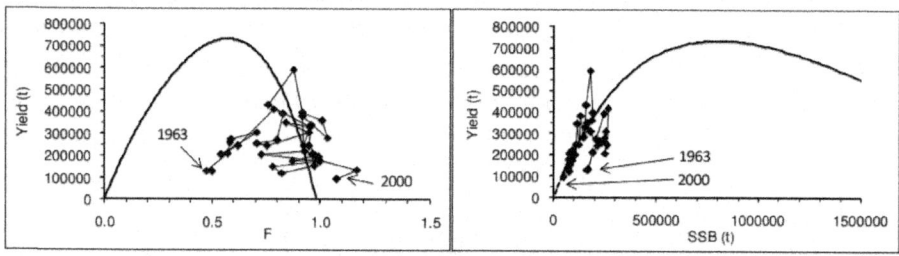

Figure. 1: North Sea cod stock. Potential annual yield as a function of fishing induced mortality F and the weight of all spawning fish (spawning stock size SSB) as estimated by means of a non-equilibrium model using data 1963-2000 from ICES North Sea working group (ICES, 2010).

Single stock fisheries, the easy case: Sustainable fisheries management requires the preagreed biological limit exploitation level based on the above outlined policies. Given that the stock of a target species can be fished without significant by-catch of other species and impact on its habitat, a simple TAC including potential discards of the target species can be used to effectively control the single species fishery induced mortality and keep it below the a-priori set limit accounting for stock specific conservation requirements. This recommendation applies to both passive gears, i.e. longlines, nets and traps, as well as actively moved gears such as seines and trawls. However, fully implemented and effective scientific monitoring and advisory as well as fishery control and enforcement systems must be in place. If the exploited stock is considered a shared resource among nations and their various fisheries, a complex framework of political commissions will be active to decide on access rights, probably to the level of individual fishermen, defined as individually transferable quotas (ITQ) and recently favored to strengthen stakeholder involvement through increased responsibility (Hauge & Wilson, 2009). Often the access rights are based on historic records on contributions of each fishery and nation to the overall exploitation. However, the prominent example of the Mediterranean blue fin tuna fisheries already demonstrates the many potential management failures starting from biased scientific knowledge and advice based on wrong catch records due to ineffective control. The simplicity of single stock/species fisheries management through catch constraints to achieve sustainable levels is enough reason to incentivize fishing industries to conduct so-called "clean" fisheries without by-catches where- and whenever possible. The incentives should force fisherman to stop effectively fishing once their allowance is exhausted. In particular, this recommendation regards fisheries for pelagic stocks/species, such as the tunas, herring, mackerel, anchovy and sardine fisheries. The pronounced schooling and migratory behavior of pelagic species over continued periods prevents relatively small areal closures or direct effort limitations being effective and safe. "Clean" catches can also be achieved by advanced selective trawl devices (Valentinsson & Ulmestrand, 2008).

Mixed fisheries, the difficult cases: We have to realize that sustainable fisheries management can be much more complex and difficult. The great majority of stocks are exploited by multi-species (mixed) fisheries, particularly the near bottom and bottom dwelling species due to their coexistence in diverse communities (Caddy & Seijo, 2005) and the poor selectivity of many gears used. Fisheries using bottom trawls and seines might severely impact the structure of the sea bottom (Kaiser & Spencer, 1996). Still, the variety of exploited stocks in mixed fisheries requires specific conservation needs based on the specific ecological role and stock status. In addition, the selection of the various mixed fisheries involved in the exploitation of certain stocks varies significantly with the gears and the fishing strategies. It is argued that the mixed fisheries are best managed by fishing effort (Kell et al., 2005; Schwach et al., 2007), if they deploy trawled (active) gears. This can be done by settings of effort constraints (TAE) in units of days at sea or the product of kilo Watt times days at sea to account for engine power (Cotter, 2010). It's noteworthy that such effort measure can be easily controlled. However, the effectiveness of such effort measures regarding passive gears has still to be proven. Fishing grounds with high stratification, e.g. along continental shelves, may force certain stocks or parts of them to occur highly aggregated and thus make pure effort measures ineffective to control fishing mortality, like in the example of pelagic fisheries (see above). However, catch constraints (TAC) estimated and set consistently with effort constraints (TAE) will help to communicate foreseen fishing possibilities to the involved stakeholders.

Now, since we've learnt that many stocks are exploited simultaneously by various mixed fisheries, we may understand that, under such circumstances, fisheries management can be very complex. While the agreed stock specific conservation requirements can be defined as F_{MSY} (1), the way towards it appears less clear when simultaneously considering all jointly exploited stocks by a variety of fisheries characterized by different selection patterns. A stochastic medium term forecast model for North Sea demersal fisheries (7 stocks, 9 fisheries) based on data from ICES (2010) and STECF (2011) provides some robust conclusions on future catch and biomass trends under various management scenarios. The major underlying dynamic concept is defined as

$$N_{y+1,a+1} = N_{y,a} \exp(-(M_{y,a} + F_{y,a})),$$
(3)

where N denotes stock size in numbers in given year y at age a, M equals natural mortality and F fishing mortality (Beverton & Holt, 1957).

The most important stock productivity parameter is the recruitment to the stock expressed as

$$R = a\, S \exp(-\beta\, S),$$
(4)

where R denotes the recruitment to the stock, S the parental stock size with α and β as stock specific parameters (Ricker, 1975). Finally, the catch equation links the observed catches taken from a given stock with the stock size and the two components of mortality, i.e. the natural and the fishing mortality as

$$C_{y,a} = F_{y,a}\, N_{y,a}\, ((1-\exp(-(F_{y,a} + M_{y,a}))) / (F_{y,a} + M_{y,a})$$
(5)

where C denotes catch in numbers in a given year y at age a (Beverton & Holt, 1957).

Stock specific production parameters required and the limit reference levels of exploitation of seven stocks are listed in Table 1 defining all stock areas as being consistent with the joint demersal fisheries management area of the Skagerrak, North Sea and Eastern Channel. It has to be noted that the stock dynamics of Norway lobster in the North Sea are largely unknown and they are assumed to be a short lived species with one age group only during its exploitation phase. The matrix of actual contributions in terms of fishing mortalities by stock for each of the nine fisheries is given in Table 2. The fisheries definitions are in accordance with the fleets defined in the cod management plans (EU, 2008b), one of the major concerns in European fisheries management. It can be taken from Table 2 that each of the nine defined fisheries contributes to the exploitation of each of the seven stocks with different intensity. While the trawlers are catching all gadoids, Norway lobster and flatfishes except sole, the beam trawlers are mainly targeting the flatfish plaice and sole. The major interest of the passive gillnets and trammel nets focuses on sole with some cod shares of gillnets as well. Longlines do not play an important role in the evaluated system at all and other fisheries catch a rather small share of cod and whiting.

Table 1: Stock specific parameters of seven stocks as used in the stochastic medium term forecast model of catch and biomass under various management scenarios. Cod in ICES divisions 3an, 4 and 7d (Gadus morhua, COD 3an47d), haddock in ICES divisions 3an and 4 (Melanogrammus aeglefinus, HAD 3an4), whiting in ICES divisions 4 and 7d (Merlangius merlangus WHG 47d), saithe in ICES divisions 3a, 4 and 6 (Pollachius virens, POL 3a46), plaice in ICES division 4 (Pleronectes platessa, PLE 4), common sole in ICES division 4 (Solea solea, SOL 4) and Norway lobster in ICES divisions 3a and 4 (Nephrops norvegicus, NEP 3a4). Note that n.a. assigns not available

	COD 3an47d	HAD 3an4	WHG 47d	POK 3a46	PLE 4	SOL 4	NEP 3a4
Ricker coefficient a	3.5	20	17	1.5	9	6	77
Ricker coefficient k (t)	1700000	300000	300000	300000	290000	50000	250000
first age group	1	1	1	3	1	1	1
last age group	7	7	8	8	9	9	1
recruitment relative variation CV	0.8	0.9	0.4	0.5	0.7	0.9	0.1
precautionary biomass Bpa (t)	150000	140000	200000	200000	230000	35000	150000
Fref range (fishing mortality)	age 2-4	age 2-4	age 2-6	age 3-6	age 2-6	age 2-6	age 1
F in 2010 (fishing mortality)	0.86	0.25	0.35	0.30	0.25	0.37	0.17
F limit or FMSY proxy (fishing mortality)	0.40	0.30	n.a.	0.30	0.20	0.22	n.a.
relative max. annual change Fref	0.1	0.1	0.1	0.1	0.1	0.1	0.1
relative max. annual change TAC +-	100	100	100	100	100	100	100

As we start from an overexploited situation for some stocks, the overarching rule applied is an annual reduction in fishing mortality by 10% for each stock whenever the exploitation exceeds the pre-agreed reference point. This appears close to the existing multi-annual plans for the North Sea stocks (EU, 2007b; EU, 2008b). A limitation regarding the annual variation of TACs as often requested by the fishing industry and implemented in the stock specific multiannual plans is not considered in the following simulations as such rules imply conflicts among the plans in the likely case that the stock dynamics differ. Let's start with the current situation in European mixed fisheries management, i.e. only the exploitation status of one individual stock is decisive for the regulation of the fishing mortalities induced by multi-species fisheries.

There is a good chance that any time one of the exploited stocks is in a good environmental status, and this becomes the decisive stock for the management and the fisheries continue until their last quota shares are exhausted. All other by-caught stocks, for which the limit exploitations and the respective TACs are exceeded through ongoing fisheries, have then to be discarded. Often such catches are black landed due to their economic value and ineffective control. In cases that discarding of marketable fish is not prohibited, high-grading of the landed catch proportions is a common response by the fishing industry. This strategy intends to maximize the economic value of the catches by means of discarding of low-priced catch components while keeping the landing and revenue option valid throughout the management periods.

Table 2: Nine European fisheries active in the Skagerrak, North Sea and Eastern Channel and their contributions to the overall stock specific exploitation rates expressed as partial fishing mortalities. Data are adopted from ICES (2010) and STECF (2011)

Gear	Mesh size (mm)	Fishery code	COD 3an47d	HAD 3an4	WHG 47d	POK 3a46	PLE 4	SOL 4	NEP 3a4
Trawls other than beam trawls	≥100	TR1	0.496	0.15	0.173	0.253	0.031	0.002	0.002
Trawls other than beam trawls	≥70 <100	TR2	0.192	0.076	0.057	0.033	0.021	0.006	0.151
Trawls other than beam trawls	≥16 <32	TR3	0.001	0.002	0.002	0.002	0.002	0.002	0.002
Beam trawl	≥120	BT1	0.005	0.002	0.002	0.002	0.006	0.001	0.002
Beam trawl	≥80 <120	BT2	0.012	0.002	0.002	0.002	0.177	0.299	0.002
Gillnets	all	GN1	0.048	0.002	0.002	0.002	0.003	0.018	0.002
Trammel nets	all	GT1	0.01	0.002	0.002	0.002	0.004	0.033	0.002
Bottom longline	n.a.	LL1	0.003	0.002	0.002	0.002	0.002	0.002	0.002
OTHER	n.a.	OTHER	0.095	0.009	0.103	0.002	0.002	0.002	0.002

The consequences of the management of mixed fisheries based on only one decisive stock are illustrated in Figure 2. While the exploitation of the most productive stock in the system, in this case the North Sea cod, is reduced stepwise towards the limit management reference with the logic and positive recovery of its stock size, the exploitation rates of the other stocks increase rapidly as their stock sizes diminish to very unproductive levels, in particular plaice and saithe. Only the stock size of Norway lobster remains without feedback to increased exploitation as the stock dynamics are specified as unknown in the model. As mentioned above, the simulated management scenario will allow major discarding of haddock, plaice and saithe while the discarding of cod is declining. All fleets except the trawlers with a mesh size of 70-99 mm will increase their efforts based on opportunistic catch possibilities. In summary, the effort management of mixed fisheries based on a single stock's reference point puts the goal of a sustainable exploitation at an unacceptable level of risk. The high amount of catches exceeding the TACs (overquota catches) contributes significantly to the management risk.

There will be immediate agreement among the conservative interests in the prescribed goal of the MSFD that not one but all exploited stocks shall be in a good environmental state, at least as far as the fisheries impact is concerned. Such conditioned simulations are illustrated in Figure 3, with the same annual reduction in fishing mortality by 10% if the exploitation exceeds the any of the limit reference points set. Under such circumstances all the stocks are quickly recovering to highly productive states and their exploitation rates are consistently reduced. Maintaining at and below or reaching such goals simultaneously for all exploited stocks implies renouncement of catches in short term from more productive stocks which are by-caught in the various fisheries. However, the previous overall catch reduction will be compensated after about 6 years with some changes in the contributions by the various stocks, there will be more cod and saithe while sole landings will remain unchanged.

Haddock and plaice landings will be significantly reduced. The projection of increased Norway lobster landings must be interpreted with care due to the largely unknown stock dynamics. Discarding will be largely reduced after a short period of few years, as all catches can be landed without further restrictions and minimum landing sizes will have a reduced effect on the amount discarded as higher individual survival will result in higher abundance of large fish.

As such this management scenario supports the idea of a discard ban. The results of the management scenario suggest that all fisheries will reduce their effort proportionally by more than 60%. Although this reduction across the board offers a huge potential to economically safe investments and thus increase the economic viability of the fisheries, it equally requires the need to adequately cover the social consequences of such a drastic effort reduction. However, the winning argument for a similar management of mixed fisheries is the gain in stock size with the related high security against fisheries collapses. Mixed fisheries management based on specific limit reference points of all stocks may require the option of disproportional weighing of specific fisheries, e.g. by favoring fisheries avoiding overfished stocks or selecting less stocks from the ecosystem (Rätz et al., 2007). In this way fisheries management can adaptively benefit from stock specific fishing possibilities. Focusing exclusively on the exerted fleet specific impact expressed as the ratio between fishing mortality in relation to the sustainable management limit on a stock by stock basis one could assign the fisheries a specific relative factor according to the formula

$$\text{fac}_{\text{fishery}} = (P \, / \, \Sigma \, (F_{\text{fishery}} \, / \, F_{\text{MSY}})) \, / \, (\Sigma \, (P \, / \, \Sigma \, (F_{\text{fishery}} \, / \, F_{\text{MSY}})) \, / \, L), \quad (6)$$

where $\text{fac}_{\text{fishery}}$ denotes a fisheries specific weighing factor, P the number of stocks caught by a given fishery and L the number of fisheries. F_{fishery} quantifies the fishing mortality exerted by specific fishery, known as the partial fishing mortality. Such factor $\text{fac}_{\text{fishery}}$ would be relatively low if a given fishery contributes more to overfishing than other fisheries. Contrarily, fisheries contributing less to the risk of overfishing would be assigned a relatively high factor which could be then applied to allow for an increased impact of such fisheries, i.e. their partial fishing mortality determining the specific fishing possibilities of future years.

The fisheries specific management scenario applying the above outlined algorithm of a specific factor to consistently estimate landings, discards, fishing mortality and specific relative fishing effort is illustrated in Figure 4. In comparison with the proportional fisheries management scenario illustrated in the preceding Figure 3, the arbitrary choice to privilege certain fisheries at the expenses of more problematic ones results in almost unchanged stock dynamics but increased landings, which are still taken consistently with the sustainable management goals. The possibility of continued and incentivized fishing strategies if considered less problematic is demonstrated by their relative effort trends in Figure 4, i.e. constant or increasing trends. Such potential solutions for conflicting interests among various fisheries and the predefined regulatory frameworks shall be discussed and agreed among stakeholders and managers in advance and be implemented in multi-annual plans of the fisheries.

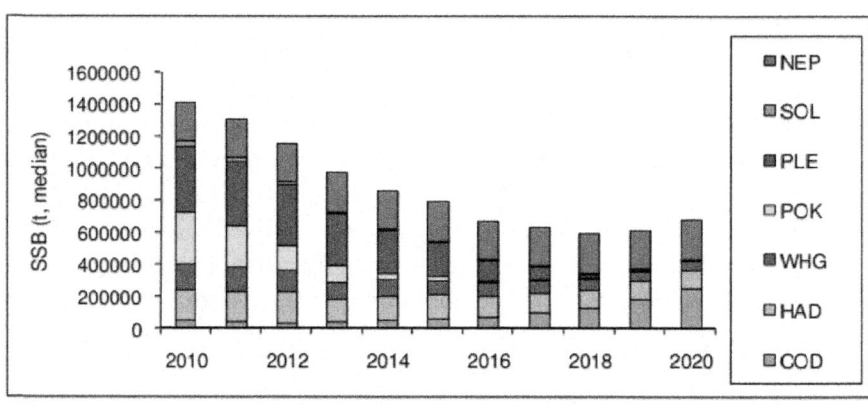

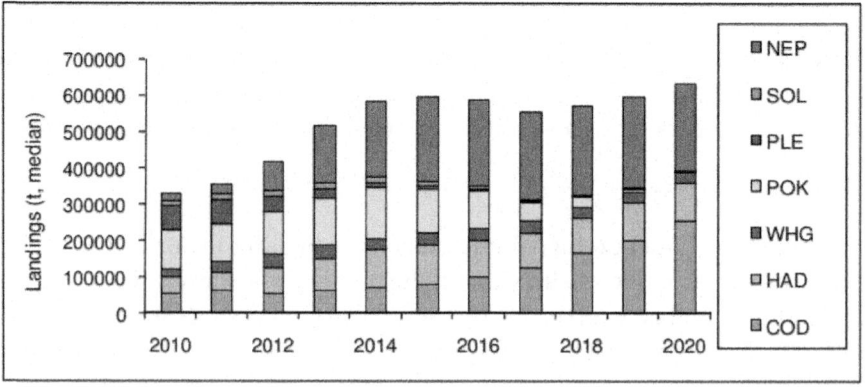

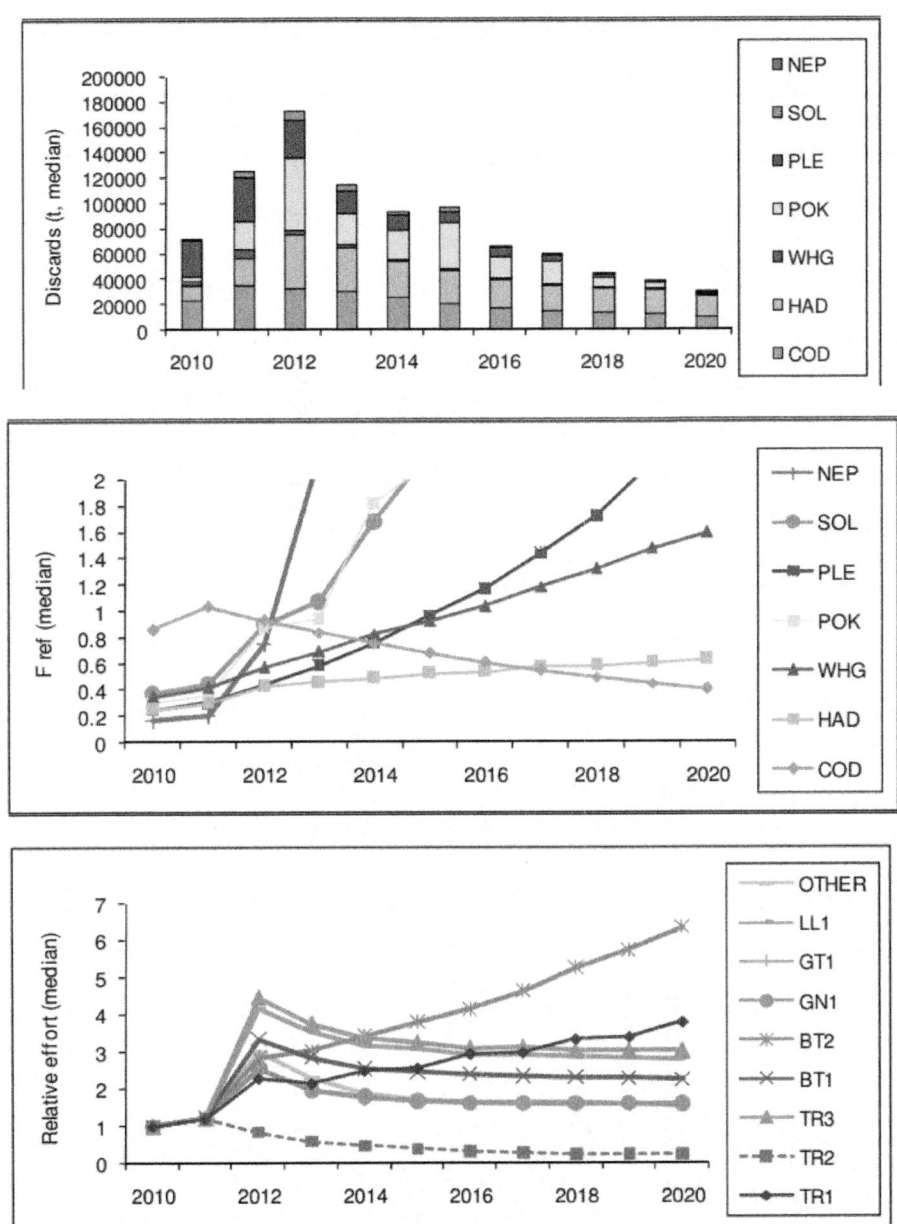

Figure 2: Decadal trends of median stock (SSB=spawning stock biomass, Fref=fishing mortality) and fisheries parameters (landings, discards and relative fishing effort) based on 100 iterations obtained from a stochastic forecast model to simulate mixed fisheries effects for 7 stocks and 9 fisheries in the Skagerrak, North Sea and Eastern Channel.

Stocks and fisheries are defined in Tables 1 and 2, respectively. A harvest control rule to reduce exploitation below or to maintain exploitation at the agreed limit reference point (Table 1) by means of an annual variation in fishing mortality constrained to a maximum of 10% is applied. Only one (the highest) stock specific and sustainable limit reference point is decisive for the control of fishing mortality.

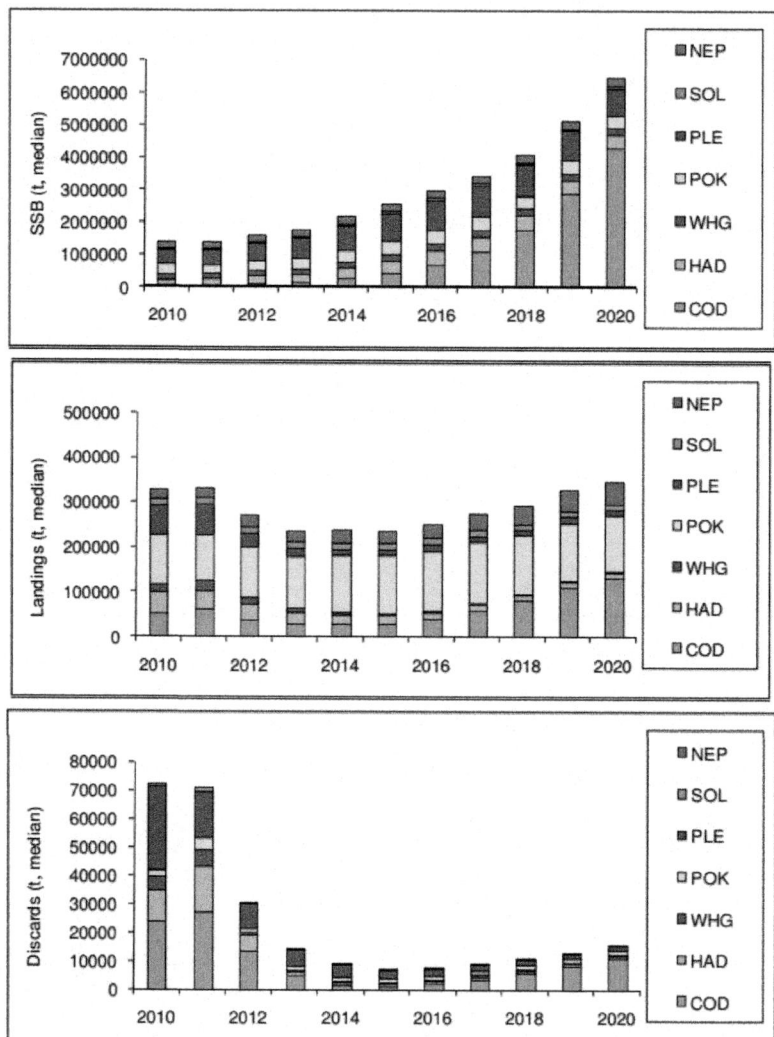

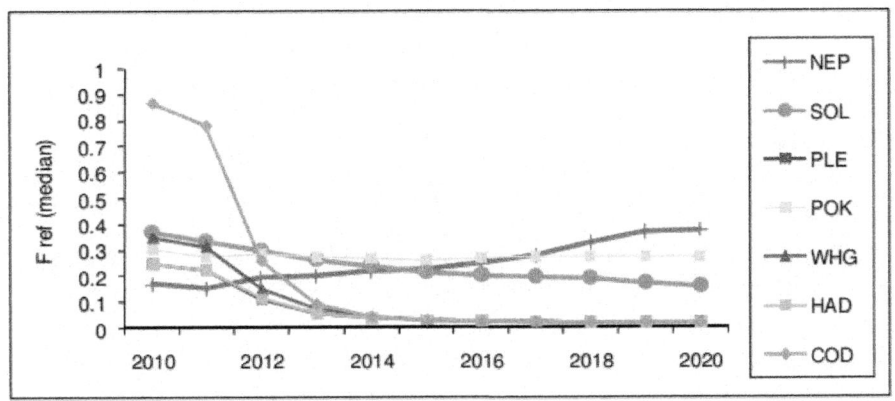

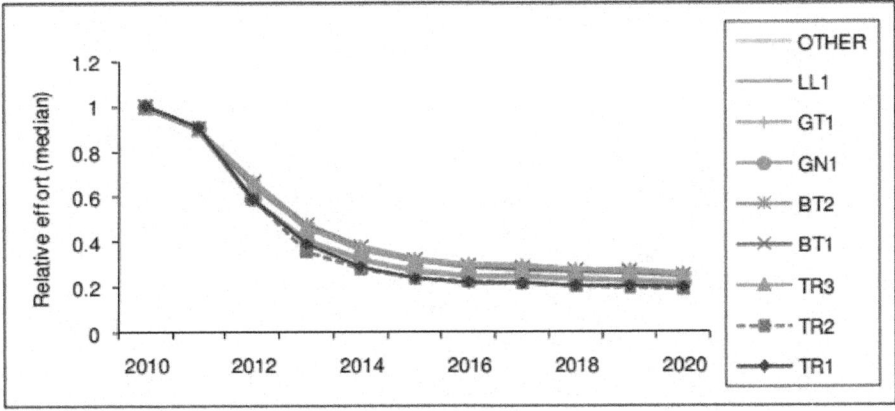

Figure 3: Decadal trends of median stock (SSB=spawning stock biomass, Fref=fishing mortality) and fisheries parameters (landings, discards and relative fishing effort) based on 100 iterations obtained from a stochastic forecast model to simulate mixed fisheries effects for 7 stocks and 9 fisheries in the Skagerrak, North Sea and Eastern Channel. Stocks and fisheries are defined in Tables 1 and 2, respectively. A harvest control rule to reduce exploitation below or to maintain exploitation at the agreed limit reference point (Table 1) by means of an annual variation in fishing mortality constrained to a maximum of 10% is applied. All defined stock specific and sustainable limit reference points are simultaneously decisive for the control of fishing mortality.

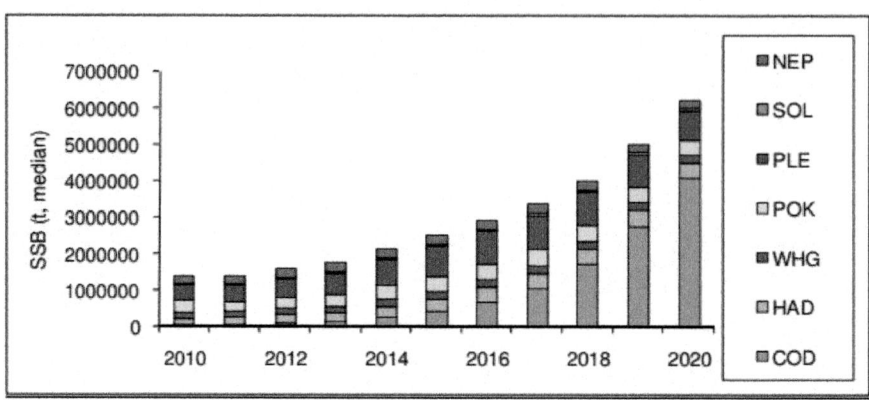

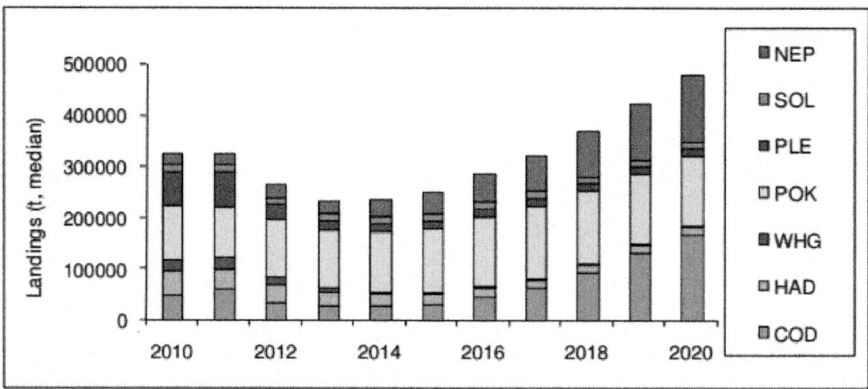

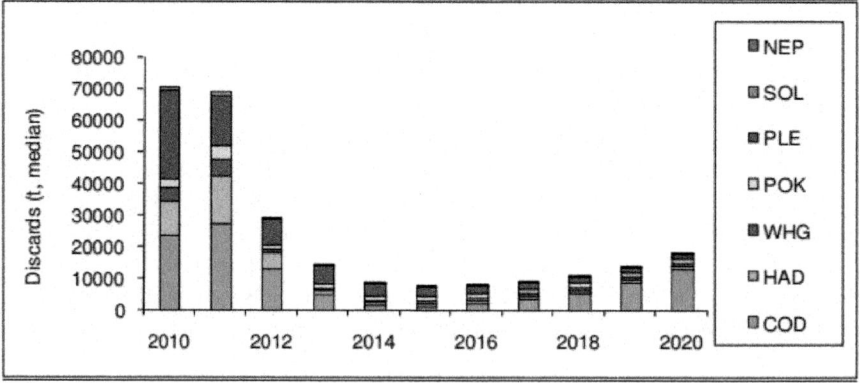

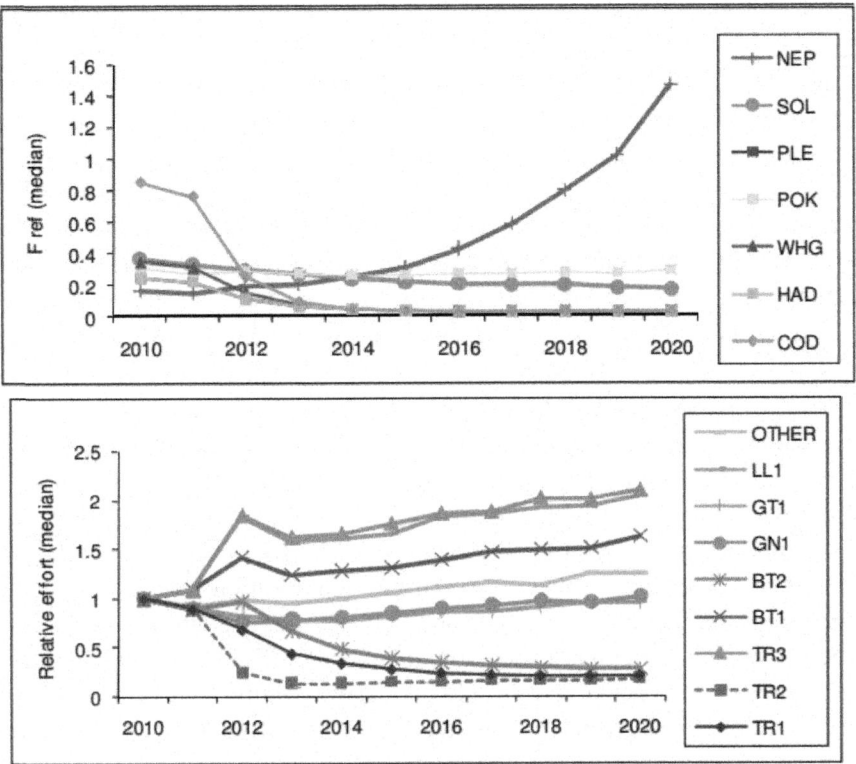

Figure 4: Decadal trends of median stock (SSB=spawning stock biomass, Fref=fishing mortality) and fisheries parameters (landings, discards and relative fishing effort) based on 100 iterations obtained from a stochastic forecast model to simulate mixed fisheries effects for 7 stocks and 9 fisheries in the Skagerrak, North Sea and Eastern Channel. Stocks and fisheries are defined in Tables 1 and 2, respectively. A harvest control rule to reduce exploitation below or to maintain exploitation at the agreed limit reference point (Table 1) by means of an annual variation in fishing mortality constrained to a maximum of 10% is applied. All defined stock specific and sustainable limit reference points are simultaneously decisive for the control of fishing mortality with a non-proportional fisheries specific management scheme.

MONEY DOESN'T MAKE THE FISHERIES GO-ROUND: SUSTAINABLE NUTRITION AND ETHICAL RESPONSI-BILITY TO THE BENEFITS OF ALL!

We have realized that the oceans are not empty but overfishing occurs frequently at an unacceptable level with significant disadvantages for the entire society including industry and consumers. While the objectives of sustainable fisheries management are internationally agreed (UN indicators for exploitation of marine stocks defined as F_{MSY}), the road to implement them remains long and bumpy, also because coastal nations of a marine region have to be consulted, come to an agreement, implement it to their national legislation, enforce and finally control it. Given the improved information available from scientific assessments of exploited stocks and their fisheries impacts and in line with the responsible fisheries management, there shall be no further reason to postpone necessary actions regarding evaluations, decisions and measures implemented to achieve high long term yields at reduced ecological and economic risk within a reasonable time (OECD, 2011).

Unfortunately, slow decision-making and implementation has been identified to delay or even prevent a sustainable approach – once decided lately, many decisions appear outdated and their implementation often turn counterproductive. As a consequence, lost value through forgone future opportunities caused by depleted and non-rebuilt fisheries are seldom accurately accounted for in arguing to delay implementation of sustainable fisheries management (Shelton, 2009). The responsible parties shall immediately develop policies aimed at sustainable stewardship of the biosphere; in easy words: how our oceans shall look like in 50 years and how Neptune's garden shall be used. By doing this, the human role needs to be re-identified and respected; yet we are players in and not controlling managers of the ecosystems. Although the global modeling including climate and other ecological effects are rapidly improving and leading to a better understanding, the sustainable management of ecosystems appears rather ideological and shall be approached by adaptive regional steps while considering the existing gaps in knowledge and political power (Norton, 2005). Gladwin et al. (1995) were calling for re-integration of humanity into nature and truth to morality. The policy makers already raised the need for economic information to assess and consider the socio-economic consequences and the potential conflict with confidentiality of individual data. Socio-economic consequences are commonly presented in so-called "impact assessments" to verify the social welfare.

However, it is of vital importance that ecological and economic goals are harmonized as functional ecosystems are seen as the natural capital, i.e. there must first be something you can harvest, and secondly the economy deals with the strategy of the investments and revenues. The evolution of ecological economics as an extended "ecological regime" is both qualitatively and quantitatively dependent on an adequate understanding of the behavior of living systems (Jansson et al., 1994). Economists have long argued that a fishery that maximizes its economic potential usually will also satisfy its conservation objectives. To add, it is well acknowledged that subsidies to fisheries that contribute to overcapacity and overfishing will turn the effectively strong relation between ecology and economy to perversity (Meyers & Kent, 2001).

Recently, maximum economic yield (MEY) has been identified as a primary management objective for Australian fisheries and is under consideration elsewhere. However, the avoidance of significant tradeoffs is complex and to develop an implementable management strategy in an adaptive management framework, a set of assumptions must be agreed among scientists, economists, and industry and managers, indicating strong industry commitment and involvement (Dichmont et al., 2010). The optimum structure of regional decisive power and whether the fishing industry is willing and able to assume greater responsibility for its actions remain key questions (Lassen et al., 2008). We conclude that fisheries management has to be fisheries specific to be acceptable and effective. Furthermore, it is obvious that smaller systems are easier to manage than starting top down on large and complex fisheries at a global or continental scale. The mutual educational processes between scientists and decision makers, from scientific monitoring, modeling, understanding to the complementary advisory role of global political frameworks, is exemplarily documented for air quality targets by Hordijk & Amann (2007). However, such demanding process will certainly benefit if the various parties involved keep their cooperation strictly constrained towards exchange of relevant information and their mandates, i.e. scientists shall undertake accurate science and advice and policy makers shall undertake and defend the sustainable decisions. Given the global poor status of many exploited stocks and their fisheries, which appear depleted in many cases, the realization of the political goal towards sustainable fisheries would require stringent or even brutal management actions. For obvious reasons, multi-annual management plans accompanied by impact assessments offer preferable solutions to avoid irrational responses (Symes & Hoefnagel, 2010). Needless to emphasize that specific multi-annual management plans and their outcomes depend on a full implementation into the fisheries management schemes including monitoring, enforcement and control without any tolerance against violations.

The vision is that once fishing capacity and deployed fishing effort are adapted to the production of the exploited marine stocks, the required investments into control and enforcement could be minimized. Fisheries science is to support the achievement of the sustainable use of the oceans by sound scientific advice based on accurate data from monitoring, and in close relation with economy and social sciences (Symes & Hoefnagel, 2010). Nature is everything else but stable, instead high and increasing variability appears the normal under the climate changes we are recently facing (Walther et al., 2002). Fish production is considered highly variable even without fishing. However, status classification of exploited marine stocks and ecosystems requires more consistent frameworks worldwide to further develop and review integrated scientific advice to sustainable fisheries management considering also economic impacts in the format of an integrated advice. Many of the regional fishery organizations need to consider such needs and their advisory bodies need to be reformed regarding their structure and mandate towards integrated advice. In particular, sustainable fisheries shall support the food production from sustainable aquaand agriculture.

The aquaculture, as closest sector to fisheries, is boosting as it maintained an average annual growth rate of 8.3 percent worldwide between 1970 and 2008 (FAO, 2010 a), peaking at 52.5 million tons reported for 2008. In common with all other food production practices, aquaculture is facing challenges for sustainable development, including genetic conservation and environmental risk of genetically altered aquatic organisms (NACA/FAO, 2001). The continued efforts in optimizing production practices, including food supply and pollution, have to be assessed, regulated and controlled to avoid environmental problems. Like capture fisheries, aquaculture will contribute to food security only after full compliance with long term sustainability criteria.

But do we need all this fish and shellfish that can potentially be produced once we are fishing and rearing sustainably? Annual per capita fish consumption grew from an average of 9.9 kg in the 1960s to 11.5 kg in the 1970s, 12.6 kg in the 1980s, 14.4 kg in the 1990s and reached 17.0 kg in 2007. In 2007, fish accounted for 15.7 percent of the global population's intake of animal protein (FAO, 2010 a). There is of course significant regional variation in the dietary. Given the recent status of marine fisheries and resources, their relevance in protein supply has to be significantly improved, which is considered necessary to effectively increase food security and thus combat world's hunger. The world population is expected to grow from the present 6.8 billion people to about 9 billion by 2050. The growing need for nutritious and healthy food will increase the demand for fisheries products from marine sources, whose productivity is already highly stressed by excessive fishing pressure, growing organic pollution, toxic contamination, coastal degradation and climate change (Garcia & Rosenberg, 2010). The number and the proportion of undernourished people have declined, but they remain unacceptably high. Although the number and proportion of hungry people have declined in 2010 as the global economy recovers and food prices remain below their peak levels, hunger remains higher than before (FAO, 2010 b). Fisheries and their management shall adopt the challenge of sustainable food production and adjust their goals and actions at global as well as regional scales in accordance with the view to safeguard biodiversity, bio-production and thus increase livelihood of humanity.

REFERENCES

1. Bertalanffy, L. von (1938). A quantitative theory of organic growth (Inquiries on growth laws. II). Human Biol. 10, pp. 181-213

2. Beverton, R. J. H. & Holt, S. J. (1957). On the dynamics of exploited fish populations. Fishery Investigations. London, HMSO, Ser. 2 (19), ISBN 0412 54960 3, pp. 541

3. Braham, H. W. (1984). Marine Fisheries Review Vol. 46(4), pp. 45-53

4. Caddy, J. F. & Seijo, J. C. (2005). This is more difficult than we thought! The responsibility of scientists, managers and stakeholders to mitigate the unsustainability of marine fisheries. Phil. Trans. R. Soc. B 360, pp. 59–75

5. Cochrane, K.; de Young, C.; Soto, D. & Bahri, T. (20090. Climate change implications for fisheries and aquaculture. FAO, Fisheries and Aquaculture Technical Paper 530, Food and Agriculture Organization of the United Nations, Rome, ISBN 978-92-5- 106347-7, pp. 221

6. Costanza, R. & Patten, B. C. (1995). Defining and predicting sustainability. Ecological Economics Vol. 15, pp. 193-196

7. Cotter, J. (2010). Reforming the European Common Fisheries Policy: a case for capacity and effort controls. Fish and Fisheries, Vol. 11, pp. 210–219

8. Dichmont, C. M., Pascoe, S.; Kompas, T.; Punt, A. E. & Deng, R. (2010). On implementing maximum economic yield in commercial fisheries. Proceeding of the National Academy of Science of the United States of America, Vol. 107 (No. 1), pp. 16–21

9. EU (2002). European Council Regulation (EC) No.2371/2002 of 31 December 2002 on the conservation and sustainable exploitation of fisheries resources under the Common Fisheries Policy, pp. 22

10. EU (2007a). Treaty of Lisbon amending the Treaty on European Union and the Treaty establishing the European Community, signed at Lisbon, 13 December 2007. http://eur-lex.europa.eu/JOHtml.do?uri=OJ:C:2007: 306:SOM:EN:HTML

11. EU (2007b). Council Regulation (EC) No 676/2007 of 11 June 2007 establishing a multiannual plan for fisheries exploiting stocks of plaice and sole in the North Sea. Official Journal of the European Union, pp. 6

12. EU (2008a). Directive 2008/56/EC of the European Parliament and of the Council of 17 June 2008 establishing a framework for community action in the field of environmental policy (Marine Strategy Framework Directive), pp. 22

13. EU (2008b). Council Regulation (EC) No 1342/2008 of 18 December 2008 establishing a longterm plan for cod stocks and the fisheries exploiting those stocks and repealing Regulation (EC) No 423/2004. Official Journal of the European Union, 14 pp.

14. EU (2009). European Commission 2009. Green Paper: Reform of the Common Fisheries Policy. COM(2009) 163 final, pp. 28

15. EU (2010). Commission Decision of 1 September 2010 on criteria and methodological standards on good environmental status of marine waters (2010/577/EU), pp. 11

16. FAO (1995). Code of Conduct for Responsible Fisheries. Food and Agriculture Organization of the United Nations (FAO), Rome, Italy 1945, ISBN 92-5-103834-1, pp. 41

17. FAO (2008). Report of the FAO Workshop on Vulnerable Ecosystems and Destructive Fishing in Deep-Sea Fisheries. Rome, 26–29 June 2007. FAO Fisheries Report No. 829 FIEP/R829, ISBN 978-92-5-105994-4, pp. 27

18. FAO (2010a). The State of World Fisheries and Aquaculture 2010. Food and Agriculture Organization of the United Nations (FAO), Rome, Italy 2010, ISBN 978-92-5-106675- 1, pp. 218

19. FAO (2010b). State of Food Insecurity in the World 2010: Addressing Food Insecurity in Protracted Crises Food and Agriculture Organization of the United Nations (FAO), Rome, Italy 2010, ISBN 978-92-5-106610-2, pp. 62

20. Garcia, S. M. & Rosenberg, A. A. (2010). Food security and marine capture fisheries: characteristics, trends, drivers and future perspectives. Phil. Trans. R. Soc. B Vol. 365, pp. 2869-2880

21. Garstang, W. (1900). The impoverishment of the sea –a critical summary of the experimental and statistical evidence bearing upon the alleged depletion of the trawling grounds. Journal of the Marine Biological Association 6, pp. 1-69

22. Gladwin, T. N.; Kennelly, J. J. & Krause, T.-S. (1995). Shifting paradigms for sustainable development: Implications for management theory and research. Academy of Management Review 1995 Vol. 20(4), pp. 874-902

23. Graham, M. (1943). The Fish Gate. London: Faber and Faber, pp. 196

24. Gulland, J. A. (1965). Population of mortality rates. ICES CM mimeographed, Gadoid Fish Committee, No. 3 (Annex to Arctic fisheries working group report)

25. Hansen, J. W. (1996). Is agricultural sustainability a useful concept? Agricultural Systems, Vol. 50(2), pp. 117-143

26. Hardin, G. (1968). The Tragedy of the Commons. Science, Vol. 162, pp. 1243–1248

27. Hart, P. J. B. & Reynolds, J. D. (2002). Handbook of Fish Biology and Fisheries. Blackwell Publishing, ISBN 0-632-06483-8

28. Hauge, K. H. & Wilson, D. C. (2009). Comparative Evaluations of Innovative Fisheries Management. Global Experiences and European Prospects. Springer Dordrecht Heidelberg London New York, ISBN 978-90-481-2663-7, pp. 272 pp

29. Hilborn, R. (2010). Apocalypse Forestalled: Why All the World's Fisheries Aren't Collapsing. The science Chronicles, November 2010, pp. 5-9

30. Hilborn, R. (2011). Future directions in ecosystem based fisheries management: A personal perspective. Fisheries Research 108, pp. 235–239

31. Hordijk, L. & Amann, M. (2007). How science and policy combined to combat air pollution problems. Environmental Policy and Law 37, pp. 336-340

32. ICES (2010). Report of the Working Group on the Assessment of Demersal Stocks in the North Sea and Skagerrak (WGNSSK), 5 -11 May 2010, ICES Headquarters, Copenhagen. ICES CM 2010/ACOM:13, pp. 1058

33. Jansson, A.-M.; Hammer, M.; Folke, C. & Costanza, R. (1994). Investing in Natural Capital. The Ecolocigal Economics Approach to Sustainability. Island Press, 1718 Connecticut Avenue, N.W., Suite 300, Washington DC 20009. ISBN 1-55963-316-6, technical editor S. Koskoff with foreword by O. Johansson, pp. 511

34. Jennings, S.; Kaiser, M. J. & Reynolds, J. D. (2001). Marine Fisheries Ecology. Blackwell Publishing, ISBN 0-632-05098-5, pp. 417

35. Kaiser, M. J., & Spencer, B. E. (1996). The effects of beam trawl disturbance on infaunal communities in different habitats. Journal of Animal Ecology, Vol. 65, pp. 348-358

36. Kell, L. T.; Pilling, G.M.; Kirkwood, G.P.; Pastoors, M.; Mesnil, B.; Korsbrekke, K.; Abaunza, P.; Aps, R.; Biseau, A.; Kunzlik, P.; Needle,

C.; Roel, B. A. & Ulrich-Rescan, C. (2005). An evaluation of the implicit management procedure used for some ICES roundfish stocks. ICES J. Mar. Sci., Vol. 62 (4), pp. 750-759

37. Kurlansky, M. (1997). Cod: a biography of the fish that changed the world. Pinguin Books Ltd, London, England, ISBN 0-14-027501-0, pp. 295

38. Lassen, H.; Sissenwine, M.; Symes, D. & Thulin, J. (2008). Reversing the burden of proof for fisheries management—managing commercial fisheries within sustainable limits. Report of a SAFMAMS workshop. Copenhagen: ICES; 4–5 March 2008, pp. 18

39. Lévêque, C. (2003). Ecology from ecosystem to biosphere. Science Publishers, Inc., P.O. Box 699, Enfield, New Hampshire 03748, USA, ISBN 1-57808-294-3, pp. 428

40. Ludwig, D.; Walker, B. & Holling, C. S. (1997). Sustainability, stability, and resilience. Conservation Ecology [online]1(1): 7. Available from the Internet. URL: http://www.consecol.org/vol1/iss1/art7/

41. MacArthur, R. & Wilson, E. O. (1967). The Theory of Island Biogeography, Princeton University Press (2001 reprint), ISBN 0-691-08836-5M, pp. 205

42. Meyers, N. & Kent, J. 2001. Perverse subsidies: how tax dollars can undercut the environment and the economy. Island Press, 1718 Connecticut Avenue, N.W., Washington DC 20009, ISBN 1-55963-834-6, pp. 279

43. NACA/FAO (2001). Aquaculture in the Third Millennium. Technical Proceedings of the Conference on Aquaculture in the Third Millennium, Bangkok, Thailand, 20-25 February 2000. Subasinghe, R.P., Bueno, P., Phillips, M.J., Hough, C., McGladdery, S.E., & Arthur, J.E. (Eds.). NACA, Bangkok and FAO, Rome, ISBN: 974-7313-55-3, pp. 471

44. Norton, B. G. (2005). Sustainability: a philosophy of adaptive ecosystem management. The University of Chicago Press, Chicago 60637, Ltd., London, ISBN 0-226-59519-6, pp. 607

45. OECD (2011). Fisheries Policy Reform. National Experiences. OECD Publishing, ISBN 9789264074958, pp. 120

46. Patterson, K.; Cook, R.; Darby, C.; Gavaris, S.; Kell, L.; Lewy, P., Mesnil, B.; Punt, A.; Restrepo, V.; Skagen, D. W. & Stefánsson, G. (2001). Estimating uncertainty in fish stock assessment and forecasting. Fish and

Fisheries (2), pp. 125–157

47. Pauly, D. & Palomares, M. L. (2005). Fishing down marine food web: it is far more pervasive than we thought. Bulletin of Marine Science, Vol. 76(2), pp. 197-211

48. Pauly, D.; Christensen, V.; Dalsgaard, J.; Froese, R. & Torres Jr., F. (1998). Fishing Down Marine Food Webs. Science, Vol. 279, pp. 860-863

49. Pickett, S. T. A.; Kolasa, J. & Jones, C. G. (2007). Ecological Understanding. 2nd ed., Elsevier Inc., Oxford UK, ISBN 978-0-12-554522-8, pp. 233

50. Pope, J. G. (1972). An investigation of the accuracy of virtual population analysis using cohort analysis. ICNAF Research Bulletin 9, pp. 65-74

51. Prager, M. H. (1994). A suite of extensions to a non-equilibrium surplus-production model. Fishery Bulletin, Vol 92, pp. 374-389

52. Rätz, H.-J. & Lloret, J. (2003). Variation in fish condition between Atlantic cod (Gadus morhua) stocks, the effect on their productivity and management implications. Fisheries Research, Vol. 60, pp. 369-380

53. Rätz, H.-J.; Bethke, E.; Dörner, H.; Beare, D. & Gröger, J. (2007). Sustainable management of mixed demersal fisheries in the North Sea through fleet-based management—a proposal from a biological perspective. – ICES Journal of Marine Science, Vol. 64, pp. 652–660

54. Rätz, H.-J.; Dörner, H.; Scott, R. & Barbas, T. (2010). Complementary roles of European and national institutions under the Common Fisheries Policy and the Marine Strategy Framework Directive. Marine Policy Vol. 34, pp. 1028-1035

55. Ricker, W. (1975). Computation and Interpretation of biological statistics of fish populations. Bull. Fish. Res. Bd. Canada 191, pp. 382

56. Rijnsdorp, A. D.; Daan, N. & Dekker, W. (2006). Partial fishing mortality per fishing trip: a useful indicator of effective fishing effort in mixed demersal fisheries. ICES J. Mar. Sci. Vol. 63 (3), pp. 556-566

57. Schaefer, M. B. (1954). Some aspects of the dynamics of populations important to the management of commercial marine fisheries. Bulletin of the Inter-American tropical tuna commission 1, pp. 25-56

58. Schwach, V.; Bailly, D.; Christensen, A.-S.; Delaney, A. E.; Degnbol, P.; van Densen, W. L. T.; Holm, P.; McLay, H. A.; Nolde Nielsen, K.; Pastoors, M. A.; Reeves, S. A., & Wilson, D. C. (2007). Policy and knowledge in fisheries management: a policy brief. ICES J. Mar. Sci., Vol. 64 (4), pp. 798-803

59. Shelton, P. A. (2009). Eco-certification of sustainably managed fisheries—

Redundancy or synergy? Fisheries Research Vol. 100, pp. 185–190

60. STECF (2011). Scientific, Technical and Economic Committee for Fisheries (STECF) – Report of the SGMOS-10-05 Working Group on Fishing Effort Regimes Regarding Annexes IIA, IIB and IIC of TAC & Quota Regulations, Celtic Sea and Bay of Biscay. (ed. Bailey, N. and Rätz, H.-J.). 2011. Publication Office of the European Union, Luxembourg, EUR 24809 EN, JRC64928, pp. 323

61. Symes, D. & Hoefnagel, E. (2010). Fisheries policy, research and the social sciences in Europe: Challenges for the 21st century. Marine Policy, Vol. 34, pp. 268–275

62. UN (1982). United Nations Convention on the Law of the Sea. Montego Bay, Jamaica, 10 December 1982. http://www.un.org/Depts/los/convention_agreements/convention_overview_co nvention.htm

63. UN (1992). Report of the United Nations Conference on Environment and Development. Rio Declaration on Environment and Development. Rio de Janeiro, Brazil, 3-14 June 1992. A/CONF.151/26 (Vol. I). http://www.un.org/documents/ga/conf151/aconf15126-1annex1.htm

64. UN (1995). Report of the United Nations Conference on Straddling Fish Stocks and Highly Migratory Fish Stocks. Sixth session, New York, USA, 24 July-4 August 1995. A/CONF.164/37. http://daccess-ddsny.un.org/doc/UNDOC/GEN/N95/274/67/PDF/N9527467.pdf?OpenElement

65. UN (2002). Report of the World Summiton Sustainable Development. Johannesburg, South Africa, 26 August–4 September 2002. A/CONF.199/20, United Nations Publication, Sales No. E.03.II.A.1, ISBN92-1-104521-5, pp. 173

66. Valentinsson, D. & Ulmestrand, M. (2008). Species-selective Nephrops trawling: Swedish grid experiments. Fisheries Research, Vol. 90 (1-3), pp. 109-117.

67. Walther, G.-R.; Post, E.; Convey, P.; Menzel, A.; Parmesank, C.; Beebee, T. J. C.; Fromentin, J.-M.; Hoegh-Guldberg, O. & Bairlein, F. (2002). Ecological responses to recent climate change. Nature, Vol. 416, pp. 389-395

68. Worm, B.; Barbier, E. B.; Beaumont, N.; Duffy, J. E.; Folke, C.; Halpern, B. S.; Jackson, J. B. C.; Lotze, H. K.; Micheli, F.; Palumbi, S. R.; Sala, E.; Selkoe, K. A.; Stachowicz, J. J. & Watson, R. (2006). Impacts of Biodiversity Loss on Ocean Ecosystem Services. Science, Vol. 314, pp. 787-790

69. Worm, B.; Hilborn, R.; Baum, J. K.; Branch, T. A.; Collie, J. S.; Costello,

C.; Fogarty, M. J.; Fulton, E. A.; Hutchings, J. A.; Jennings, S.; Jensen, O. P.; Lotze, H. K.; Mace, P. M.; McClanahan, T. R.; Minto, C.; Palumbi, S. R.; Parma, A. M.; Ricard, D.; Rosenberg, A. A.; Watson, R. & Zeller, D. (2009). Rebuilding Global Fisheries. Science, Vol. 325, pp. 578-585

Chapter 2

ONTOGENETIC DIETARY SHIFTS IN A PREDATORY FRESHWATER FISH SPECIES: THE BROWN TROUT AS AN EXAMPLE OF A DYNAMIC FISH SPECIES

Javier Sánchez-Hernández[1, 2], María J. Servia[3], Rufino Vieira-Lanero[2] and Fernando Cobo[1, 2]

[1]Department of Zoology and Physical Anthropology, Faculty of Biology, University of Santiago de Compostela, Spain

[2]Station of Hydrobiology "Encoro do Con", Castroagudín s/n, Vilagarcía de Arousa, Pontevedra, Spain

[3]Department of Animal Biology, Vegetal Biology and Ecology, Faculty of Science, University of A Coruña, Spain

INTRODUCTION

The brown trout (*Salmo trutta*) is a species of Eurasian origin, but at has become naturalized in many other parts of the world. It has an outstanding socio-economic importance, both in commercial and sport fisheries, and it is frequently used as tourist attraction [1,2]. The case of brown trout is a clear example of a 'dynamic' fish species, as its diet and feeding behaviour can vary greatly among individuals, age classes, seasons and rivers. The composition of brown trout diet is strongly influenced by environmental and biotic factors. For example, water temperature plays an important role, as it influences food intake and the activity of fishes [3], but also the emergence and activity of aquatic insects or other potential prey items. Also water flow rate can be extremely important for drifting feeders such as brown trout, as they regulate food availability. The are many abiotic factors that influence feeding behaviour, but in general, biotic factors such as the locomotor ability of fishes, accessibility, abundance and antipredator behaviour of prey are thought to be the most important factors in the determination of the diet and feeding strategies in fishes. Usually not all the available prey is consumed by the predator, a feature that allows biologists to distinguish between trophic base and trophic niche (Figure 1). The trophic base consists of all potential prey items that the brown trout is able to consume and it is determined by the feeding habits of the fish, the size of the mouth and the anatomical characteristics of its digestive tract. However, the trophic niche is the variety of organisms that are really consumed by the predator, which depends on different factors that play an important role when choosing criteria prey items as, for example, prey abundance, including site-specific prey accessibility, prey size, energetic selection criteria and prey preference. In this context, the trophic niche for brown trout is very flexible and is usually broader in adults than juveniles.

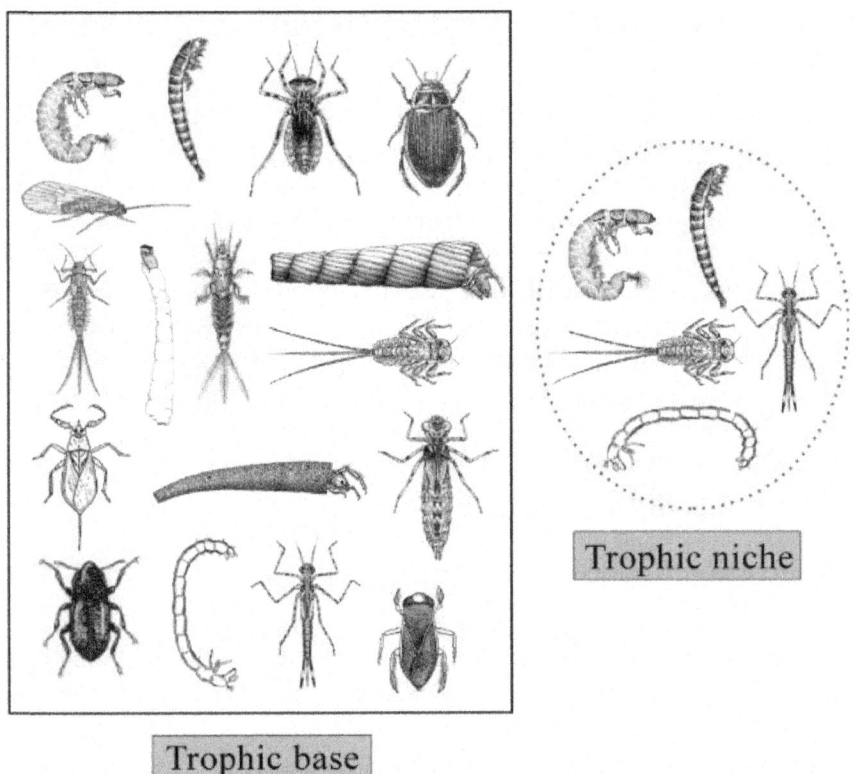

Trophic niche

Trophic base

Figure 1: Graphical example of the trophic base and niche for brown trout.

A knowledge of the foraging ecology of fishes is fundamental to understanding the processes that function at the individual, population and community levels since the factors that influence the acquisition and assimilation of food can have significant consequences for the condition, growth, survival and recruitment of fishes [4]. In this context, the development of effective conservation programmes requires a clear understanding of fish ecological requirements, so the knowledge of its feeding habits is essential to achieve this objective. For example, the knowledge on how food is shared among individuals of the same population is critical for understanding its functioning. Hence, conclusions of field studies on feeding could help wildlife managers to take measures to preserve fish populations, especially for threatened and exploited species. In this chapter we will briefly discuss the variables that are involved in the feeding behaviour of brown trout as an example of a predatory freshwater fish species.

METHODOLOGY AND TYPES OF ANALYSIS EMPLOYED IN FEEDING AND ONTOGENETIC DIETARY SHIFTS STUDIES

The majority of researchers have conducted feeding studies in feral fish populations based on diet descriptions of the stomach contents, using occurrence, numerical, gravimetric and volumetric methods. The main disadvantage of feeding studies is that fish are systematically killed in order to study their stomach contents. However, due to the decline of many natural fish populations, the studies that use non-lethal methods are now more frequent. Different techniques have been used to collect stomach contents without harming the fish such as gastric lavage, emetics or forceps [5-7]. The effectiveness of the gastric lavage is not related to the size of the trouts, but rather to the prey's own morphological characteristics, the degree of repletion of the stomach and the extent of digestion of the food [8]. The effectiveness of this method is inversely related to the degree of repletion [8].

Prey Selection Analysis

Prey selection is an important part of fish feeding ecology. In order to study prey selection of fishes, several indices have been employed, such as the Savage index [9] and Ivlev's selectivity index [10]. The Savage index varies from zero (maximum negative selection) to infinity (maximum positive selection), whereas possible values of Ivlev's selectivity index range from −1 to +1, with negative values indicating rejection or inaccessibility of the prey, zero indicating random feeding, and positive values indicating active selection. Moreover, several researchers have demonstrated that studies based on food selection provide insight into factors involved in prey choice of brown trout [e.g. 11-13].

Stomach Content Analysis

In the early 80s Hyslop reviewed the methods used to study the feeding behaviour of fishes and their application to stomach content analyses [14]. Hyslop pointed out the difficulties in the application of these methods and, where appropriate, proposed alternative approaches. Food overlap between age classes can be assessed with Schoener's overlap index [15]. The overlap index has a minimum of 0 (no prey overlap), and a maximum of 1 (all prey items in equal proportions), and diet overlap is usually considered significant when value of the index exceeds 60% [16]. A chi-square (χ^2) test can be used to test for significant differences in the diet composition between age classes [e.g. 17].

Graphical Methods

Graphical methods proposed by Costello [18] and Tokeshi [19] were used to illustrate the relative importance of prey species and to assess the feeding strategy of fish species. Amundsen and collaborators designed an alternative method of Costello graphical method, by plotting prey abundance (Ai) (y - axis) against the frequency of occurrence in diet (Fi) (x - axis) for each prey species. Information on prey importance, feeding strategy and niche breadth can be obtained by examining the distribution of points along the diagonals and axes of the graph [20] (Figure 5 and Section 4.1).

Niche Breadth Indexes

Marshall and Elliott compared univariate and multivariate numerical and graphical techniques for determining inter- and intraspecific feeding relationships in estuarine fish [21] and on the basis of this study, different indices have been employed by ichthyologists to study niche breadth and diet specialisation. Generally, the Shannon diversity index was combined with the Levin's index to assess niche breadth [21] and the evenness index was used to evaluate diet specialisation, these being indices employed to study feeding habits in brown trout populations [22,23]. However, stable isotope analysis is a potentially powerful method of measuring trophic niche width, particularly when combined with conventional approaches [24]. For this reason, over the past two decades this methodology has been employed to study the trophic interactions and dietary niche in different fish species, and it has been recently used to study ontogeny and dietary specialization in brown trout [25,26].

Multivariate Approaches

Recently prey trait analysis has been proposed as a functional approach to understand mechanisms involved in predator–prey relationships [27,28]. Despite the disadvantages of this methodology [29 and references therein], it has been used in order to get a deeper insight into the mechanisms that regulate diet composition and feeding habits of fishes, providing extremely valuable ecological information and complementing traditional diet analysis [23,29,30].

For the application of prey trait analysis, researchers have to use the same trait database and trait analyses as de Crespin de Billy [27]. To evaluate the potential vulnerability of invertebrates to fish predation, de Crespin de Billy and Usseglio-Polatera created a total of 71 different categories for 17 invertebrate traits [(1) macrohabitat, (2) current velocity, (3) substratum, (4) flow exposure, (5) mobility/attachment to substratum, (6) tendency to drift in the water column, (7) tendency to drift at the water surface, (8) trajectory on the bottom substratum or in the drift, (9) movement frequency, (10) diel drift behaviour, (11) agility, (12) aggregation tendency, (13) potential size, (14) concealment, (15) body shape (including cases/tubes), (16) body flexibility (including cases/tubes) and (17) morphological defences] [28]. The information of this trait database is structured using a 'fuzzy coding' procedure; thus, a score is assigned to each taxon describing its affinity for each category of each trait, with '0' indicating 'no affinity' to '5' indicating 'high affinity'. The taxonomic resolution (order, family and genus) use in the classification process corresponds to the lowest possible level of determination of taxa in fish gut contents. When identification to genus is not possible or in the case of missing information for a certain genus, the value assign for a trait is that of the family level, using the average profile of all other genus of the same family. Additionally, all the taxa and their assigned scores for each category can be found in previous works [27,28]. Prey trait analysis should be carried out with the software R (version 2.11.1), its ADE4 library for the analysis in R is free and downloadable at http://cran.es.r-projet.org/. Finally, the analysis of prey traits has provided ichthyologists with important clues for understanding the ontogenetic dietary shifts in freshwater fish species. As shown in sections 4 and 5, it is an important tool to disentangle the food resource partitioning among both sympatric age classes and fish species.

DIET COMPOSITION OF NEWLY EMERGED BROWN TROUT FRY

In brown trout populations there is strong evidence for a critical period with high mortality in the first few weeks after fry emergence [3]. Furthermore, the most critical stage for population regulation in the whole life cycle is the density-dependent mortality of young trout in the first few weeks of the life cycle soon after the young fish start to feed [3 and references therein]. Thus, first feeding of newly emerged fry is very important for brown trout survival in this phase of the life cycle, and in newborns of brown trout first feeding can occur even prior to emergence [13,31,32]. In this sense, the feeding behaviour of newly emerged brown trout fry has been studied in both laboratory conditions and in natural spawning areas. Results of those studies show that feeding in recently emerged fry can be initiated before complete yolk exhaustion [13,31,32]. Zimmerman and Mosegaard observed that alevins of brown trout began feeding in experimental conditions when yolk constituted approximately 40% of the total alevin dry weight [31]. Other researchers have indicated that brown trout fry under natural conditions start feeding when having almost 30% of yolk sac remaining compared to the presumed original size of the yolk sac at hatching [32], while in a recent study no food particles have been found in the stomachs of fry having >10% of the yolk sac remaining [13].

The optimal foraging theory (OFT) explains adaptation via natural selection through quantitative models, which led to a better understanding of foraging behaviour. Hence, OFT predicts that predators should select prey that maximise the energetic gains available in relation to the energetic costs of capturing, ingesting and digesting the prey [33,34]. In this context, Many researchers have found that chironomid larvae and baetid nymphs seem to be the most important food items for newborns in different geographical areas [e.g. 13,32,35]. These are probably the most accessible invertebrates living in the gravel interstices on nesting grounds at the moment of emergence, providing over 80% of the energetic input [13].

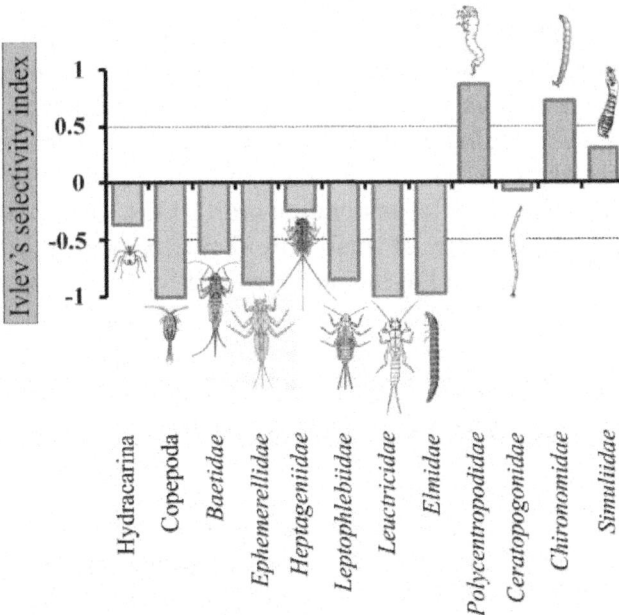

Figure 2: Prey selectivity according to Ivlev's selectivity index of newly emerged brown trout fry in the River Iso (NW Spain) (modified from [13]).

However, although chironomid larvae and baetid nymphs seem to be the most important food items for newborns, newly emerged brown trout fry can show differences in the selection of these prey items. Although Baetidae is abundant in the benthos, this taxon is negatively selected according to Ivlev's selectivity index, whereas Chironomidae remains positively selected (Figure 2), demonstrating that abundance of prey items in the benthos is not the only factor explaining the complex mechanism that operates in the food selection during this phase of the ontogeny. Thus, prey size may affect the prey ingestion in early fish larvae, and much literature focuses on the relationship between prey size and mouth size as the primary factor of prey selection [e.g. 36]; but in general, other factors apart from size, such as locomotor skills of fish or accessibility and antipredator behaviour of prey items play an important role in feeding behaviour. These hypotheses that could explain the absence of some items in the stomachs in spite of their abundance in the benthos [13].

The feeding diversity of juvenile fishes is generally greater than during the larval period, and there is often an increase in the importance of species-specific dietary traits [4]. However, recent studies have demonstrated that at the moment of complete yolk absorption, the fry shows a dramatic shift in niche breadth, which might be related to the improvement of swimming and handling ability of fry for capturing and ingesting both aquatic invertebrates and aerial imagoes [13].

A common practice in many countries associated with river restoration is the rehabilitation of spawning sites with different techniques [37], but recently different authors have emphasized the importance of the complete recolonization of spawning grounds by benthic macroinvertebrates, including first instars, in order to assure the presence of the required amount of prey for the feeding of young fry after restoration works [13]. Hence, at the moment of hatching, a certain density of small prey should be present in the gravel, as searching for food is limited to the nest area and fry forage on available prey [13].

DIET CHANGES WITH AGE: FOOD RESOURCE PARTITIONING AND CHANGE IN PISCIVOROUS BEHAVIOUR

In Salmonid populations, dominant fishes may exclude less aggressive individuals, limiting their access to resources within patches. For example, dominant Atlantic salmon *Salmo salar* Linnaeus, 1758 may exclude subordinates from high-quality patches by intimidation or direct aggression [38,39]. However, subordinate fish may gain access to food by using high-quality patches when dominants are absent [40] or may be constrained to foraging in marginal areas [41]. When brown trout and Atlantic salmon co-occurred, trout has been observed to be dominant over salmon, holding feeding stations by swimming actively in the central regions of food patches, whereas salmon occupied the margins, generally remaining stationary on the stream bed [41]. Moreover, in habitats in which food is patchily distributed in time and space, fishes can benefit by moving between patches [42,43], with subordinate animals moving little in comparison with dominant fishes [39].

During their life history brown trout undergo ontogenetic habitat shifts [44 and references therein] due to changes in habitat selection operating at multiple spatial scales [44]. These shifts during fish life stage transitions may be accompanied by a marked reduction in intra-specific competition in the fish population, facilitating the partitioning of resources [e.g. 45,46]. Moreover, dietary analyses usually show high values of diet overlap among age classes, but the differences in the use of feeding habitat and behavioural feeding habits are important adaptive features that may reduce the intra-specific competition in the population [23]. Thus, although the diet comparison among age classes can show a remarkable similarity in their prey utilization patterns, sometimes the high overlap values may not indicate competition, since fishes can adopt different strategies to overcome competence, i.e. resource partitioning among age classes can occur at five different levels: (1) diet composition; (2) prey selection; (3) prey size; (4) habitat utilization for feeding; and (5) niche breadth. Also stomach fullness can vary among age classes as shown in section 4.6.

Changes in Diet Composition with Age

In brown trout, as in many other fish species, there is normally a change in the diet composition during the life of the fish. Thus, juveniles mainly consume prey items linked to the bottom of the river, many of them interstitial, i.e. living among grains of sand or gravel. Opposite, terrestrial invertebrates and fishes are important resources for large trouts. The contribution of these food items to fish diets increases with predator size or age because larger fish can feed on a wider range of preys as shown inFigures 3 and 4. Within a population, the percentage of the most important prey items change with age. In one study of a river in Italy the percentage of plecopteran nymphs in the diet tended to increase with the individual's age [17]. In another study, *Baetis* spp. dominated in all age classes in different proportions, whereas the percentage of caddisflies with cases, (*Allogamus* sp.) and mayflies (*Ecdyonurus* spp.) tended to increase with age [23]. Thus, as shown in Table 1, each age class consumes significantly different prey items, Chironomidae being the most frequently consumed prey item in age 0+ and age 3+ (44.41% and 29.84% respectively) and aerial imagoes of Ephemeroptera in age 1+ and age 2+ (77.49% and 49.08% respectively).

Regarding the changes in the proportion of terrestrial invertebrates consumed by trouts during the ontogeny, previous studies have demonstrated that aquatic invertebrates dominated the diet in all age classes [23,47], but it seems clear that terrestrial invertebrates were more frequently consumed by older trout [17,23,48], terrestrial organisms being important prey during warmer seasons in salmonids [49,50].

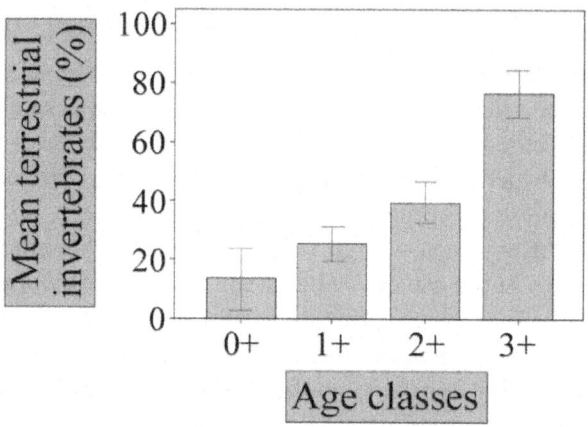

Figure 3: Percentage of terrestrial invertebrates (in terms of relative abundance) consumed by each age class of*Salmo trutta* in the River Anllóns (NW Spain) during summer. Error bars represent 95% confidence intervals.

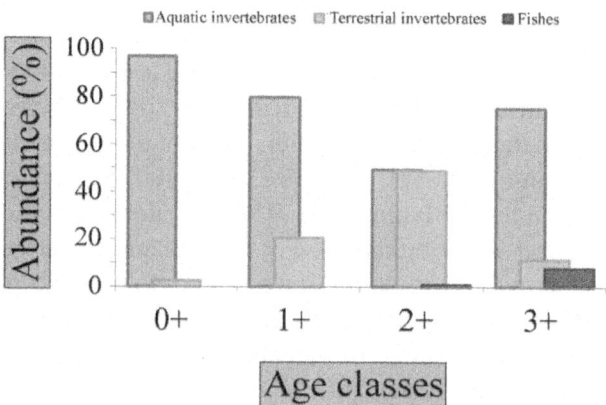

Figure 4: Diet composition consumed by each age class of *Salmo trutta* in the River Lengüelle (NW Spain) during summer.

Table 1: Diet composition and prey selection according to Ivlev's selectivity index in each age class of *Salmo trutta* in the River Furelos (NW Spain) during summer

	Age 0+			Age 1+			Age 2+			Age 3+		
	Diet (%)	Ivlev		Diet (%)	Ivlev		Diet (%)	Ivlev		Diet (%)	Ivlev	
		Ben-thos	Drift		Benthos	Drift		Ben-thos	Drift		Ben-thos	Drift
Aquatic invertebrates												
Oligochaeta gen. sp.	0.14	-0.68	1	0	-1	—	0	-1	—	0	-1	—
Ancylidae	2.07	0.04	1	0.15	-0.85	1	0	-1	—	0	-1	—
Hydrobiidae	2.48	0.90	0.82	2.11	0.88	0.79	0.92	0.74	0.58	4.84	0.95	0.90
Lymnaeidae	1.66	-0.18	-0.01	1.96	-0.10	0.07	20.64	0.79	0.85	29.03	0.85	0.89
Sphaeriidae	0	-1	—	0.53	0.59	1	1.38	0.82	1	0	-1	—
Hydracarina gen. sp.	0.41	0.64	-0.61	0.23	0.43	-0.76	0	-1	-1	0	-1	-1
Ostracoda gen. sp.	0.55	1	1	0.23	1	1	0	—	—	0	—	—
Baetidae	1.93	0.49	-0.06	0.15	-0.63	-0.87	8.72	0.86	0.60	0	-1	-1
Caenidae	5.10	0.94	0.50	0.60	0.59	-0.48	0.46	0.49	-0.57	0	-1	-1
Ephemerellidae	0	-1	-1	0.23	-0.77	-0.86	2.29	0.14	-0.12	0	-1	-1
Ephemeridae	0	—	—	0.08	1	1	0	—	—	0	—	—
Leuctridae	0.55	0.85	1	0.08	0.25	1	0.46	0.82	1	0	-1	—
Aeshnidae	0	—	—	0.08	1	1	0	—	—	0	—	—
Calopterygidae	6.07	1	1	3.78	0.45	0.17	8.72	0.72	0.53	0	-1	-1
Coenagrionidae	0.28	0.61	1	0	-1	—	0	-1	—	0	-1	—
Gomphidae	0.14	-0.13	1	0	-1	—	0	-1	—	0	-1	—
Aphelocheiridae	0.41	-0.27	-0.28	0.08	-0.81	-0.81	0	-1	-1	0.81	0.05	0.05
Gerridae	0.14	1	-0.28	0	—	-1	0	—	-1	0.81	1	0.54
Sialidae	0.14	1	1	0.23	1	1	0	—	—	0	—	—
Dytiscidae	0	—	—	0	—	—	0.46	1	1	0	—	—
Elmidae	0.41	-0.58	-0.56	0.15	-0.83	-0.81	0.46	-0.55	-0.52	0	-1	-1
Brachycentridae	0	-1	—	0.08	-0.64	1	0	-1	—	0	-1	—
Hydropsychidae	2.76	-0.61	0.17	0.45	-0.92	-0.62	0.92	-0.85	-0.36	0.81	-0.87	-0.41
Leptoceridae	0.28	1	1	0	—	—	0	—	—	0	—	—
Limnephilidae	0.55	0.92	1	0.45	0.91	1	0.46	0.91	1	0.81	0.95	1
Philopotamidae	0.14	0.51	1	0	-1	—	0	-1	—	0	-1	—
Polycentropodidae	0	-1	—	0.30	0.25	1	0	-1	—	0	-1	—
Rhyacophilidae	0	-1	—	0	-1	—	0.46	0.49	1	0	-1	—
Sericostomatidae	0.14	1	1	0.15	1	1	0.46	1	1	0	—	—
Chironomidae	44.41	0.31	0.18	4.08	-0.71	-0.76	1.83	-0.86	-0.89	29.84	0.12	-0.01
Simuliidae	15.17	-0.51	-0.18	1.89	-0.92	-0.84	0.92	-0.96	-0.92	12.90	-0.56	-0.26
Tipulidae	0	-1	—	0.08	0.54	1	0	-1	—	0	-1	—

Terrestrial invertebrates											
Ephemeroptera gen. sp.	12.14	—	1	77.49	—	1	49.08	—	1	2.42	—	1
Trichoptera gen. sp.	0.14	—	-0.28	0.45	—	0.30	0	—	-1	0	—	-1
Chironomidae	0.69	—	-0.93	2.04	—	-0.82	0	—	-1	13.71	—	-0.20
Simuliidae	0.69	—	1	0.68	—	1	0	—	—	2.42	—	1
Arachnida gen. sp.	0	—	—	0.08	—	—	0	—	—	0	—	—
Acanthosomatidae	0	—	—	0.08	—	1	0.46	—	1	0	—	—
Psyllidae	0	—	—	0.15	—	1	0	—	—	0	—	—
Chloropidae	0.14	—	-0.28	0	—	-1	0	—	-1	0	—	-1
Diptera gen. sp.	0	—	-1	0.08	—	-1	0	—	-1	0	—	-1
Syrphidae	0	—	-1	0	—	-1	0	—	-1	0	—	-1
Xylomyidae	0	—	—	0.15	—	1	0	—	—	0	—	—
Cynipidae	0	—	-1	0	—	-1	0	—	-1	0.81	—	0.25
Formicidae	0	—	-1	0	—	-1	0	—	-1	0.81	—	0.54
Carabidae	0	—	—	0.08	—	1	0	—	—	0	—	—
Chrysomelidae	0.28	—	-0.28	0	—	-1	0	—	-1	0	—	-1
Coleoptera gen. sp.	0	—	—	0	—	—	0.46	—	1	0	—	—
Other prey items											
Pseudochondrostoma duriense	0	—	—	0.45	—	—	0.46	—	—	0	—	—
Eggs	0	—	—	0.15	—	1	0	—	—	0	—	—

The maximum and mean prey size eaten generally increases with size in predatory fish species [51]. Piscivorous behaviour is most frequent in large brown trout, and studies show that it occurs in older individuals with a size of 20–30 cm. Trout in smaller size classes have rarely/never been recorded eating other fish [23,26,52,53]. In contrast, Sánchez-Hernández and collaborators have recorded piscivorous behaviour in an age-0 trout (the individual found was 8.5 cm in fork length) [29]. This behaviour could be related with the hypothesis of Mittelbach and Persson, who stated that fish species that had larger mouth gapes became piscivorous at younger ages and at smaller sizes [51], and demonstrating that mouth gape is not a limitation to use fishes by small trouts [29]. One possible advantage for small trout in diversifying into eating fish is a reduction in competition with other individuals in the same size classes.

The feeding strategy among age classes may be illustrated with graphical methods, such as the modified Costello graphical method [20] based on the relative importance of prey species. Figure 5, shows plots of prey abundance (*Ai*) against frequency of occurrence of prey in the diet (*Fi*) for three different age classes. The plots show feeding strategy differs between age classes, varying in degrees of specialization and generalization on different prey types. For all age classes, *Baetis* spp. were the most important prey, always being eaten by more than eighty-five percent of the individuals ($Fi = 87.5\%$ in age-1+ to 100% in age-2+) and represented a high contribution in specific abundance ($Ai = 23.93\%$ in age-1+ to 68.95% in age-0+). However, the majority of the prey items presented low values for both Fi and Ai (lower left quadrant) for all age classes (Figure 5), displaying evidence of a generalist strategy for these prey (e.g. *Chimarra marginata*, Simuliidae (adult), Chironomidae (adult) and *Ophiogomphus* sp.).

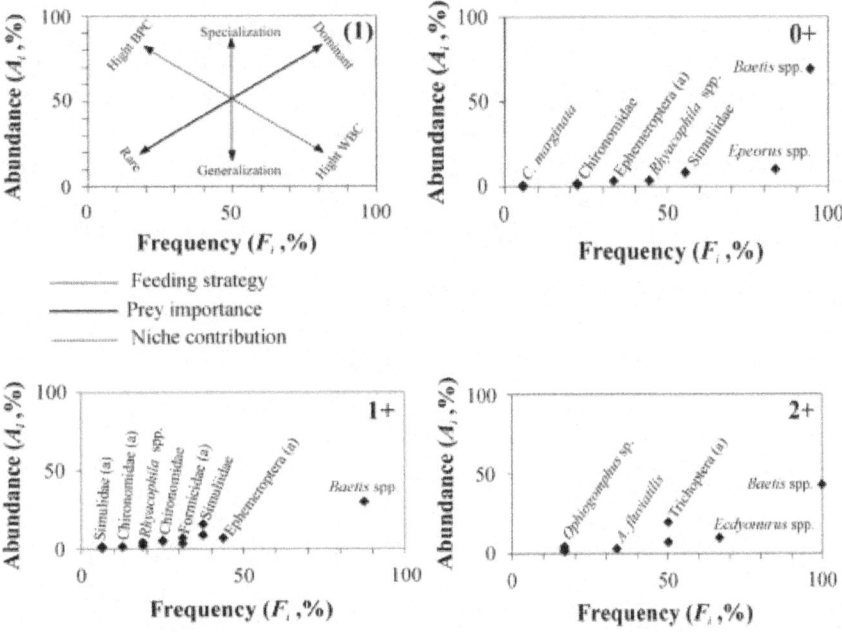

Figure 5: Feeding strategy diagram. (1) Explanatory diagram of the modified Costello method according to Amundsen and collaborators [20]. Data are presented for each age class.

Finally, although in some occasions the diet comparison among age classes can show a remarkable similarity in their prey utilization patterns, sometimes the high overlap values may not indicate competition, since the differences in the behavioural feeding habits are important adaptive features that may reduce the intra-specific competition in the population [23]. Previously, the differences in the behavioural feeding habits among age classes had been studied using prey trait analysis [23], obtaining important advances to disentangle food resource partitioning among cohorts. Details and information needed for the elaboration of prey trait analysis can be found in the introduction section and bibliography [27-29]. Thus, in the reference [23] 'diel drift behaviour' trait showed that age-2+ is clearly separated from the other age classes. Age-2+ tended to feed on prey with no tendency in diel drift behaviour as shown by the presence of *Ancylus fluviatilis*, Gerridae, Coleoptera and Formicidae in their stomachs. On the contrary, age-0+ preferred to feed on prey with nocturnal diel drift behaviour tendency due to the presence of Philopotamidae (*C. marginata*). Age-0+ showed a wider distribution of values in the fuzzy principal component analysis (FPCA) of the 'trajectory' trait; age-0+ tended to feed on prey with oscillatory and by random trajectory (Lumbriculidae, Simuliidae and Chironomidae), whereas age-2+ tended to feed on prey with lineal trajectory due to the presence of Gerridae [23]. On the other hand, no clear differences have been found in the traits 'tendency to drift in the water column' and 'tendency to drift at the water surface' for prey among age classes [23], however as shown in Figure 6 (fuzzy principal component analysis calculated with the values reported in Table 1), age-3+ is clearly separated from the other age classes, showing that the differences ability to feed at different depths of the water column is possible among cohorts.

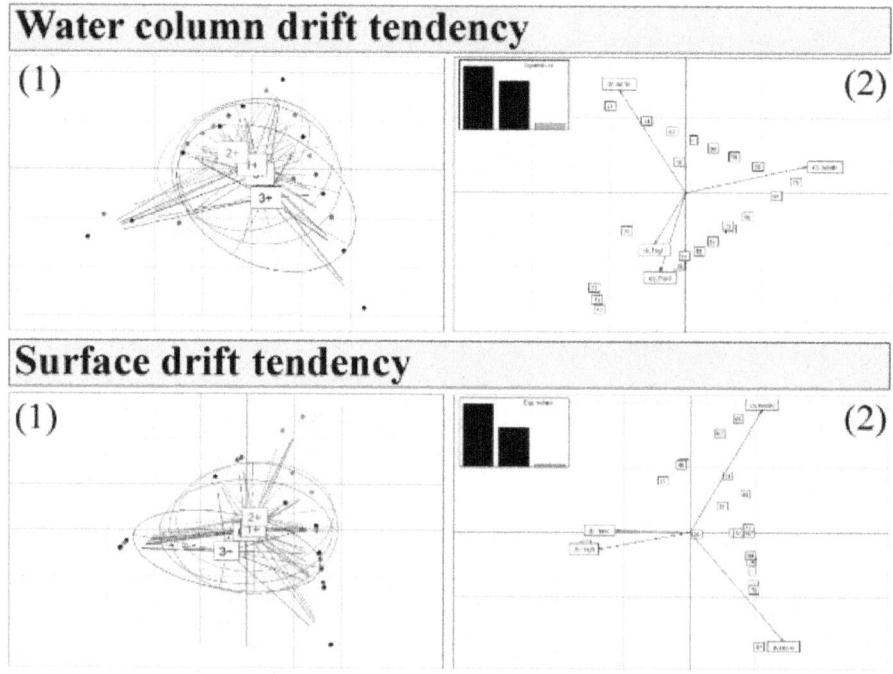

Figure 6: Biplot of gut contents obtained from a fuzzy principal component analysis (FPCA) based on behavioural feeding habits of the four age classes. (1) Similarity results among age classes according to the gut contents of Table 1. Data are presented for

each age class. 0+: age-0+, 1+: age-1+, 2+: age-2+ and 3+: age-3+. Ellipses envelop weighted average of prey taxa positions consumed by age classes: Labels (0+, 1+, 2+ and 3+) indicate the gravity centre of the ellipses (2) Factorial correspondence analysis between both traits and prey items and their spatial distribution with histogram of eigenvalues. Details needed for the elaboration of these graphics can be found in the introduction section and bibliography [27-29].

Changes in Prey Selection with Age

Prey abundance should be an important factor involved in prey choice. However, fishes do not always consume the most abundant taxa available in the environment [13,29,54]. The total abundance and biomass of invertebrates drifting during the day describe the potential prey available to juvenile brown trout better than abundance and biomass of benthic invertebrates do [55]. Different authors have shown that active choice guided by energetic optimization criteria appeared to be of limited importance in determining the size composition of prey eaten by trouts [11]. These authors also stated that the operating mechanisms of prey-size selection are probably not independent of the characteristics of the size-frequency distribution of the available prey. Moreover, in brown trout spatial and temporal variations in prey selection are possible, due to the preference for the different prey items related to the site-specific prey accessibility [12]. As shown in section 3, the size-frequency distribution of potential prey items in the benthos was different to that of prey in the stomachs of newly emerged brown trout [13], a result that agrees with observations made in other fish species [e.g. 54] and observations made in age-0+ individuals of *S. trutta* during the autumn [29]. In spite of the high abundance of Elmidae in the benthos, this prey item was negatively selected [29]. The rejection of the elmid beetle may be due to their low energetic value, as they have an intense sclerotisation, but it may also be due to their bad taste [56-58]. Hence, other factors, besides prey abundance, including site-specific prey accessibility, prey size, energetic selection criteria and prey preference of fishes, play an important role in the feeding behaviour of freshwater fishes [54].

Fochetti and collaborators found a high preference for species of Trichoptera by trout younger than 3+, a preference for plecopteran species by those older than three years, and a general negative avoidance for species of Ephemeroptera by all age classes [17]. As shown in Table 1, prey selection is clearly different among age classes and important prey items such as Chironomidae in age 0+ and age 3+ and aerial imagoes of Ephemeroptera in age-1+ and age-2+ are positively selected in the benthos and drift respectively. Overall, as different researchers have demonstrated, the mechanisms involved in prey selection among age classes in the same population are complex and may be related at different levels: (1) prey selection in fishes is related to prey characteristics, such as size, locomotor skills, accessibility or anti-predator behaviour, (2) prey selection in fishes is related to fish characteristics, like prior experience, locomotor skills, stomach fullness, mouth gape, sensory capabilities and fish size and (3) prey selection in fishes is related to physical habitat characteristics, as flow patterns and structural complexity of habitat [12,13,34,54].

Changes in Prey Size with Age

Ontogenetic dietary shifts may also occur at the level of prey size [23,54]. Steingrímsson and Gíslason showed a consistent, but moderate, shift towards larger prey with increased body size in brown trout [59]. Several researchers have found that mean prey size increases as predator size increases [e.g. 23,48,54,60]. The size-frequency distributions of the available terrestrial prey were always greatly dominated (75–90%) by the two smallest size classes (1–2 and 2–3 mm long), prey over 4 mm long being extremely scarce, while size distributions of aquatic prey were less skewed [11]. Thus, in general it is correct to state that trout fed mainly on prey within 1–4 mm size range [11,48], with 2–3 mm prey being the most commonly consumed [48]. On the contrary, Rincón and Lobón-Cerviá showed that organisms 1–2 mm long were generally the most numerous [11]. By age classes the average prey size consumption is different with higher values in age-2+ (8.4 mm ± 1.62) than age-0+ (4.2 mm ± 0.25) and age-1+ (5.9 mm ± 0.51), but there are no significant differences between age-0+ and age-1+ [23]. Figure 7 illustrates the age-related variation in prey size, showing that mean prey size tends to increase with age.

In conclusion, prey-size selection is probably dependent on the characteristics of the size-frequency distribution of the available prey [11], and the size-related differences in the diet of trout can be related to gape-limitations, increasing mean prey size and maximum prey-size with trout size [48].

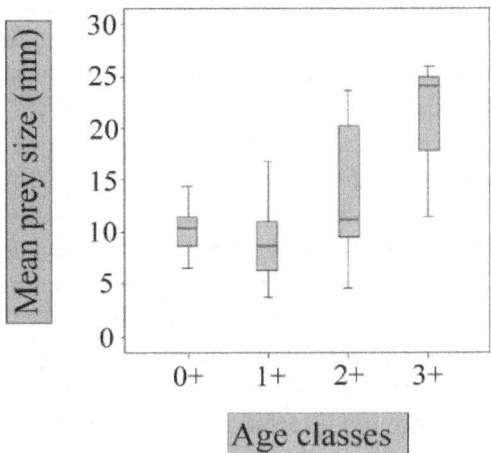

Figure 7: Box plots of the age-related variation in prey size of *Salmo trutta* in the River Furelos (NW Spain) during summer. The solid line within each box represents the median, the bottom and top borders indicate the 25th and 75th percentiles, the notches represent the 95% confidence intervals.

Changes in the Habitat Used For Feeding With Age

In fishes, patches used for feeding and refuges are normally different, as shown by several researchers [61-63], due to brown trout being a habitat generalist. Also, patterns in habitat selection have been shown to be driven by physical and environmental factors operating at multiple spatial scales [44].

Many organisms exhibit ontogenetic shifts in their diet and habitat use, which often exert a large influence on the structure and expected dynamics of food webs and ecological communities [64]. Special attention has been given to ontogenetic shift in habitat preference in brown trout populations [e.g. 44,65-67]. It is well-known that in brown trout populations habitat use changes during ontogeny, preferring deeper and slower flowing water as they increased in size [e.g. 44,65].

Habitat patches used by brown trout can be monitored by radio telemetry [e.g. 68] and although these studies have shown that brown trout feed on young white suckers *Catostomus commersonii* (Lacepède, 1803) at night in shallow habitats, little information was obtained about the habitat used for feeding. Also, microhabitat use of freshwater fishes has been studied by snorkel observations in previous studies [e.g. 69]. This methodology could be used to study feeding habitat requirements of fish species. However, there are still gaps to be filled before snorkel surveys can be fully adopted in fish diet studies.

In fact, one of the main disadvantages of this approach is the need for good visibility. This has been one of the main handicaps because, although brown trout normally yields a bimodal (crepuscular) pattern of activity with a major peak at dawn and a lower one around dusk [70], brown trout can also feed at night [71]. Another limiting factor for the application of snorkel surveys to study feeding habitat requirements is related to the physical characteristic of the river such as current, depth or turbidity. It is well-known that different macroinvertebrates have different preferences for habitats [72] and so prey trait analysis has been proposed as a functional approach to understand mechanisms involved in predator-prey relationships [27-29]. Consequently it may be useful for understanding inter-species interactions and the mechanisms that determine food partitioning between them [29,30]. Nowadays, they have been recently used to provide interesting results about differences in feeding habitat requirements among age classes in brown trout [23], for example, young of the year (0+) tend to capture prey living in moderate current velocities, whereas other age classes (1+ and 2+) tend to feed on prey living in fast current velocities [23].

A previous study on habitat choice in a littoral zone of Lake Tesse (Norway) showed that small trout had a strong association with the bottom and larger trout occurred more frequently higher up in the water column, and suggested that this difference in vertical distribution was also reflected in food choice [73]. In streams, competition among fish species may also be reduced by vertical segregation [e.g. 41,69,74]. In spite that terrestrial invertebrates, as component of the diet, are more important in adults than juveniles of brown trout [23], these same authors found that the ability to feed at different depths of the water is similar among age classes. Nevertheless, it is possible that different age classes of fish may become vertically segregated by concentrating on different prey types living in different parts of the water column as shown in section 4.1. Hence, additional studies are needed in order to clarify whether vertical segregation among cohorts is related to the ability to feed at different depths of the water column.

Finally differences in the use of feeding habitat are important adaptive features that may reduce the intra-specific competition in the population. In this context, fuzzy principal component analysis (FPCA) has shown that age-0+ tended to feed on prey living in moderate current velocities, although overlap was higher between age-1+ and age-2+, preferring to feed on prey living in fast current velocities [23]. Moreover, these researchers have found that age-0+ showed a higher spectrum of prey, which revealed a greater ability to prey on different macrohabitats, whereas age-1+ and age-2+ preferred to feed on epibenthic prey living in erosional macrohabitats. In our case with the values reported in Table 1 and as shown in Figure 8, 'current velocity' trait shows no clear differences for prey of the four age classes. On the contrary, 'macrohabitat' trait shows that age-1+ has the most ample spectrum for this trait, tending to feed on epibenthic prey living in erosional macrohabitats, whereas age-3+ tends to feed on prey items available in the water column (Figure 8).

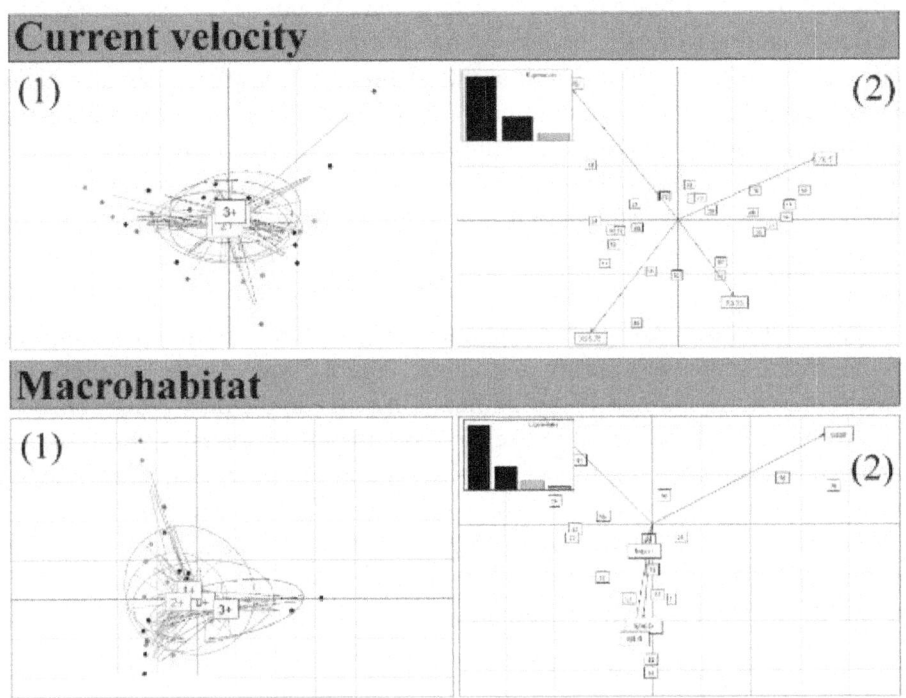

Figure 8: Biplot of gut contents obtained from a fuzzy principal component analysis (FPCA) based on preferential habitat utilization for feeding of the four age classes. (1) Similarity results among age classes according to the gut contents of Table 1. Data are presented for each age class. 0+: age-0+, 1+: age-1+, 2+: age-2+ and 3+: age-3+. Ellipses envelop weighted average of prey taxa positions consumed by age classes: Labels (0+, 1+, 2+ and 3+) indicate the gravity centre of the ellipses. (2) Factorial correspondence analysis between both traits and prey items and their spatial distribution with histogram of eigenvalues. Details needed for the elaboration of these graphics can be found in the introduction section and bibliography [27-29].

Changes in the Niche Breadth with Age

Deady and Fives showed that niche breadth decreases with fish length in corkwing wrasse, *Symphodus (Crenilabrus) melops* (Linnaeus, 1758), indicating an increase in dietary specialization with increasing length [75].

Magalhães found dietary shifts throughout the ontogeny in an endemic cyprinid of the Iberian Peninsula (*Squalius pyrenaicus* (Günther, 1868)), including shifts from soft-bodied to hard-shelled prey and decreased animal prey breadth [76]. In contrast, several other researchers have found that niche breadth increases with body size [e.g. 23,77]. Oscoz and collaborators found that larger fish have a higher number of potential prey items available and a wider niche breadth, as indicated by their higher trophic diversity index values [77]. As mentioned in section 3, analysis of diet changes on newly emerged brown trout fry suggests a dramatic shift in niche breadth at the moment of complete yolk absorption, which might be related to the improvement the fry's swimming and handling ability in capturing and ingesting prey [13]. In a recent study on brown trout populations, it has been demonstrated that niche breadth, measured as Levin's index, increases with fish length in *Salmo trutta*. However, no differences were found in the Shannon diversity and evenness indices of prey eaten among age classes [23].

As can be seen in Figure 9; the number of different prey types consumed by S. trutta increases during ontogeny. Age-0+ shows the smallest prey spectrum, whereas in age-1+ a significant increase in the prey types consumed by juveniles is observed. However, no significant differences in the dietary niche among ages-1+, 2+ and 3+ are observed in Figure 9. Hence, although no clear results have been observed in the variation of the diversity indices of prey among age classes [23], it could be argued that there is a tendency to increase dietary niche with increasing length or age, at least when the niche breadth of juveniles (0+), subadults (1+) and adults (≥2+) is compared.

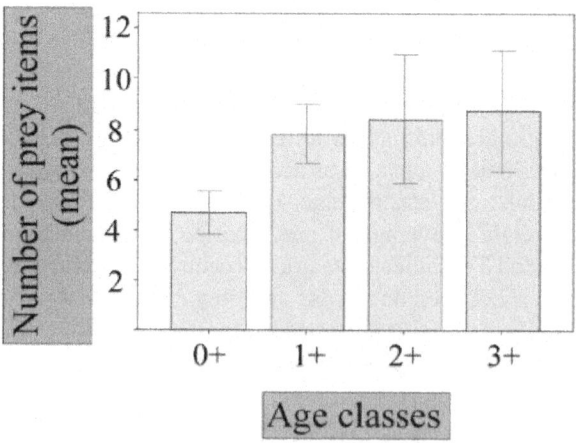

Figure 9: Age-related variation in the number of different prey types consumed by *Salmo trutta* in the River Lengüelle (NW Spain) during summer. Error bars represent the 95% confidence intervals.

Changes in the Fullness Index with Age

Several researchers have found in different fish species that the stomach fullness index (defined as the weight of the stomach contents (in grams) divided by the weight of the predator (in grams) and multiplied by 100) varies during ontogeny [78,79]. In salmonids the results are contradictory: in brook charr *Salvelinus fontinalis* (Mitchill, 1814) no differences have been found in the stomach fullness [80], whilst other researchers have demonstrated that the stomach fullness of brown trout varies among size classes [47]. Brown trout between the size of 40 mm and 320 mm fed more intensively, whilst the intensity declined above 320 mm length [47]. In the River Furelos (NW Spain), we have found that stomach fullness during the summer is different among age classes (Kruskal-Wallis test; $p < 0.001$), being higher in age-0+ (9% ± 0.64) than age-1+ (1.1% ± 0.14), age-2+ (1% ± 0.25) and age-3+ (1.1% ± 0.24) (all Mann–Whitney U test, $p < 0.001$) but no differences have been found between ages-1+, 2+ and 3+ (all Mann–Whitney U test, $p > 0.05$). Moreover, stomach fullness decreases with fish size ($r = -0.72$; $p < 0.001$) (unpublished data). Hence, stomach fullness can vary among age classes; however additional studies are needed in order to clarify whether stomach fullness varies during ontogeny in brown trout.

COMPETION FOR FOOD BETWEEN BROWN TROUT AND OTHER SYMPATRIC FISH SPECIES

Trophic interactions between species are important factors structuring animal communities. Brown trout are top-consumers in freshwater habitats and play an important role as carriers of energy from lower to higher trophic levels (i.e. predators). Many freshwater fish species tend to occupy a specific type of habitat but there are lots of exceptions. For example, spatial niche overlap is considerable where Atlantic salmon and brown trout co-occur, although young Atlantic salmon tend to occupy faster flowing and shallower habitats [e.g. 74]. Moreover, when both fish species co-occur, the habitat used by Atlantic salmon is restricted through interspecific competition by the more aggressive brown trout, indicating an interactive segregation between fish species [e.g. 41,74]. Indeed, it is well known that brown trout is a territorial drift feeder [3,81], and several authors have reported the behavioural dominance of trout over cyprinids in streams [82-84].

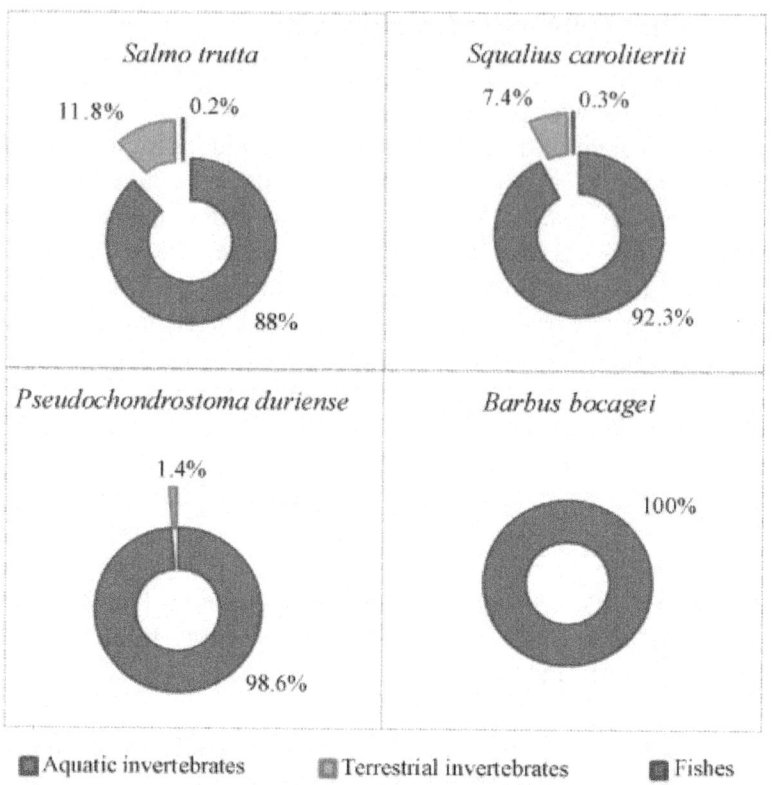

Figure 10: Diet composition consumed by each fish species in the Tormes River (Central Spain) during summer.

The competitive coexistence between species occupying similar niches may be facilitated by a generalisation of niche width as predicted by the optimal foraging theory (OFT), rather than the specialised niche width predicted by the classic niche theory as a response to interspecific competition [85]. However, studies on food partitioning in fish communities have obtained contradictory results. Whereas several authors have found differences in diet composition among sympatric fish species [e.g. 86,87], other researchers concluded that the same food resource can be shared by several species [29,30,85]. In these cases, the differences in behavioural feeding habits, handling efficiency and feeding habitat utilization are important adaptive features that may reduce the inter-specific competition in the fish community and permit the partitioning of food that allows coexistence [29,30]. Thus, sympatric fish species can adopt different strategies to overcome competion and food resource partitioning can occur at different levels.

Firstly, the use of microhabitats is often different between species, due to segregation of microhabitats, an important factor in reducing the effects of competition for food [69,88,89]. For example, *Barbus bocagei* Steindachner, 1865 occupied deeper habitats and selected lower positions in the water column than *Pseudochondrostoma polylepis* (Steindachner, 1865), and *Squalius pyrenaicus* (Günther, 1868), *P. polylepis* occupied microhabitats with greater velocities than the other two species and *S. pyrenaicus* selected shallower habitats than the other two species [69]. In another study, *S. trutta* showed wider diversity in the habitat used for feeding than *Squalius carolitertii* (Doadrio, 1988), *Pseudochondrostoma duriense* (Coelho, 1985) and *B. bocagei* [30]. Hence, differences were found among species in their ability to feed at different depths of the water column [29,30] as shown in snorkelling studies into microhabitat use in fish [69].

Secondly, different species may specialise in different resources. For example, many cyprinid fish sympatric with trout feed on a significant amount of detritus and plant material not used by trout, leading to reduced inter-specific competition [29,30]. Moreover, resource partitioning may also occur at the level of prey size [29,30,90], although it is not clear whether this size selective strategy is adopted to reduce interspecific competition or it is the result of foraging behaviour and/or morphological constraints such as gape size [29,91]. Also, terrestrial prey are present primarily on the stream surface and although tend to be absent from the diets of benthic feeders such as *B. bocagei* (Figure 10), terrestrial inputs may constitute an important food resource for freshwater fish species and especially for brown trout. Thus, the utilization of allochthonous food resources such as terrestrial invertebrates by fishes may reduce competition facilitating the partitioning of resources [30].

Thirdly, diel segregation is possible among fish species, and this may also lead to reduced interspecific competition between fish [29,92,93]. According to macroinvertebrate trait analyses, sticklebacks (*Gasterosteus aculeatus* Linnaeus, 1758) and *P. duriense* show a slight preference for prey that drift during the day, whilst age-0 *S. trutta* seem to prefer to feed at dusk, whereas *Achondrostoma arcasii* (Steindachner, 1866) differs from the other three species due to its preference to feed on prey on organisms with weak or no tendency to drift [29]. However, the "diel drift behaviour" of macroinvertebrate prey of brown trout and three sympatric cyprinids is similar [30]. Hence, the differences in the diel feeding behaviour among sympatric fish species might only be adopted in highly competitive communities, where food is a more limiting resource.

CONCLUSION

To summarize, the present study supports the hypothesis differences in the feeding habits and habitat utilization of different age classes of trout could reduce competition for food, by allowing food resource partitioning. Hence, age-related diet shifts occur at five different levels: (1) diet composition changes with fish age; (2) prey selection varies with fish age, probably due to prey-size selection which is in turn dependent on the size-frequency distribution of the available prey; (3) mean prey size increases with fish size and age; (4) habitat utilization for feeding may be different among age classes; (5) niche breadth tends to increase with age and fish size. Finally, also the stomach fullness can vary among age classes. However, additional studies are needed in order to clarify whether stomach fullness varies during the ontogeny in brown trout.

ACKNOWLEDGEMENTS

Dr. Adrian Seymour and Josué Sánchez are acknowledged for valuable comments and grammar corrections on the manuscript.

REFERENCES

1.   Aas O, Haider W, Hunt L. Angler responses to potential harvest regulations in a Norwegian sport fishery: a conjoint-based choice modeling approach. North American Journal of Fisheries Management 2000;20(4) 940-950. DOI: 10.1577/1548-8675(2000)020<0940:ARTPHR>2.0.CO;2

2.   Butler JRA, Radford A, Riddington G, Laughton R. Evaluating an ecosystem service provided by Atlantic salmon, sea trout and other fish species in the River Spey, Scotland: the economic impact of recreational rod fisheries. Fisheries Research 2009;96(2-3) 259-266. DOI: 10.1016/j. fishres.2008.12.006

3.   Elliott JM. Quantitative Ecology and the Brown Trout. Oxford: Oxford University Press; 1994.

4.   Nunn AD, Tewson LH, Cowx IG. The foraging ecology of larval and juvenile fishes. Reviews in Fish Biology and Fisheries 2012;22(2) 377-408. DOI: 10.1007/s11160-011-9240-8

5.   Seaburg KG. A stomach sampler for live fish. The Progressive Fish-Culturist. 1957;19(3)137-139. DOI:10.1577/1548-8659(1957)19[137:AS SFLF]2.0.CO;2

6.   Wales JH. Forceps for removal of trout stomach content. The Progressive Fish-Culturist 1962;24(4)171. DOI:10.1577/1548-8659(1962)24[171:FF ROTS]2.0.CO;2

7. Jernejcic F. Use of emetics to collect stomach contents of Walleye and Large mouth Bass. Transactions of the American Fisheries Society 1969;98(4) 698-702. DOI: 10.1577/1548-8659(1969)98[698:UOETCS]2.0.CO;2

8. Sánchez-Hernández J, Servia MJ, Vieira-Lanero R, Cobo F. Evaluación del lavado gástrico como herramienta para el análisis de la dieta en trucha común. Limnetica 2010;29(2) 369-378.

9. Savage RE. The relation between the feeding of the herring off the cast coast of England and the plankton of the surrounding waters. Fishery Investigation, Ministry of Agriculture, Food and Fisheries, Series 2, 1931;12 1-88.

10. Ivlev VS. Experimental ecology of the feeding of fishes. Translated from the Russian by Douglas Scott. New Haven: Yale University Press; 1961.

11. Rincón PA, Lobón-Cerviá J. Prey-size selection by brown trout (Salmo trutta L.) in a stream in northern Spain. Canadian Journal of Zoology 1999;77(5) 755-765. DOI: 10.1139/z99-031

12. Johnson RL, Coghlan SM, Harmon T. Spatial and temporal variation in prey selection of brown trout in a cold Arkansas tailwater. Ecology of Freshwater Fish 2007;16(3) 373-384. DOI: 10.1111/j.1600-0633.2007.00230.x

13. Sánchez-Hernández J, Vieira-Lanero R, Servia MJ, Cobo F. First feeding diet of young brown trout fry in a temperate area: disentangling constraints and food selection. Hydrobiologia 2011a;663(1) 109-119. DOI: 10.1007/s10750-010-0582-3

14. Hyslop EJ. Stomach contents analysis—a review of methods and their application. Journal of Fish Biology 1980;17(4) 411-429. DOI: 10.1111/j.1095-8649.1980.tb02775.x

15. Schoener TW. Nonsynchronous spatial overlap of lizards in patchy habitats. Ecology 1970;51(3) 408-418. DOI: 10.2307/1935376

16. Wallace RK Jr. An assessment of diet overlap indexes. Transactions of the American Fisheries Society 1981;110(1) 72-76. DOI: 10.1577/1548-8659(1981)110<72:AAODI>2.0.CO;2

17. Fochetti R, Argano R, Tierno de Figueroa JM. Feeding ecology of various age-classes of brown trout in River Nera, Central Italy. Belgian Journal of Zoology 2008;138(2) 128-131.

18. Costello MJ. Predator feeding strategy and prey importance: a new graphical analysis. Journal of Fish Biology 1990;36(2) 261-263. DOI: 10.1111/j.1095-8649.1990.tb05601.x

19. Tokeshi M. Graphical analysis of predator feeding strategy and prey importance. Freshwater Forum 1991;1 179-183.

20. Amundsen P-A, Gabler HM, Staldvik FJ. A new approach to graphical analysis of feeding strategy from stomach contents data – modification of the Costello (1990) method. Journal of Fish Biology 1996;48(4) 607-614. DOI: 10.1111/j.1095-8649.1996.tb01455.x

21. Marshall S, Elliott M. A comparison of univariate and multivariate numerical and graphical techniques for determining inter- and intraspecific feeding relationships in estuarine fish. Journal of Fish Biology 1997;51(3) 526-545. DOI: 10.1111/j.1095-8649.1997.tb01510.x

22. Oscoz J, Leunda PM, Campos F, Escala MC, Miranda R. Diet of 0+ brown trout (*Salmo trutta* L., 1758) from the river Erro (Navarra, North of Spain). Limnetica 2005;24(3-4) 319-326.

23. Sánchez-Hernández J, Cobo F. Summer differences in behavioural feeding habits and use of feeding habitat among brown trout (Pisces) age classes in a temperate area. Italian Journal of Zoology 2012a;79(3) 468-478. DOI: 10.1080/11250003.2012.670274

24. Bearhop S, Adams CE, Waldron S, Fuller RA, MacLeod H. Determining trophic niche width: a novel approach using stable isotope analysis. Journal of Animal Ecology 2004;73(5) 1007-1012. DOI:10.1111/j.0021-8790.2004.00861.x

25. Grey J. Ontogeny and dietary specialization in brown trout (*Salmo trutta* L.) from Loch Ness, Scotland, examined using stable isotopes of carbon and nitrogen. Ecology of Freshwater Fish 2001;10(3) 168-176. DOI: 10.1034/j.1600-0633.2001.100306.x

26. Jensen H, Kiljunen M, Amundsen, P-A. Dietary ontogeny and niche shift to piscivory in lacustrine brown trout *Salmo trutta* revealed by stomach content and stable isotope analyses. Journal of Fish Biology 2012;80(7) 2448-2462. DOI: 10.1111/j.1095-8649.2012.03294.x

27. de Crespin de Billy V. Régime alimentaire de la truite (*Salmo trutta L.*) en eaux courantes: rôles de l'habitat physique des traits des macroinvertébrés. PhD thesis. Université Claude Bernard, Lyon; 2001

28. de Crespin de Billy V, Usseglio-Polatera P. Traits of brown trout prey in relation to habitat characteristics and benthic invertebrate communities. Journal of Fish Biology 2002;60(3) 687-714. DOI: 10.1111/j.1095-8649.2002.tb01694.x

29. Sánchez-Hernández J, Vieira-Lanero R, Servia MJ, Cobo F. Feeding habits of four sympatric fish species in the Iberian Peninsula: keys to understanding coexistence using prey trais. Hydrobiologia 2011b;667(1) 119-132. DOI: 10.1007/s10750-011-0643-2

30. Sánchez-Hernández J, Cobo F. Summer food resource partitioning between four sympatric fish species in Central Spain (River Tormes). Folia Zoologica 2011;60(3) 189-202.

31. Zimmerman CE, Mosegaard H. Initial feeding in migratory brown trout (*Salmo trutta* L.) alevins. Journal of Fish Biology 1992;40(4) 647-650. DOI: 10.1111/j.1095-8649.1992.tb02612.x

32. Skoglund H, Barlaup BT. Feeding pattern and diet of first feeding brown trout fry under natural conditions. Journal of Fish Biology 2006;68(2) 507-521. DOI: 10.1111/j.0022-1112.2006.00938.x

33. Pyke GH, Pulliam HR, Charnov EL. Optimal foraging: a selective review of theory and tests. Quarterly Review of Biology 1977;52(2) 137-154.

34. Gerking SD. Feeding ecology of fish. San Diego: Academic Press; 1994.

35. McCormack JC. The food young trout (*Salmo trutta*) in two different necks. Journal of Animal Ecology 1962;31(2) 305-316.

36. Cunha I, Planas M. Optimal prey size for early turbot larvae (*Scophthalmus maximus* L.) based on mouth and ingested prey size. Aquaculture 1999;175(1-2) 103-110. DOI: 10.1016/S0044-8486(99)00040-X

37. Hunter CJ. Better trout habitat: a guide to stream restoration and management. Washington DC: Island Press; 1991.

38. Gotceitas G, Godin J-GJ. Foraging under the risk of predation in juvenile Atlantic salmon (*Salmo salar* L.): effects of social status and hunger. Behavioral Ecology and Sociobiology 1991;29(4) 255-261. DOI: 10.1007/BF00163982

39. Griffiths SW, Armstrong JD. Kin-biased territory overlap and food sharing among Atlantic salmon juveniles. Journal of Animal Ecology 2002;71(3) 480-486. DOI: 10.1046/j.1365-2656.2002.00614.x

40. Alanära A, Burns MD, Metcalfe NB. Intraspecific resource partitioning in brown trout: the temporal distribution of foraging is determined by social rank. Journal of Animal Ecology 2001;70(6) 980-986. DOI: 10.1046/j.0021-8790.2001.00550.x

41. Höjesjö J, Armstrong J, Griffiths S. Sneaky feeding by salmon in sympatry with dominant brown trout. Animal Behaviour 2005;69(5) 1037-104. DOI: 10.1016/j.anbehav.2004.09.007

42. Armstrong JD, Braithwaite VA, Huntingford FA. Spatial strategies of

wild Atlantic salmon parr: exploration and settlement in unfamiliar areas. Journal of Animal Ecology 1997;66(2) 203-211.

43. Ruxton GD, Armstrong JA, Humphries S. Modelling territorial behaviour of animals in variable environments. Animal Behaviour 1999;58(1) 113-120. DOI: 10.1006/anbe.1999.1114

44. Ayllón D, Almodóvar A, Nicola GG, Elvira B. Ontogenetic and spatial variations in brown trout habitat selection. Ecology of Freshwater Fish 2010;19(3) 420-432. DOI: 10.1111/j.1600-0633.2010.00426.x

45. Elliott JM. The food of trout (*Salmo trutta*) in a Dartmoor stream. Journal of Applied Ecology 1967;4(1) 59-71.

46. Amundsen P-A, Bøhn T, Popova OA, Staldvik FJ, Reshetnikov YS, Kashulin N, Lukin A. Ontogenetic niche shifts and resource partitioning in a subarctic piscivore fish guild. Hydrobiologia 2003;497(1-3) 109-119. DOI: 10.1023/A:1025465705717

47. Kara C, Alp A. Feeding habits and diet composition of brown trout (*Salmo trutta*) in the upper streams of river Ceyhan and river Euphrates in Turkey. Turkish Journal of Veterinary and Animal Sciences 2005;29 417-428.

48. Montori A, Tierno de Figueroa JM, Santos X. The diet of the brown trout *Salmo trutta* (L.) during the reproductive period: size-related and sexual effects. International Review of Hydrobiology 2006;91(5) 438-450. DOI: 10.1002/iroh.200510899

49. Nakano S, Kawaguchi Y, Taniguchi Y, Miyasaka H, Shibata Y, Urabe H, Buhara N. Selective foraging on terrestrial invertebrates by rainbow trout in a forested headwater stream in northern Japan. Ecological Research 1999;14(4) 351-360. DOI: 10.1046/j.1440-1703.1999.00315.x

50. Utz RM, Hartman KJ. Identification of critical prey items to Appalachian brook trout (*Salvelinus fontinalis*) with emphasis on terrestrial organisms. Hydrobiologia 2007;575(1) 259-270. DOI: 10.1007/s10750-006-0372-0

51. Mittelbach GG, Persson L. The ontogeny of piscivory and its ecological consequences. Canadian Journal of Fisheries and Aquatic Sciences 1998;55(6) 1454-1465. DOI: 10.1139/f98-041

52. Kahilainen K, Lehtonen H. Brown trout (*Salmo trutta* L.) and Arctic charr (*Salvelinus alpinus* (L.)) as predators on three sympatric whitefish (*Coregonus lavaretus* (L.)) forms in the subarctic Lake Muddusjärvi. Ecology of Freshwater Fish 2002;11(3) 158-167. DOI: 10.1034/j.1600-0633.2002.t01-2-00001.x

53. Jensen H, Bøhn T, Amundsen P-A, Aspholm PE. Feeding ecology of piscivorous brown trout (*Salmo trutta* L.) in a subarctic watercourse. Annales Zoologici Fennici 2004;41(1) 319-328.

54. Sánchez-Hernández J, Cobo F. Ontogenetic dietary shifts and food selection of endemic *Squalius carolitertii* (Actinopterygii: Cypriniformes: Cyprinidae) in River Tormes, Central Spain, in summer. Acta Ichthyologica et Piscatoria 2012b;42(2) 101-111. DOI: 10.3750/AIP2011.42.2.03

55. Sagar PM, Glova GJ. Prey availability and diet of juvenile brown trout (*Salmo trutta*) in relation to riparian willows (*Salix* spp.) in three New Zealand streams. New Zealand Journal of Marine & Freshwater Research 1995;29(4) 527-537. DOI: 10.1080/00288330.1995.9516685

56. Ochs G. The ecology and ethology of whirligig beetles. Archiv für Hydrobiologie 1969;37 375-404.

57. Power G. Seasonal growth and diet of juvenile Chinook salmon (*Oncorhynchus tshawytscha*) in demonstration channels and the main channel of the Waitaki river, New Zealand, 1982–1983. Ecology of Freshwater Fish 1992;1(1) 12-25. DOI: 10.1111/j.1600-0633.1992.tb00003.x

58. Oscoz J, Escala MC, Campos F. La alimentación de la trucha común (*Salmo trutta* L., 1758) en un río de Navarra (N. España). Limnetica 2000;18(1) 29-35.

59. Steingrímsson SÓ, Gíslason GM. Body size, diet and growth of landlocked brown trout, Salmo trutta, in the subarctic river Laxá, North-East Iceland. Environmental Biology of Fishes 2002;63(4) 417-426. DOI: 10.1023/A:1014976612970

60. Blanco-Garrido F, Sánchez-Polaina FJ, Prenda J. Summer diet of the Iberian chub (*Squalius pyrenaicus*) in a Mediterranean stream in Sierra Morena (Yeguas Stream, Córdoba, Spain). Limnetica 2003;22(3-4) 99-106.

61. Clapp DF, Clark RD, Diana JS. Range, activity, and habitat of large, free-ranging brown trout in a Michigan stream. Transactions of the American Fisheries Society 1990;119(6) 1022-1034. DOI: 10.1577/1548-8659(1990)119<1022:RAAHOL>2.3.CO;2

62. Metcalfe NB, Fraser NHC, Burns MD. State-dependent shifts between nocturnal and diurnal activity in salmon. Proceedings of the Royal Society of London, Series B 1998;265(1405) 1503-1507. DOI: 10.1098/rspb.1998.0464

63. Meyer CG, Holland KN. Movement patterns, home range size and habitat utilization of the bluespine unicornfish, *Naso unicornis* (Acanthuridae) in a Hawaiian marine reserve. Environmental Biology of Fishes 2005;73(2) 201-210. DOI: 10.1007/s10641-005-0559-7

64. Ramos-Jiliberto R, Valdovinos FS, Arias J, Alcaraz C, García-Berthou E. A network-based approach to the analysis of ontogenetic diet shifts: An example with an endangered, small-sized fish. Ecological Complexity 2011;8(1) 123-129. DOI: 10.1016/j.ecocom.2010.11.005

65. Haury J, Ombredane D, Baglinière JL. L'habitat de la truite commune (*Salmo trutta* L.) en cours d'eau. In: Baglinière JL, Maisse G. (eds.) La trutie: biologie et écologie. Paris: INRA éditions; 1991. p47-96.

66. Parra I, Almodóvar A, Ayllón D, Nicola GG, Elvira B. Ontogenetic variation in density-dependent growth of brown trout through habitat competition. Freshwater Biology 2011;56(3) 530-540. DOI: 10.1111/j.1365-2427.2010.02520.x

67. Roussel J, Bardonnet A. Ontogeny of diel pattern of stream-margin habitat use by emerging brown trout, *Salmo trutta*, in experimental channels: influence of food and predator presence. Environmental Biology of Fishes 1999;56(1-2) 253-262. DOI: 10.1023/A:1007504402613

68. O'Connor RR, Rahel FJ. A patch perspective on summer habitat use by brown trout *Salmo trutta* in a high plains stream in Wyoming, USA. Ecology of Freshwater Fish 2009;18(3) 473-480. DOI: 10.1111/j.1600-0633.2009.00364.x

69. Martínez-Capel F, García de Jalón D, Werenitzky D, Baeza D, Rodilla-Alamá M. Microhabitat use by three endemic Iberian cyprinids in Mediterranean rivers (Tagus River Basin, Spain). Fisheries Management and Ecology 2009;16(1) 52-60. DOI: 10.1111/j.1365-2400.2008.00645.x

70. Bachman RA, Reynolds WW, Casterlin ME. Diel locomotor activity patterns of wild brown trout (*Salmo trutta* L.) in an electronic shuttlebox. Hydrobiologia 1979;66(1) 45-47. DOI: 10.1007/BF00019138

71. Railsback SF, Harvey BC, Hayse JW, LaGory KE. Tests of theory for diel variation in salmonid feeding activity and habitat use. Ecology 2005;86(4) 947-959. DOI: 10.1890/04-1178

72. Tachet H, Richoux P, Bournaud M, Usseglio-Polaterra P. Invertébrés d'eau douce. Paris: CNRS Éditions; 2002.

73. Hegge O, Hesthagen T, Skurdal J. Vertical distribution and substrate preference of brown trout in a littoral zone. Environmental Biology of Fishes 1993;36(1) 17-24. DOI: 10.1007/BF00005975

74. Heggenes J, Baglinière JL, Cunjak RA. Spatial niche variability for young Atlantic salmon (*Salmo salar*) and brown trout (*S. trutta*) in heterogeneous streams. Ecology of Freshwater Fish 1999;8(1) 1-21. DOI: 10.1111/j.1600-0633.1999.tb00048.x

75. Deady D, Fives JM. The diet of corkwing wrasse, *Crenilabrus melops*, in Galway Bay, Ireland, and in Dinard, France. Journal of the Marine Biological Association of the United Kingdom 1995;75(3) 635-649. DOI: 10.1017/S0025315400039060

76. Magalhães MF. Effects of season and body-size on the distribution and diet of the Iberian chub *Leuciscus pyrenaicus* in a lowland catchment. Journal of Fish Biology 1993;42(6) 875-888. DOI: 10.1111/j.1095-8649.1993.tb00397.x

77. Oscoz J, Leunda PM, Miranda R, Escala MC. Summer feeding relationships of the co-occurring *Phoxinus phoxinus* and *Gobio lozanoi* (Cyprinidae) in an Iberian river. Folia Zoologica 2006;5(4) 418-432.

78. Akpan BE. Ontogenetic and monthly feeding behaviour of *Liza falcipinnis* (Mugilidae) from Cross River Estuary, Nigeria. World Journal of Applied Science and Technology 2011;3(2) 48-56.

79. Magnussen E. Food and feeding habits of cod (*Gadus morhua*) on the Faroe Bank. ICES Journal of Marine Science 2011;68(9) 1909-1917. DOI: 10.1093/icesjms/fsr104

80. Sánchez-Hernández J, Cobo F, González MA. Biología y la alimentación del salvelino, *Salveliml fontinalis* (Mitchill, 1814), en cinco lagunas glaciares de la Sierra de Gredos. Nova Acta Científica Compostelana 2007;16 129-144.

81. Friberg N, Andersen TM, Hansen HO, Iversen TM, Jacobsen D, Krojgaard L, Larsen, E. The effect of brown trout (*Salmo trutta* L.) on stream invertebrate drift, with special reference to *Gammarus pulex* L. Hydrobiologia 1994;294(2) 105-110. DOI: 10.1007/BF00016850

82. Grossman GD, Boulé V. An experimental study of competition for space between rainbow trout (*Oncorhynchus mykiss*) and rosyside dace (*Clinostomus funduloides*). Canadian Journal of Fisheries and Aquatic Sciences 1991;48(7) 1235-1243. DOI: 10.1139/f91-149

83. Facey DE, Grossman GD. The relationship between water velocity, energetic costs and microhabitat use in four North American stream fishes. Hydrobiologia 1992;239(1) 1-6. DOI: 10.1007/BF00027524

84. Hill J, Grossman GD. An energetic model of microhabitat use for rainbow trout and rosyside dace. Ecology 1993;74(3): 685-698. DOI: 10.2307/1940796

85. Gabler H-M, Amundsen P-A. Feeding strategies, resource utilisation and potential mechanisms for competitive coexistence of Atlantic salmon and alpine bullhead in a sub-Arctic river. *Aquatic Ecology* 2010;44(2) 325-336. DOI: 10.1007/s10452-009-9243-x

86. Encina L, Rodríguez-Ruiz A, Granado-Lorencio C. Trophic habits of the fish assemblage in an artificial freshwater ecosystem: the Joaquin Costa reservoir, Spain. *Folia* Zoologica 2004;53(4) 437-449.

87. Novakowski GC, Hahn NS, Fugi R. Diet seasonality and food overlap of the fish assemblage in a pantanal pond. Neotropical Ichthyology 2008;6(4) 567-576. DOI: 10.1590/S1679-62252008000400004

88. Grossman GD, de Sostoa A, Freeman MC, Lobón-Cerviá J. Microhabitat use in a Mediterranean riverine fish assemblage. Fishes of the lower Matarraña. Oecologia 1987a;73(4) 490-500. DOI: 10.1007/BF00379406

89. Grossman GD, de Sostoa A, Freeman MC, Lobón-Cerviá J. Microhabitat use in a Mediterranean riverine fish assemblage. Fishes of the upper Matarraña. Oecologia 1987b;73(4) 501-512. DOI: 10.1007/BF00379407

90. Jepsen DB, Winemiller KO, Taphorn DC. Temporal patterns of resource partitioning among Cichla species in a Venezuela blackwater River. Journal of Fish Biology 1997;51(6) 1085-1108. DOI: 10.1111/j.1095-8649.1997.tb01129.x

91. Stevens M, Maes J, Ollevier F. Taking potluck: trophic guild structure and feeding strategy of an intertidal fish assemblage. In: Stevens M. (ed.) Intertidal and basin-wide habitat use of fishes in the Scheldt estuary. Heverlee (Leuven): Katholieke Universiteit Leuven; 2006. p37-59.

92. Alanärä A, Burns MD, Metcalfe NB. Intraspecific resource partitioning in brown trout: the temporal distribution of foraging is determined by social rank. Journal of Animal Ecology 2001;70(6) 980-986. DOI: 10.1046/j.0021-8790.2001.00550.x

93. David BO, Closs GP, Crow SK, Hansen EA. Is diel activity determined by social rank in a drift-feeding stream fish dominance hierarchy? Animal Behaviour 2007;74(2) 259-263. DOI: 10.1016/j.anbehav.2006.08.015

Chapter 3

A REVIEW OF THE EFFECTS OF HYDROLOGIC ALTERATION ON FISHERIES AND BIODIVERSITY AND THE MANAGEMENT AND CONSERVATION OF NATURAL RESOURCES IN REGULATED RIVER SYSTEMS

Peter C. Sakaris[1]

[1]Department of Biology and Chemistry, Southern Polytechnic State University, Marietta, GA, USA

INTRODUCTION

Hydrologic alterations resulting from dam construction and other human activities have negatively impacted the biodiversity and ecological integrity of rivers worldwide (Dudgeon 2000, Pringle et al. 2000). These alterations have included habitat fragmentation, conversion of lotic to lentic habitat, variable flow and thermal regimes, degraded water quality, altered sediment transport processes, and changes in timing and duration of floodplain inundation (Cushman 1985, Pringle 2000). The negative impacts of altered hydrologic regimes on aquatic organisms are well documented. For example, dam construction has blocked the migratory routes of diadromous and potamodromous species (e.g., salmonids and white sturgeon, *Acipenser transmontanus*), which has severely reduced their spawning and overall reproductive success (Wunderlich et al. 1994, Beamesderfer et al. 1995). In the Alabama River system (USA), flow-modification in regulated reaches has resulted in losses of river-dependent ("fluvial") fish species, and distributions of federally listed species have been restricted by main stem impoundment (Freeman et al. 2004). Several researchers have documented major changes in fish assemblage structure following dam construction (Paragamian 2002, Quinn and Kwak 2003; Gillete et al. 2005). For example, Quinn and Kwak (2003) reported that long-term changes in the fish assemblage after dam construction on the White River (Arkansas, USA) included a shift from warmwater to coldwater species, a substantial decrease in fluvial specialists, and dramatic reductions in species richness. The negative effects of dam construction are not only limited to fishes. Freshwater mussel diversity has declined substantially, particularly in the southeast USA, as a consequence of hydrologic alteration (Watters 2000).

This chapter begins with an examination of the hydrologic alterations that may be caused by dam construction. Several examples are presented for different continents, emphasizing that hydrologic alteration is an issue of global concern. Next, the effects of altered hydrologic regimes on the growth, recruitment, and survival of organisms and on the overall biodiversity and community structure in regulated river systems are reviewed. Subsequently, tools and strategies to manage and conserve aquatic fauna in regulated river systems are discussed. In the past several decades, a wealth of information has been published on these topics. This chapter provides a general overview of the impacts of hydrologic alteration and presents several management approaches, which have been developed to address it. More detailed information may be found in the review papers referenced in this chapter.

ALTERATION OF THE FLOW REGIME

In their highly impactful paper, Poff et al. (1997) outlined five important characteristics of a flow regime: *magnitude, frequency, duration, timing* (or predictability), and the *rate of change* (or flashiness). These major components of the flow regime are ecologically relevant to the system. For example, the magnitude of flow (e.g., mean monthly discharge) may define habitat characteristics such as wetted area or habitat volume in a stream or river (Richter et al. 1996). The frequency of episodic flows (e.g., high or low pulse frequencies) may lend insight on how often drought or flood conditions occur within a system (Richter et al. 1996). Each of these flow attributes may be altered by dam construction and hydropeaking operations. For example, flows have rapidly fluctuated between extremely low and high discharges as a result of hydropeaking operations downstream of Harris Dam on the Tallapoosa River, USA (Irwin and Freeman 2002). Extreme discharge fluctuations during a period of only four to six hours have generated a highly variable flow regime that has potentially threatened the persistence of several native fishes (i.e., fluvial specialists) below the dam (Irwin and Freeman 2002). Irwin and Freeman (2002) reported significant changes in hydrology after construction of Harris Dam in 1982, which included increases in high-pulse frequency, low-pulse frequency, fall rate, and the number of flow reversals. Irwin and Freeman (2002) documented release-driven, diel temperature fluctuations as high as 10°C, producing highly stressful conditions for resident organisms.

Hydrologic alterations that occur as a result of dam construction have been well documented (Galat and Lipkin 2000, Maingi and Marsh 2002, Yang et al. 2008). For example, Yang et al. (2008) applied the Range of Variability Approach (RVA discussed later; Richter et al. 1997) to evaluate the effects of dam construction on the hydrologic regimes of middle and lower river networks in Yellow River, China. The authors stressed that assessments of hydrologic alteration are extremely complex, particularly in systems that are impounded by more than one dam (Yang et al. 2008). In addition, both pre- and post-impact discharge data must be sufficient to effectively assess the effects of dams on hydrologic processes. The Yellow River in China was impounded by the Sanmenxia and Xiaolangdi dams to meet several objectives: flood control and electricity generation, to reduce downstream sediment deposition, and to provide water for irrigation. Unfortunately, the natural flow regime has become significantly altered in the system, with the lower Yellow River recently experiencing zero flow conditions as a result of increased water consumption (Yang et al. 2008). Significantly reduced flows will have negative effects on biodiversity and the persistence of viable wetlands and fisheries in the Yellow River Delta (Yang et al. 2008). The analysis of Yang et al. (2008) indicated that Xiaolangdi dam significantly altered the natural flow regime of the lower Yellow River in the following ways: decreased median of monthly flow, decreased medians of annual 1-, 3-, 7-, 30-, and 90-day minimum and maximum flows, higher low pulse and high pulse counts, and decreased medians of fall rate, rise rate, and number of reversals in the post-impact period.

Maingi and Marsh (2002) studied the effects of dam construction on hydrologic conditions in the Tana River, Kenya. Kenya has a growing population, and water needs have increased as populations are forced to expand into semi-arid regions (Maingi and Marsh 2002). Five dams were constructed from 1968 to 1988 along the upper Tana, the largest river in Kenya (Maingi and Marsh 2002). The largest dam (Masinga Dam) was built to provide hydropower, to increase irrigation potential in the lower basin, and to increase use of dry season flows in the upper Tana (Maingi and Marsh 2002). Of special concern is a tract of riverine tropical forest along the mid- to lower Tana River. This forest extends 0.5 to 3 km from the bank of the river, and the forest largely depends on regular flooding and sufficient groundwater. With decreased peak flows and a declining water table due to river regulation, preservation and regeneration of this riverine forest has become a challenge (Maingi and Marsh 2002). Analyses revealed that major changes in the flow regime occurred after Masinga Dam was constructed, including a significant reduction in May flows, reduced variability in monthly discharges, reduced 7-d, 30-d, and 90-d maximum annual discharges, decreased mean low pulse duration from 14.6 to 7.9 days, increased annual rises and annual falls of the river, and increased mean fall rates from 15.1 m/s to 21.6 m/s (Maingi and Marsh 2002). Experiments with vegetation sample plots indicated that vegetation located above 1.80 m of dry season river level has experienced an average 67.7% reduction in days flooded after construction of Masinga Dam. Experiments also revealed that flood pulse duration declined significantly for all vegetation plots by an average of 87.6% (Maingi and Marsh 2002). These reductions in flood frequency and duration have negative implications for the preservation of riverine forest in the Tana River Basin.

Along the Missouri River (USA), a series dams were constructed for improved navigation, irrigation, and flood control, as well as hydropower generation (Galat and Lipkin 2000). Galat and Lipkin (2000) attributed the listing of the Missouri River as North America's most endangered river in 1997 (American Rivers, 1997) to the numerous alterations that have significantly impacted the ecosystem. In their study, Galat and Lipkin (2000) divided the Missouri River into three sections and classified them as 1) an upper, least-altered section with four dams, 2) a middle highly impacted section with six large mainstem dams, and 3) a regulated and channelized lower section. The authors also used Richter's RVA approach (Richter et al. 1997) to compare pre- and post-impoundment hydrologic conditions at the upper (least-altered), middle, and lower (channelized) sections (Galat and Lipkin 2000). Their analyses indicated that numerous hydrologic changes occurred after mainstem impoundment of the Missouri River including: 1) increased mean annual discharges (i.e., 30 to 38% higher at channelized locations), 2) a stabilization of mean monthly discharges with higher flows from August through February and a reduction in June and July high flows, 3) loss of a natural bimodal flood pulse, 4) lower flow variability at most stations, 5) higher 1-, 7-, 30-day annual minimum flows at all stations, 6) altered timing of annual peak and minimum flows (particularly at middle, inter-reservoir sites), 7) increased frequency of high pulses at two middle, inter-reservoir locations and three of four lower-basin channelized stations, 8) decreased frequency of high pulses at one middle, inter-reservoir and two channelized stations, 9) changes in mean duration of high-flow pulses, 10) decrease in the number of low-flow pulses, and 11) reduction in mean rise and fall rates. Some changes, such as stabilization of mean monthly discharges and variation in the number of high pulses, were absent or mild at least-altered locations and more pronounced at middle and channelized locations (Galat and Lipkin 2000).Pegg et al. (2003) used an alternative, time-series approach for assessing the impacts of hydrologic alteration on the Missouri River, and their findings generally corroborated those of Galat and Lipkin (2000). Modification of the landscape, or watershed, through land-use activities has also influenced hydrologic processes and has complicated our understanding of how hydrologic alteration affects the ecological integrity of freshwater ecosystems. The ability to return to some semblance of a natural flow regime will require knowledge of how land-use (e.g., residential, industrial, agriculture, etc.) also impacts hydrologic conditions in a regulated river system (Poff et al. 1997, Poff et al. 2006). Streams and rivers are four-dimensional systems (i.e., including their temporal dimension) that are intimately linked with the groundwater and landscape, and their lateral interactions with the landscape are vital to maintaining the integrity and function of these ecosystems (Fausch et al. 2002). Stream and riverine ecosystems exchange sediments, nutrients, and energy with the landscape, and, therefore, are part of a larger "riverscape" (Fausch et al. 2002), open to and affected by external processes.

Poff et al. (2006) sought to understand how geomorphological alterations due to land-use changes interact with hydrologic alterations from dam construction to influence the hydrogeomorphic integrity of streams in the USA. The interaction between the natural flow regime and geomorphology of a stream is important in defining the habitat that is available to organisms in a system (Poff et al. 2006). Therefore, alterations of the flow and/or geomorphic properties of a system will likely induce changes in resident fauna. Poff et al. (2006) also emphasized the importance of understanding how the underlying natural variation in physiography across major regions can influence the impact of land-use changes on hydrogeomorphic processes in a stream. The authors explained that the natural topographical, geological, and climatic features of a region will influence how deforestation and agricultural activities, for example, might affect the rates of sediment and nutrient input and water flow in local streams (Poff et al. 2006). In their study, four U.S. regions, distinct in their natural vegetation, climate, geology, and physiography, were identified and examined: 1) Pacific Northwest, with Pacific Lowland Mixed Forest and Cascade Mixed Forest-Coniferous Forest-Alpine Meadow provinces, 2) Southwest, with the Colorado Semi-Arid, American Semi-desert and Desert provinces, 3) Central, consisting of the Prairie Parkland Temperate Forest, and 4) Southwest Region, consisting of South Eastern Mixed Forest (Poff et al. 2006). Within each region, areas were classified as *Least Disturbed*, *Agriculture*, or *Urban*, and the percent of each class was calculated (Poff et al. 2006). Poff et al. (2006) discovered numerous regional differences in hydrologic responses to flow alteration along gradients of increasing urban and agricultural land cover. For example, with increased urban cover, peak flows increased in the Southeast and Northwest regions, minimum flows increased in the Central Region and decreased in the Northwest, duration of near-bankfull flows declined in the Southeast and the Northwest, and flow variability increased in three regions (Southeast, Central, and Northwest; Poff et al. 2006). Poff et al.'s (2006) study highlighted the importance of accounting for regionally specific, landscape-level effects in the assessment of local hydrologic conditions in stream ecosystems.

CASE STUDY

In the Alabama River system (USA), four hydropower dams were constructed on the main stem of the Tallapoosa River (Boschung and Mayden 2004). In the Northern Piedmont, flows have rapidly fluctuated between extremely low and high flows as a result of hydropeaking operations downstream of Harris Dam on the Tallapoosa River (Irwin and Freeman 2002). Hydrologic data were retrieved from United States Geological Survey (USGS) stream gauge stations at three locations downstream of Harris Dam: 1) Wadley (USGS 02414500), a regulated site downstream of Harris Dam, 2) Horseshoe Bend (USGS 02414715), a regulated site downstream of the Wadley location, and 3) Hillabee Creek (USGS 02415000), an unregulated tributary of the Tallapoosa River (Figure 1, website: waterdata.usgs.gov/al/nwis).

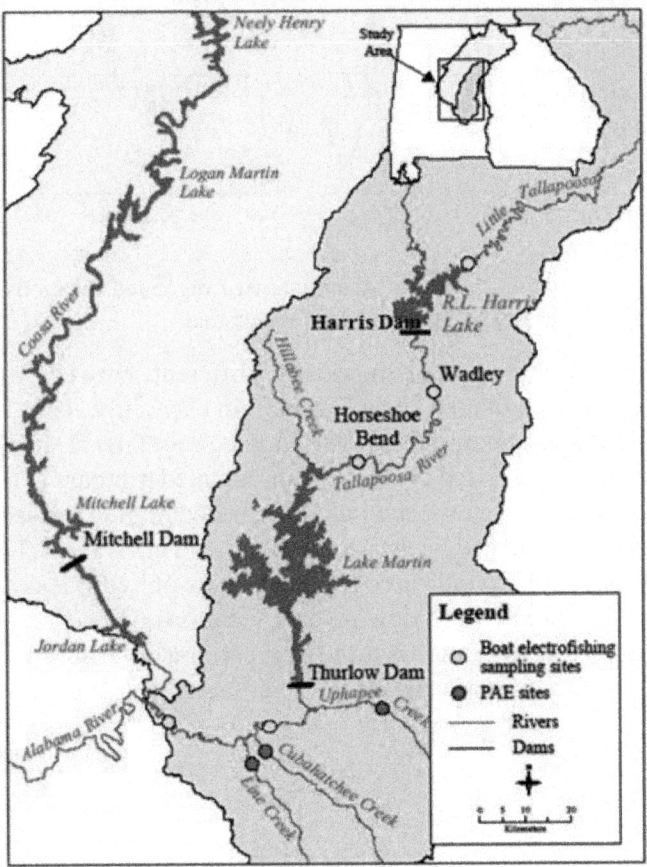

Figure 1: Hydrologic data were retrieved from USGS stream gauge stations at two regulated sites downstream of Harris Dam, Wadley and Horseshoe Bend, and at one unregulated site, Hillabee Creek (fromSakaris 2006).

Hourly variation in stream discharge (m³/s) was compared among the three sites during the first week of July 2012. Ecologically relevant hydrologic variables were also calculated in the Indicators of Hydrologic Alteration Program (IHA, Sustainable Waters Program, The Nature Conservancy, Boulder, CO) for each location from water years 1987 to 2012. Water years were started on October 1 of each year and ended on September 30 of the following year (e.g., water year 1987 = 10/01/86 – 9/30/87). Analyses focused on annual high and low pulse frequencies, number of reversals, and rise rates. Annual hydrologic conditions were compared between the two regulated sites and the unregulated site, which was treated as a "reference site."

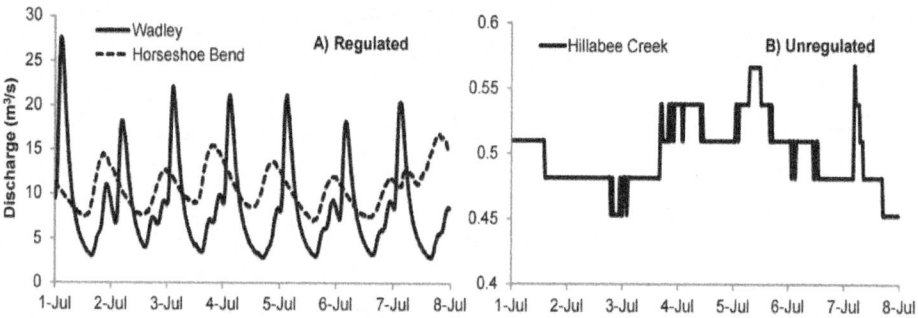

Figure 2: Daily variation in river discharge at two regulated locations (A) and at an unregulated site (B) in the Tallapoosa River Watershed.

Hydrologic regimes were markedly different between the regulated locations and the unregulated site (Figure 2). In early July, daily hydropeaking operations produced unnatural flow variation below Harris Dam, while a more natural flow regime persisted in the local unregulated tributary (Figure 2). Daily variation in discharge was substantially dampened at the Horseshoe Bend site (Figure 2), which is located farther downstream of Harris Dam (Figure 1). This reduced variation in flow indicated that the effects of hydropeaking operations may not be as severe at more downstream locations. However, unnatural and rapidly fluctuating flows, such as those observed below Harris Dam, generally produce a stressful environment for the river fauna that reside there.

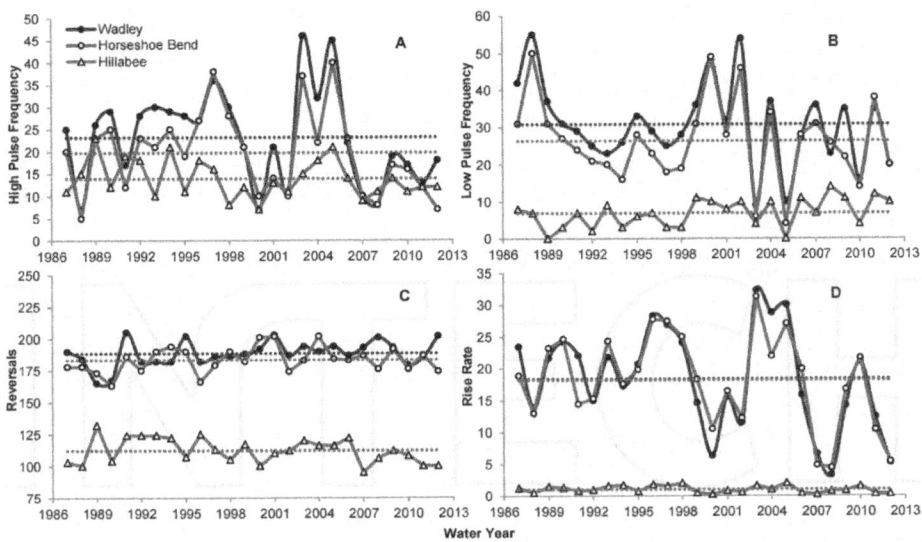

Figure 3: Annual variation in high pulse frequencies (A), low pulse frequencies (B), number of reversals (C), and rise rates (D) at two regulated locations and at an unregulated site in the Tallapoosa River Watershed. Dotted lines represent the statistical mean for each location. Mean rise rate was nearly equal at the two regulated sites.

As expected, high pulse and low pulse frequencies, the number of reversals, and rise rates were similar between the two regulated sites, as well as the overall annual variation in these hydrologic parameters (Figure 3). All four hydrologic parameters were substantially lower at the unregulated site (Figure 3). Fewer high pulses, low pulses, and reversals at Hillabee Creek indicated that the flow regime was much less variable and may be more representative of natural flow conditions in this region. Higher rise rates at the regulated sites are likely due, in part, to the rapidly increasing flows during hydropeaking events.]

HYDROLOGIC EFFECTS ON RECRUITMENT, GROWTH, SURVIVAL

Altered flow regimes below dams have typically produced unfavorable conditions for the recruitment of fishes (Fraley et al. 1986; Brouder 2001, Freeman et al. 2001, Wildhaber et al. 2000, Propst and Gido 2004). Freeman et al. (2001) reported that juvenile fish abundances were strongly related to the persistence of shallow habitats in a regulated reach of the Tallapoosa River, Alabama. However, the persistence of these habitats was severely reduced by rapid flow fluctuations resulting from hydropeaking operations (Freeman et al. 2001). In a regulated section of the Neosho River (Kansas), the reduction of minimum flows below John Redmond Dam reduced the availability of riffle habitats that were suitable for Neosho madtoms (*Noturus placidus*, Wildhaber et al. 2000). Moreover, hypolimnial-release of coldwater from dams will generally slow growth and development and alter physiology of fish during early life stages; whereas, the release of warm water from small, surface release dams may result in reduced densities of coldwater fish species (Clarkson and Childs 2000,Lessard and Hayes 2003).

Recruitment of fishes has been related to hydrology in freshwater ecosystems; however, most studies have been conducted in reservoirs (see, for example: Maceina and Stimpert 1998, Buynak et al. 1999,Sammons and Bettoli 2000, Schultz et al. 2002). Few studies have directly examined relations between hydrology and recruitment of fishes in regulated river sections. In a regulated section of the Roanoke River (North Carolina), Rulifson and Manooch (1990) reported that striped bass *Morone saxatilis*recruitment was highest when river flows were low to moderate (142 – 283 m³/s) during the spawning season. During the years when recruitment was highest, *flows typically resembled pre-impoundment flow conditions* (Rulifson and Manooch 1990). Striped bass require a specific flow regime for successful transport of eggs and larvae to nursery habitats (Rulifson and Manooch 1990). Irwin et al. (1999) reported that riffle habitats (i.e., shallow-fast and shallow coarse) were utilized by juvenile channel catfish *Ictalurus punctatus* and flathead catfish *Pylodictis olivaris*. However, persistence of these habitats may decrease in highly regulated systems (Bowen at al. 1998), thereby negatively influencing the recruitment of catfishes. Furthermore, Holland-Bartels and Duval (1988) suggested that variation in channel catfish productivity was related to river discharges. A decrease in age-0 channel catfish abundance was attributed to a sharp increase in river discharge that likely disrupted spawning activity and flushed young from nests (Holland-Bartels and Duval 1988). Therefore, one would suspect that highly variable flows (i.e., high rise and fall rates) during hydropeaking operations would negatively affect the spawning success and recruitment of channel catfish. In middle reaches of the regulated Missouri River, Pegg et al. (2003) identified a significant reduction in spring spawning flows as a major impairment of fish spawning and recruitment.

Studies have indicated that reduced flooding, or a diminished flood pulse, has contributed to low fish recruitment in river systems. Bonvechio and Allen (2005) studied recruitment of sunfishes *Lepomis*spp. and black basses *Micropterus* spp. in relation to hydrology in four Florida rivers. The authors suggested that high flows in the fall would increase access to floodplain habitats, thereby increasing prey availability (i.e., invertebrates) for adult sunfishes before the spawning season (Bonvechio and Allen 2005). Sunfishes would likely consume more prey and allocate more energy towards reproduction (i.e., fecundity), producing a stronger year class. In the inter-reservoir and lower channelized sections of the Missouri River, changes in the magnitude, frequency, timing, and duration of the annual flood pulse (i.e., inundation of the floodplain) was indicated as the likely cause of reduced recruitment and production of floodplain fishes (Galat and Lipkin 2000). The elimination of a fall flood pulse in the lower Missouri River has limited fish and wildlife access to floodplain habitats (Galat and Lipkin 2000). Brouder (2001) found a strong positive relationship between maximum mean daily discharge and recruitment of the roundtail chub, *Gila robusta*, in the upper Verde River, Arizona. Brouder (2001) explained that a reduction or elimination of flooding through hydrologic alteration would be deleterious to the recruitment of native roundtail chub. Flooding helps to prepare spawning substrate by clearing interstitial spaces for eggs, potentially reduces population sizes of nonnative species (thereby reducing competition), and possibly dilutes contaminants that would negatively affect reproductive success (Brouder 2001).

Alteration of the thermal regime as a consequence of dam construction can also have negative effects on fish recruitment (Clarkson and Childs 2000, Horne et al. 2004). Hydroelectric dams on the Manistee River, a tributary of Lake Michigan, negatively impact steelhead (anadromous rainbow trout,*Oncoryhnchus mykiss*) recruitment, by preventing steelhead access to potential upstream spawning habitats (Horne et al. 2004). Furthermore, the release of warm surface water from the reservoir results in increased summer temperatures that reduce the survival of age-0 steelhead in the river (Horne et al. 2004). Clarkson and Childs (2000) proposed that declines of native big-river fishes of the Colorado River Basin were partly due to the release of cold, hypolimnial water from dams. In laboratory experiments, growth rates of four species (razorback sucker, flannelmouth sucker, humpback chub, and Colorado squawfish) were slower and their development was delayed at colder temperatures (Clarkson and Childs 2000). Larval fish also lost equilibrium when transferred from 20°C to 10°C (Clarkson and Childs 2000). Slow growth, delayed development, and loss of equilibrium at early life history stages all likely contribute to reduced recruitment in a system.

Growth of fishes may also be related to hydrology in river systems. Quist and Guy (1998) suggested that increased growth of channel catfish in the Kansas River (USA) resulted from floodplain inundation. Inundation of the floodplain typically provides shallower, prey-rich habitats for fishes (Welcomme 1979). Mayo and Schramm (1999) hypothesized that growth of flathead catfish was also influenced by water temperature during the growing season, in addition to the number of flood days in the lower Mississippi River system. Unfortunately, hydrologic alterations may include changes to the timing and duration of floodplain inundation as well as thermal regimes (Cushman 1985, Pringle 2000). Rutherford et al. (1995) determined that growth of age-0 channel catfish in the Mississippi River was also related to the length of the growing season, which could theoretically be shortened with the release of cold, hypolimnetic water from a dam. Coldwater from hypolimnial-release dams may dramatically lower spring and summer tailwater temperatures, which may slow the growth and development of fishes (Clarkson and Childs 2000).

Hydrologic alteration has also strongly impacted the growth and recruitment of riparian and wetland vegetation (Young et al. 1995, Burke et al. 2008). For example, the recruitment of cottonwoods (*Populus* spp.) is closely linked to hydrologic and geomorphic processes (Scott et al. 1997, Burke et al. 2008). Burke et al. (2008) applied a hierarchical approach to studying the impacts of dam operations on the Kootenai River Ecosystem, focusing on cottonwood recruitment as their ecological response to river regulation. This hierarchical approach assessed *first order* impacts (changes in hydrologic conditions) that led to *second order* impacts, such as altered sediment transport and channel morphology (Burke et al. 2008). Burke et al. (2008) then described *third-level* impacts as biological functions that are influenced by first and second-level impacts. *Fourth-level* impacts are those involving feedbacks between biological and physical conditions (Burke et al. 2008). Overall, this hierarchical approach studied the effects of hydrologic alteration at the ecosystem level, assessing the physical, chemical, and biological changes that occur in a system.

The Kootenai River system, located in parts of Idaho, Montana, and British Columbia, has been modified in various ways. Levees were constructed in lower floodplain sections of the river to convert floodplain for agriculture, and Corral Linn Dam was constructed in the 1930s for hydropower and flood control (Burke et al. 2008). In 1974, Libby Dam was constructed upstream of Corral Linn Dam to provide additional hydropower and flood control, impounding 145 km of the river upstream of Libby Dam (Burke et al. 2008). As a result, natural flow conditions were altered and inundation of the floodplain was limited in the system (Burke et al. 2008). Specifically, higher flows are now maintained during naturally low-flow periods (fall-winter), while much lower flows are maintained during naturally high-flow periods (spring-summer; Burke et al. 2008). Burke et al. (2008) analyzed three time periods in their study: 1) *historic*, before Corral Linn Dam was operational, 2) *pre-Libby Dam*, and 3) *post-Libby Dam*. First, second, and third-order impacts as a result of dam construction and other activities were examined in a 233-km study reach between Libby Dam and the downstream Corral Linn Dam. To examine first-order impacts, Burke et al. (2008) utilized *Indicators of Hydrologic Alteration* (IHA) software (Richter et al. 1996) to compare hydrologic regimes across the time periods. The authors used a one-dimensional hydrodynamic flow model to evaluate second-order changes in hydraulics and bed mobility. Third-order impacts were assessed by changes in the recruitment potential of black cottonwoods. Due to the significant reduction of naturally high flows during the spring snowmelt, cottonwood seed germination sites are no longer prepared through the mobilization and redistribution of sediments. In addition, germination sites no longer experience the slow and gradual recession of these natural spring flows that would help maintain adequate soil moisture for the establishment of seedling roots (Burke et al. 2008). Burke et al.'s (2008) analyses revealed major changes in the hydrologic regime below Libby Dam, which included significantly greater median monthly flows during winter and the near elimination of a spring snowmelt peak (i.e., an "inverted annual hydrograph"). Notable second order, temporal and spatial alterations were also detected in stage fluctuation, stream power, shear stress, and bed mobility. The authors determined that the activities of both dams have contributed to lower cottonwood recruitment, by increasing stage recession rates during the seedling establishment period in the lower study reach and changing the timing, magnitude, and duration of flow in the upper and middle sections of the study reach (Burke et al. 2008).

Young et al. (1995) conducted a study examining how the growth of the Baldcypress, a tree common to wetlands in the southeastern United States, may respond to an altered flow regime. This tree is often found in wetlands or swamps that are subject to frequent or permanent flooding (Young et al. 1995). Before their study site was impounded by road construction in 1973, it existed as a floodplain swamp with permanent shallow flooding. The impoundment resulted in increased water levels on the upstream side of the road and apparently no effects on water levels on the downstream side (Young et al. 1995). Young et al. (1995) discovered that trees at the impacted, upstream site initially exhibited a significant growth surge due to increased flooding, but an overall decline in growth followed for 16 years. The initial surge in growth was possibly due to a pulse of sediment nutrient deposition with flooding, while the later decline in growth may have been a result of increased anoxic conditions in the rooting zone (Young et al. 1995).

The recruitment and survival of native riparian vegetation may not only depend on the restoration of more natural flow conditions, but also the removal of invasive riparian species that have better success in altered flow conditions

(Merritt and Poff 2010). In western North America, the cottonwood,*Populus deltoides*, has declined substantially, while the invasive saltcedar, *Tamarix*, has become well established (Merritt and Poff 2010). Merritt and Poff (2010) determined that recruitment potential of*Tamarix* was highest along unregulated reaches in the Southwestern United States, *but remained high across a gradient of regulated flows*. In contrast, recruitment of cottonwoods was highest under natural flow conditions and declined abruptly with even slight flow modification (Merrit and Poff 2010). In addition, *Tamarix* was most dominant along the most altered river reaches, whereas *Tamarix* and*Populus* were equally dominant at the least regulated sites. The authors concluded that altered flow regimes have further enhanced the dominance of *Tamarix* over native plant species (Merritt and Poff 2010).

River regulation can also affect the reproductive success of marine fishes and invertebrates (Drinkwater and Frank 1994). For example, the recruitment of marine fish and invertebrates appears to be highly correlated with freshwater input (Drinkwater and Frank 1994). In most cases, increased river runoff into coastal oceans positively influences fish and invertebrate production, by increasing nutrient inputs that enhance primary production (Drinkwater and Frank 1994). The impoundment of rivers by dams, the diversion of water for agricultural purposes (irrigation), and regulated release of water from dams can modify the amount and/or timing of freshwater released to coastal estuaries. See Drinkwater and Frank (1994) for a thorough review of the effects of river regulation on marine fish and invertebrates.

HYDROLOGIC EFFECTS ON COMMUNITY STRUCTURE AND BIODIVERSITY

The effects of hydrologic alteration on community structure of aquatic organisms have been well documented (Dudgeon 2000, Pringle et al. 2000, Marchetti and Moyle 2001, Humphries et al. 2008).Marchetti and Moyle (2001) reported that most native fish species of the regulated Putah Creek, a tributary of the Sacramento River, were more often found in habitats that were characteristic of the natural flow regime (i.e., increased canopy, higher streamflow, decreased conductivity, cooler temperatures, and fewer pools). In contrast, most nonnative species appeared to be adapted to conditions opposite to those of the natural flow regime (Marchetti and Moyle 2001). Therefore, restoration of a native-dominated fish assemblage would require a return to natural flow conditions. In the regulated Campaspe River (Australia), only four of ten native species were consistently documented from 1995 to 2003, while historically an estimated 18 to 20 native fishes once inhabited the river (Humphries et al. 2008). The authors also documented the presence of six exotic fishes, with common carp and European perch being the most abundant fishes (i.e., 36% of the overall fish abundance, Humphries et al. 2008).

Dam construction and river regulation has threatened aquatic and terrestrial biodiversity worldwide (Dudgeon 2000, Pringle et al. 2000). Pringle et al. (2000) provided a thorough review of the effects of hydrologic alterations on riverine biota in temperate and neotropical regions. The authors explain that, although construction of new dams in the United States has declined, large dam construction has occurred more recently in tropical regions of South America (Pringle et al. 2000). In temperate regions, migratory diadromous fishes, such as salmon, sturgeon, American shad, and American eel, have been extirpated from much of their native ranges due to dams that block their spawning migrations (Drinkwater and Frank 1994, Pringle 2000). Movements of potamodromous fishes have also been impeded, restricting their reproductive success and overall distributions (Pringle et al. 2000). Habitat fragmentation and the conversion of lotic, free-flowing habitat to more lentic conditions, in the form of impoundments, reservoirs, etc., have resulted in the imperilment of small-bodied fishes, particularly fluvial-dependent species (Pringle et al. 2000). This increase in availability of lentic habitat has also allowed for the expansion of lentic fishes and the introduction of lentic fishes into systems beyond their native range (Pringle et al. 2000). As mentioned in the previous section, Pringle (2000) also identified a reduction or alteration in the timing and/or duration of floodplain inundation as a major factor contributing to the decline of flood-dependent taxa. Reduced freshwater flows into estuarine habitats have also threatened species, such as the delta smelt in San Francisco Bay, California (Pringle et al. 2000). In Pringle et al.'s (2000) case study of the Mobile River Basin in the southeast USA, the authors report from the literature that at least 16 endemic mussels and 38 gastropods are thought to be extinct.

Although the negative impacts of hydrologic alterations on biota are well documented in North America, less is known about the effects of river regulation on South American tropical systems (Pringle et al. 2000). Pringle et al. (2000) explained that biota of Neotropical rivers are highly vulnerable to hydrologic alterations for several reasons. The habitat heterogeneity of tropical ecosystems has produced highly diverse communities with high rates of endemism. Many South American fishes are highly migratory and have complex life cycles and depend on seasonal floodplain inundation that provides food, refuge, and nursery habitat for young fishes. The accumulation of organic material in reservoirs of low-gradient Amazonian streams has led to undesirable water quality conditions (e.g., hypoxia) that can result in fish kills. Reduced freshwater input to estuaries has also led to an increased presence of marine fishes in lower sections of rivers, replacing native freshwater species (Pringle et al. 2000). Agostinho et al. (2008) provided a more recent, extensive review of the effects of dams on fish fauna in the Neotropical Region. The authors focused on the highly regulated Paraná River, which flows through the most highly populated region in Brazil. Data are presented illustrating the negative impacts of dams on fish diversity and fisheries in the region. Agostinho et al. (2008) expressed the need for improved management approaches in Brazil, such as taking a more ecosystem-level rather than reductionist approach. The authors also mention that little information exists regarding the effects of hydrologic alteration (dams), fishery exploitation, and other impacts on aquatic resources in Brazil. These data needs must be addressed so that managers can formulate and inform effective management decisions (Agostinho et al. 2008).

Asia possesses a proportionately high number of dams, with the number of large dams increasing from 1,541 in 1950 to a staggering 22,701 in 1982 (Dudgeon 2000). China has constructed the greatest number of dams in tropical Asia (Dudgeon 2000). The climate of this region alternates between a wet and dry season, with many organisms depending on the wet season and the associated flood pulse for sustenance and access to floodplain habitats (Dudgeon 2000). Dam construction, however, has focused on flood control during wet periods and storing water during dry periods, resulting in significantly altered hydrologic regimes in most major rivers (Dudgeon 2000). Dudgeon (2000) also mentions that other factors, such as pollution, deforestation, overharvesting and rapidly growing human populations in the landscape, have further exacerbated conditions in these systems. Hydrologic alteration has negatively impacted a wide diversity of taxa in this region, including crocodiles, terrestrial mammals, fishes, and river dolphins (Dudgeon 2000). The Mekong River Basin supports a high diversity of over 500 fishes. Unfortunately, the construction of large dams on the Mekong River has threatened the persistence of many species (Dudgeon 2000). See Dudgeon (2000) for a thorough overview of the ecological consequences of large dam construction on the Mekong River.

Negative impacts of hydrologic alteration are not only limited to aquatic organisms. Riverine forest along the regulated Tana River in Kenya serves as habitat for two endemic primates, the rare Tana River Red Colobus and the critically endangered Tana River Mangabey (Maingi and Marsh 2002). Regulation of the Tana River threatens the persistence of riverine forest along mid and lower sections of Tana River and, therefore, further endangers these rare primates (Maingi and Marsh 2002). Hill et al. (1998) examined the effects of dams on the shoreline vegetation of lakes and reservoirs in southern Nova Scotia, Canada. Hill et al.'s (1998) study included 37 unregulated and 13 regulated lakes, for which plant species inventories were conducted. Plant communities of regulated lakes were less diverse, contained more exotic species, and typically lacked rare shoreline herbs. The authors attributed this reduction in diversity and introduction of nonnative species to the altered hydrologic regimes of reservoirs that produce extreme fluctuations in water levels.

A significant reduction in *hydrologic connectivity*, as a result of dam construction and other anthropogenic activities in the landscape, has also threatened aquatic biodiversity in riverine systems (Pringle 2003). Hydrologic connectivity refers to "the water-mediated transfer of matter, energy, and/or organisms within or between elements of the hydrologic cycle (Pringle 2001, Pringle 2003)." Pringle's (2001, 2003) definition of hydrologic connectivity emphasizes its importance at a regional or global scale, whereas *river connectivity* refers to the continuity or linkages of a river ecosystem as it operates across its four dimensions (i.e., temporal, and longitudinal, lateral, and vertical spatial dimensions, Freeman et al. 2007). Dam construction (i.e., reduced hydrologic connectivity) has impeded the spawning migrations of anadromous fishes, preventing these fishes from returning to their natal sites. Substantial reductions in the distribution and abundance of freshwater mussel species have been attributed to reduced habitat connectivity. Fragmentation of habitats isolates local populations from others, limiting or eliminating the exchange of individuals and the potential for recolonization of habitat patches when a local extinction occurs. Reduced hydrologic connectivity also has negative impacts on broader-scale functions, such as biogeochemical cycling in ecosystems (Pringle 2003). For example, dams act as barriers to the transport of silica to coastal oceans, limiting primary production (i.e., diatom production) and the integrity of coastal food webs (Pringle 2003, Freeman et al. 2007).

FLOW MANAGEMENT AND MODELING

In 1997, Poff et al. succinctly explained that "current management approaches often fail to recognize the fundamental scientific principle that the integrity of flowing water systems depends largely on their*natural dynamic character.*" Although in today's society returning riverine systems to their "natural dynamic character" is nearly impossible, the authors indicated that conservation and management strategies should attempt to restore the ecological integrity of these regulated systems by enhancing their "natural" flow variability (Poff et al 1997). Although regulated rivers can never be fully restored to natural conditions, flows below dams should be managed to best represent natural flow conditions (Poff et al. 1997). Previous management strategies focused on improving water quality and simply implementing minimum flow requirements (Poff et al. 1997, Richter et al. 1997, Arthington et al. 2006). Richter et al. (1996) also mentioned that past management strategies focused on the flow requirements of only a few selected aquatic species and neglected the flows needed to maintain aquatic-riparian systems and broader ecosystem functions. Management of freshwater resources was also conducted in a compartmentalized or "fragmented" fashion (Poff et al. 1997, Karr 1991). Management approaches today have evolved to incorporate the prescription and implementation of natural aspects of the flow regime. In addition, a more concerted effort is applied to coordinate management activities among various resource agencies. Furthermore, current strategies attempt to apply a more holistic, ecosystem-level (rather than reductionist) approach to the management of regulated rivers and conservation of freshwater resources.

Various techniques and modeling approaches have been developed to enhance our understanding of how hydrologic alteration affects aquatic ecosystems, as well as improve our management of regulated rivers (Richter et al. 1996, Richter et al. 1997, Irwin and Freeman 2002, Olden and Poff 2003, Arthington et al. 2006 , Poff et al. 2009, Merrit and Poff 2010, Sakaris and Irwin 2010). Richter et al. (1996) emphasized the importance of selecting hydrologic parameters that are most "biologically relevant" when assessing hydrologic alteration in a regulated system. In other words, we should focus on the parameters that most influence the ecological integrity of a system. Richter et al. (1996) presented a well-structured approach for hydrologic assessment, *Indicators of Hydrologic Alteration*(IHA), which accounts for the most biologically relevant parameters. This approach defines and calculates a series of hydrologic attributes and then compares the hydrologic regime of a system before and after impact (e.g., impoundment). A total of 32 biologically relevant hydrologic parameters are calculated for each year from these five IHA statistics groups: 1) magnitude of monthly water conditions, 2) magnitude and duration of annual extreme water conditions, 3) timing of annual extreme water conditions, 4) frequency and duration of high and low pulses, and 5) rate and frequency of water condition changes (Richter et al. 1996). These parameters account for the five important and ecologically relevant characteristics of a flow regime: magnitude, frequency, duration, timing, and the rate of change (Poff et al. 1997). The four steps of Richter et al.'s (1996) approach are: 1) define the data series for pre- and post-impact periods (usually collected from USGS flow gauges), 2) calculate values of hydrologic attributes for each year in each data series (i.e., pre-impact and post-impact data series), 3) compute inter-annual statistics for the 32 parameters in each data series, specifically 32 measures of central tendency and 32 measures of dispersion, and 4) calculate values of the IHA. The fourth step involves comparing the 64 inter-annual statistics between pre- and post-impact periods, as a percent deviation of one time period to the other (Richter et al. 1996). The IHA approach can also be used to compare hydrologic regimes between regulated and "reference" sites.

Richter et al. (1997) further improved the approach to river management with the development of the "Range of Variability Approach (RVA)." Richter et al. (1997) mention that previous approaches did not provide specific flow targets to be met, focused on a limited number of features of the hydrologic regime, and/or focused on only a few target species and a limited number of their habitat requirements. In addition, research studies examining relationships between hydrologic conditions and ecological responses in a system are typically time-consuming (often taking several years) and are usually not completed within the timeframe during which flow management decisions are typically made (Richter et al. 1997). Richter et al.'s (1997) RVA assists river managers in the identification of flow-based management targets that should enhance the overall ecological integrity of a system. For systems with highly altered hydrologic regimes, the main idea is *to restore hydrologic conditions within the historical or "natural" range of variation*, particularly for streamflow characteristics that are well outside the historical range (Richter et al. 1997). Richter et al. (1997) recommended that the RVA be applied in the preliminary stages of *adaptive flow management programs* (see below), providing initial flow management targets that can be modified as more ecological information is gathered for a specific ecosystem. The RVA approach has six steps, which are briefly described here. For a more in-depth overview, see Richter et al. (1997). The six steps are as follows: 1) Characterize the natural range of streamflow variation using the IHA approach (Richter et al. 1996) described above. 2) Select management targets, one for each of the 32 hydrologic parameters, with the idea that each management target should fall within the natural range of variation. Each target may have upper and lower bounds (e.g., ± 1 standard deviation). 3) The river management team formulates a management "system" or plan, using the RVA targets as design guidelines. 4) Scientists conduct routine ecological monitoring and/or river research program to evaluate ecological effects of the management system as it is implemented. 5) Characterize actual streamflow variation using the IHA method at the end of each year and compare the values of hydrologic parameters with the RVA target values. 6) Revise either the management system or RVA targets based on new information that is collected (Richter et al. 1997).

An adaptive approach, termed *adaptive-flow management*, has been recommended for the management of regulated river systems (Irwin and Freeman 2002). In adaptive-flow management, managers attempt to restore rivers to near-natural flow regimes while accounting for societal needs (Irwin and Freeman 2002). The main goal of adaptive-flow management is to continually improve management as uncertainty about a river system is reduced. This management approach requires the cooperation and long-term commitment of natural resource personnel, private industry, landowners, and other stakeholders. Adaptive-flow management can be best described as an iterative process with a series of steps that include 1) prescription of a flow/management regime that satisfies all stakeholders, 2) monitoring and evaluation of the flow regime's effect on habitat and biota, and 3) the recommendation of a new and improved management regime. By quantifying relationships between features of the flow regime and responses in the biota and overall ecosystem, models can be developed to predict how populations, communities or the ecosystem may respond to the prescription of flow regimes, or an "environmental flow standard (Arthington et al. 2006)." These models can be continually improved as more is learned about the ecological responses to hydrologic alteration in the managed river system.

Olden and Poff (2003) addressed a major issue confronting managers in determining which of the many published approaches and hydrologic (and "ecologically relevant") parameters should be used in river management. The authors recognized that many of the hydrologic variables proposed for use in the characterization of a flow regime (e.g., 32 hydrologic parameters, Richter et al. 1996) were inter-correlated, and little guidance was provided for the selection of appropriate parameters. Olden and Poff's (2003) main goal was to provide a standardized framework for the selection of a reduced set of hydrologic indices and to minimize redundancy among the selected parameters. This reduced set of indices would still account for the majority of the statistical variation in the complete set of hydrologic indices, minimize multicollinearity among the selected hydrologic variables, and adequately represent the critical attributes of a system's flow regime. The authors also examined the effectiveness of IHA and the overall transferability of indices to facilitate comparisons across systems that differ in their streamflow characteristics (Olden and Poff 2003). See Olden and Poff (2003) for a detailed overview of the approach and a review of the 171 hydrologic indices published in the literature.

In 2006, Arthington et al. proposed a mechanism for developing regional environmental flow "standards." Their rationale was that hydrologic and ecological data are often lacking for specific streams or rivers in a region, which makes it quite difficult to prescribe system-specific flow regimes in the management of regulated rivers. Arthington et al. (2006) recommended classifying rivers and streams that share important flow attributes into "ecologically meaningful groups." Within a region, the logical assumption is that rivers that are similar in their flow variability and geomorphic properties would exhibit similar ecological responses to management regimes. Arthington et al. (2006) described their approach as grouping the systems into "practical management units." See Arthington et al. (2006) for a complete overview of their management approach, which shares common features with the ELOHA approach described below.

Poff et al. (2010) explained that a strong need exists to develop ecological goals and management standards for streams and rivers at a regional or even global scale. Water resource and environmental flow management has become highly complex, because management must account for diverse societal needs while attempting to restore the ecological integrity of degraded ecosystems. Meanwhile, rapidly growing human populations will further increase water consumption and energy demands and require increased food production. As a result, restoring systems with highly altered flow regimes to "natural" flow conditions will become even more difficult. The authors, consisting of a group of international scientists, presented a framework for evaluating environmental flow needs that could potentially form the basis for implementing flow standards at a regional scale (Poff et al. 2009). Poff et al. (2009) refer to this framework as the *Ecological limits of hydrologic alteration* (ELOHA), with the goal of presenting "*a logical approach that flexibly allows scientists, water resource managers and other stakeholders to analyze and synthesize available scientific information into coherent, ecologically based and socially acceptable goals and standards for management of environmental flows.*" Poff et al. (2010) recognize that water resource managers from different regions are often confronted with unique challenges, may operate in different social and political environments, and may be at different stages of water-resource development. The necessary scientific foundations of the ELOHA framework exist and consist of: 1) essentially years of research has been conducted examining the effects of altered hydrologic regimes on population dynamics, community structure, and ecosystem-level functions, 2) the previous application of various methods for managing environment flows, which the authors refer to as a "rich toolbox" from which methods or tools can be applied by water resource managers, 3) a conceptual foundation that facilitates regional flow assessments, 4) the development of hydrologic models, and 5) an understanding that river management is complex and adaptive and must meet both ecological and societal goals (Poff et al. 2010).

The ELOHA framework consists of four major steps that can be flexibly applied by managers from different regions (Poff et al. 2009). 1) *Building a "hydrologic foundation" for the region* involves collecting hydrologic time-series data and constructing hydrographs that represent "baseline" (minimally altered) and "developed" (altered) hydrologic conditions throughout the region, particularly for all locations that require environmental flow management and protection. 2)*Classifying rivers according to their hydrology and geomorphology* assumes that rivers with similar hydrologic regimes (e.g., snowmelt driven rivers) and geomorphic characteristics would likely respond similarly to hydrologic alteration and other disturbances, whereas rivers that are dissimilar in type (e.g., snowmelt vs. desert rivers) would likely respond differently when altered. When classifying rivers based on their hydrologic regimes, chosen hydrologic features should collectively characterize the flow regime of the system and avoid redundancy in the parameters used (Olden and Poff 2003). The selected hydrologic metrics should also be ecologically relevant and be applicable in management. River classification is important, because flow management decisions will likely vary based on river type. Furthermore, if the "hydrologic foundation" is not fully built for a region, the hydrologic models and management targets developed for one river may be extrapolated to similar systems until more system-specific data are collected. 3) *Computing flow alteration* involves estimation of the degree of hydrologic alteration for each system, for which hydrologic data are available. Any deviation in the hydrologic regime from "natural" (baseline) conditions may have an ecological impact, and this ecological impact generally becomes more severe as the disparity between developed and baseline conditions widens. Programs, such as IHA (Richter et al. 1996; Mathews and Richter 2007), can be used to calculate a set of hydrologic alteration values as a percent or absolute deviation from baseline condition for each developed site. 4) *Conducting research and monitoring programs to assess ecological responses to altered hydrologic regimes* addresses the critical need for improved understanding of biotic and ecosystem responses to flow alteration in the ELOHA framework. Flow alteration-ecological response relationships guide river managers in establishing flow management targets, or "standards," and in developing flow management plans that will most likely enhance the ecological integrity of an altered system. It is important to note that the ELOHA approach is an adaptive process. Scientists play an important role in this process, by conducting research programs that attempt to reduce uncertainty and build our understanding of ecological responses to hydrologic alteration. With new information, management flow standards can be updated and implemented over time. See Poff et al. (2010) for a detailed overview of the application of ELOHA, the various models and tools that can be used in each step of the process, and the potential challenges that may confront river managers and scientists that adopt this approach.

The ELOHA framework (Poff et al. 2010) requires the assessment of "ecologically significant" differences between baseline and developed hydrologic regimes in a region. Merritt and Poff (2010) recently developed an *index of flow modification* (IFM), which is a composite metric of the most biologically relevant hydrologic variables that essentially measures how modified an altered flow regime is compared to unregulated (or baseline) conditions. Pre-dam and post-dam flow data are collected for each location (or study reach), typically from USGS (United States Geological Survey) gauges. Biologically relevant hydrologic variables are then obtained for pre-dam and post-dam periods using IHA (Indicators of Hydrologic Alteration) software (Richter et al. 1996), and then the absolute or percent change in each variable is calculated from pre-dam to post-dam periods. In their study,Merritt and Poff (2010) calculated the percentage change in spring flow (mean of April through June), summer flow (mean of July through September), low flow (mean of October through February flows), and 2-, 10-, and 25-year recurrence interval peak flows. The change in the number of days of minimum flow and the change in maximum flow were also calculated (Merrit and Poff 2010). Principle Components Analysis (PCA) is then conducted using these calculated metrics (i.e., hydrologic metrics) of all study sites, and only significant principle component axes are used in the calculation of IFM. Merrit and Poff (2010) developed the IFM "*by calculating the Euclidean distance of each observation (study reach) from the centroid of the significant PCA axis scores of relatively unregulated rivers for the hydrologic metrics.*" The IFM can then be used, for example, to examine relationships between the recruitment, abundance, and/ or growth of organisms and the degree of flow modification (IFM) across sites ranging from relatively unregulated to regulated conditions.

Population matrix models have also been developed by scientists to evaluate and predict how riparian and aquatic populations respond to hydrologic variation in systems with altered flow regimes (Lytle and Merritt 2004, Sakaris and Irwin 2010). These models may be useful in step 4 of the ELOHA approach. Lytle and Merritt (2004) developed a stochastic, density-dependent model to predict how annual hydrologic variation affects the mortality, recruitment, and population dynamics of the riparian cottonwood and to project how altered flow regimes might affect cottonwood populations. Lytle and Merritt (2004) simulated the effects of channelization and damming in the Yampa River, Colorado, and their model suggested that the observed natural flow regime would likely produce the most abundant mature cottonwood forest. Sakaris and Irwin (2010) developed matrix models for predicting the effects of altered flow regimes on the dynamics of a flathead catfish population in a regulated section of the lower Coosa River, Alabama. Matrix construction required the collection of fertility, survival, and body growth data (for size-classified matrices) for the fish population. The authors conducted multiple regression analyses to assess the influence of hydrologic features of the altered flow regime on annual recruitment of the flathead catfish in the system. Using this information, the effects of environmental stochasticity (hydrologic variation) on the long-term growth dynamics of the catfish populations was projected. Sakaris and Irwin (2010) also used their model to predict the effects of prescribed flow regimes on fish population dynamics in the river. Sakaris and Irwin (2010) presented their model as a potential tool that could be used in the adaptive flow management of regulated rivers (e.g., below Harris Dam on the Tallapoosa River, Alabama, Irwin and Freeman 2002).

Arthington et al. (2006) mentioned that general agreement exists among scientists and most river managers that to maintain the ecological integrity and biodiversity of a system, we must attempt to restore, or "mimic," natural flow conditions. That is, all general features of the natural flow regime (magnitude, frequency, duration, timing, and the rate of change, Poff et al. 1997), to some degree, should be accounted for when prescribing a flow-management regime in a regulated system. As mentioned earlier, previous management strategies typically focused on implementing a single environmental flow standard, such as maintaining a minimum flow requirement below a dam. For example, on the Tallapoosa River in the East Gulf Coastal Plain in Alabama (USA), a minimum continuous flow (34 m³/s) was established below Thurlow Dam as part of a re-licensing agreement in 1991. Although diversity of fishes increased approximately 3 km downstream of the dam (Travnichek et al. 1995), Thurlow Dam has still exhibited high annual variability in discharge that often exceeds dam capacity, which has typically resulted in prolonged periods of high flow (> 283 m³/s). Wildhaber et al. (2000) evaluated relations between Neosho madtom densities and flows in the Neosho River Basin below John Redmond Dam and suggested that higher minimum flows be prescribed in the river to improve densities of Neosho madtoms and other ictalurids. Studies have evolved since to prescribe or, at least, model flow regimes that are more natural in character. For example, Propst and Gido (2004) attempted to partially mimic the natural flow regime in a regulated reach of the San Juan River (Colorado), by increasing reservoir releases to mimic timing and only partially mimicking the amplitude, volume, and duration of spring snowmelt discharge. Densities of native fishes typically increased in years with high spring discharges (Propst and Gido 2004). Horne et al. (2004) modeled the effects of two management scenarios, bottom withdrawal and actual dam removal, on the recruitment of steelhead in the Manistee River. The authors' models predicted that bottom withdrawal (of hypolimnetic water) would slightly cool summer water temperatures and modestly enhance steelhead recruitment in the river (Horne et al. 2004). Horne et al. (2004) mention, however, that their model for dam removal did not account for the added benefit of increased steelhead access to upstream spawning habitat, which would likely improve recruitment in the system. In the lower Kootenai River,Burke et al. (2008) discovered that recruitment potential of the back cottonwood improved in 1997 and 1999, partly due to experimental flow releases from Libby Dam. These water releases helped to mimic pre-dam hydrologic conditions during the spring snowmelt, with a sustained peak in flow during the early growing season and a subsequent gradual recession of flow (Burke et al. 2008). These experimental flow releases have been implemented since 1993 to enhance the spawning success of white sturgeon (Burke et al. 2008).

River management and the conservation of natural resources in regulated rivers will become increasingly difficult, as the ecological needs of an ecosystem must be delicately balanced with societal needs. Furthermore, the management of altered flow regimes has become quite complex, as we must also account for and understand how the interaction of local climate, land use, and the unique geological and topographical features of a region influence hydrologic processes in a river system. Future management approaches will require the involvement and cooperation of governmental agencies, scientists, non-profit organizations, and the public to develop solutions that attempt to restore features of the natural flow regime, conserving and enhancing biodiversity, while providing for the needs of society.

REFERENCES

1.   A. A Agostinho, F. M Pelicice, and L. C Gomes, 2008Dams and the fish fauna of the Neotropical Region: impacts and management related to diversity and fisheries. *Brazilian Journal of Biology* 68 1119 1132

2.   American Rivers 1997North America's most endangered and threatened rivers of 1997. American Rivers, Washington D. C.

3.   A. H Arthington, S. E Bunn, N. L Poff, and R Naiman, 2006The challenge of providing environmental flow rules to sustain river esosystems. *Ecological Applications* 16 1311 1318

4.   R. C. P Beamesderfer, T. A Rien, and A. A Nigro, 1995Differences in the dynamics and potential production of impounded and unimpounded white sturgeon populations in the lower Columbia River. *Transactions of the American Fisheries Society* 124 857 872

5.   T. F Bonvechio, and M. S Allen, 2005Relations between hydrological variables and year-class strength of sportfish in eight Florida waterbodies. *Hydrobiologia* 532 193 207

6.   H. T Boschung, and R. L Mayden, 2004Fishes of Alabama. Smithsonian Books, Washington D. C.

7. Z. H Bowen, M. C Freeman, and K. D Bovee, 1998Evaluation of generalized habitat criteria for assessing impacts of altered flow regimes on warmwater fishes. *Transactions of the American Fisheries Society* 127 455 468

8. M. J Brouder, 2001Effects of flooding on recruitment of roundtail chub, *Gila robusta*, in a Southewestern River. *The Southwestern Naturalist* 46 302 310

9. M Burke, K Jorde, and J. M Buffington, 2008Application of a hierarchical framework for assessing environmental impacts of dam operation: changes in streamflow, bed mobility, and recruitment of riparian trees in a western North American river. *Journal of Environmental Management*90 224 236

10. G. L Buynak, B Mitchell, D Bunnell, B Mclemore, and P Rister, 1999Management of largemouth bass at Kentucky and Barkley Lakes, Kentucky.*North American Journal of Fisheries Management* 19 59 66

11. R. W Clarkson, and M. R Childs, 2000Temperature effects of hypolimnial-release dams on early life stages of Colorado River Basin big-river fishes. *Copeia* 2 402 412

12. R. M Cushman, 1985Review of ecological effects of rapidly varying flows downstream from hydroelectric facilities. *North American Journal of Fisheries Management* 5 330 339

13. K. F Drinkwater, and K. T Frank, 1994Effects of river regulation and diversion on marine fish and invertebrates. *Aquatic Conservation: Freshwater and Marine Ecosystems* 4 135 151

14. D Dudgeon, 2000Large-scale hydrological changes in tropical Asia: prospects for riverine diversity. *BioScience* 50 793 806

15. K. D Fausch, C. E Torgersen, C Baxter, and H Li, 2002Landscapes to riverscapes: bridging the gap between research and conservation of stream fishes. *Bioscience* 52 1 16

16. J. J Fraley, S. L Mcmullin, and P. J Graham, 1986Effects of hydroelectric operations in the Kokanee population in the Flathead River System, Montana. *North American Journal of Fisheries Management* 6 560 568

17. M. C Freeman, Z. H Bowen, K. D Bovee, and E. R Irwin, 2001Flow and habitat effects on juvenile fish abundance in natural and altered flow regimes.*Ecological Applications* 11 179 190

18. M. C Freeman, E. R Irwin, N. M Burkhead, B. J Freeman, and H. L Bart, Jr. 2004Status and conservation of the fish fauna of the Alabama River system. *American Fisheries Society Symposium* 45 557 585

19. M. C Freeman, C. M Pringle, and C. R Jackson, 2007Hydrologic connectivity and the contribution of stream headwaters to ecological integrity at regional scales. *Journal of the American Water Resources Association* 43 5 14

20. D. L Galat, and R Lipkin, 2000Restoring ecological integrity of great rivers: historical hydrographs aid in defining reference conditions for the Missouri River. *Hydrobiologia* 422/423 29 48

21. D. P Gillette, J. S Tiemann, D. R Edds, and M. L Wildhaber, 2005Spatiotemporal patterns of fish assemblage structure in a river impounded by low-head dams. *Copeia* 2005 3 539 549

22. N. M Hill, P. A Keddy, and I. C Wisheu, 1998A hydrological model for predicting the effects of dams on the shoreline vegetation of lakes and reservoirs. *Environmental Management* 22 723 736

23. L. E Holland-bartels, and M. C Duval, 1988Variations in abundance of young-of-the-year channel catfish in a navigation pool of the Upper Mississippi River. *Transactions of the American Fisheries Society* 117 202 208

24. B. D Horne, E. S Rutherford, and K. E Wehrly, 2004Simulating effects of hydro-dam alteration on thermal regime and wild steelhead recruitment in a stable-flow Lake Michigan tributary. *River Research and Applications* 20 185 203

25. P Humphries, P Brown, J Douglas, A Pickworth, R Strongman, K Hall, and L Serafini, 2008Flow-related patterns in abundance and composition of the fish fauna of a degraded Australian lowland river. *Freshwater Biology* 53 789 813

26. E. R Irwin, M. C Freeman, and K. M Costley, 1999Habitat use by juvenile channel catfish and flathead catfish in lotic systems in Alabama. 223 230 *in*E. R. Irwin, W. A. Hubert, C. F. Rabeni, H. L. Schramm, Jr., and T. Coon, editors. Catfish 2000: proceedings of the international ictalurid symposium. *American Fisheries Society, Symposium* 24, Bethesda, Maryland.

27. E. R Irwin, and M. C Freeman, 2002Proposal for adaptive management to conserve biotic integrity in a regulated segment of the Tallapoosa River, Alabama, U.S.A. *Conservation Biology* 16 1212 1222

28. J. R Karr, 1991Biological integrity: a long-neglected aspect of water resource management. *Ecological Applications* 1 66 84

29. J. L Lessard, and D. B Hayes, 2003Effects of elevated water temperature on fish and macroinvertebrate communities below small dams. *River Research and Applications* 19 721 732

30. D. A Lytle, and D. M Merritt, 2004Hydrologic regimes and riparian forests: a structured population model for cottonwood. *Ecology* 85 2493 2503

31. M. J Maceina, and M. C Stimpert, 1998Relations between reservoir hydrology and crappie recruitment in Alabama. *North American Journal of Fisheries Management* 18 104 113

32. J. K Maingi, and S. E Marsh, 2002Quantifying hydrologic impacts following dam construction along the Tana River, Kenya. *Journal of Arid Environments* 50 53 79

33. M. P Marchetti, and P. B Moyle, 2001Effects of flow regime on fish assemblages in a regulated California stream. *Ecological Applications* 11 530 539

34. R. M Mayo, and H. L. Schramm Jr. 1999Growth of flathead catfish in the lower Mississippi River. 121 124 *in* E. R. Irwin, W. A. Hubert, C. F. Rabeni, H. L. Schramm, Jr., and T. Coon, editors. Catfish 2000: proceedings of the international ictalurid symposium. *American Fisheries Society, Symposium 24*, Bethesda, Maryland.

35. R Matthews, and B Richter, 2007Application of the Indicators of Hydrologic Alteration software in environmental flow-setting. Journal of the American Water Resources Association. 43 1 4

36. D. M Merritt, and N. L Poff, 2010Shifting dominance of riparian *Populus* and *Tamarix* along gradients of flow alteration in western North American rivers. *Ecological Applications* 20 135 152

37. J. D Olden, and N. L Poff, 2003Redundancy and the choice of hydrologic indices for characterizing streamflow regimes. *River Research and Applications* 19 101 121

38. V. L Paragamian, 2002Changes in the species composition of the fish community in a reach of the Kootenai River, Idaho, after construction of Libby Dam. *Journal of Freshwater Ecology* 17 375 383

39. M. A Pegg, C. L Pierce, and A Roy, 2003Hydrological alteration along the Missouri River Basin: a time-series approach. *Aquatic Sciences* 65 63 72

40. N. L Poff, J. D Allan, M. B Bain, J. R Karr, K. L Prestegaard, B. D Richter, R. E Sparks, and J. C Stromberg, 1997The natural flow regime: a paradigm for river conservation and restoration. *BioScience* 47 769 784

41. N. L Poff, B. P Bledsoe, and C. O Cuhaciyan, 2006Hydrologic variation with land use across the contiguous United States: geomorphic and ecological consequences for stream ecosystems. *Geomorphology* 79 264 285

42. N. L Poff, B Richter, A. H Arthington, S. E Bunn, R. J Naiman, E Kendy, M Acreman, C Apse, B. P Bledsoe, M Freeman, J Henriksen, R. B Jacobson, J Kennen, D. M Merritt, J. O Keeffe, J. D Olden, K Rogers, R. E Tharme, and A Warner, 2010The Ecological Limits of Hydrologic Alteration (ELOHA): a new framework for developing regional environmental flow standards. *Freshwater Biology* 55 147 170

43. C. M Pringle, M. C Freeman, and B. J Freeman, 2000Regional effects of hydrologic alterations on riverine macrobiota in the new world: tropical-temperate comparisons. *BioScience* 50 807 823

44. C. M Pringle, 2001Hydrologic connectivity and the management of biological reserves: a global perspective. *Ecological Applications* 11 981 998

45. C. M Pringle, 2003What is hydrologic connectivity and why is it ecologically important? *Hydrologic Processes* 17 2685 2689

46. D. L Propst, and K. B Gido, 2004Responses of native and nonnative fishes to natural flow regime mimicry in the San Juan River. *Transactions of the American Fisheries Society* 133 922 931

47. J. W Quinn, and T. J Kwak, 2003Fish assemblage changes in an Ozark river after impoundment: a long-term perspective. *Transactions of the American Fisheries Society* 132 110 119

48. M. C Quist, and C. S Guy, 1998Population characteristics of channel catfish from the Kansas River, Kansas. *Journal of Freshwater Ecology* 13 351 359

49. B. D Richter, J. V Baumgartner, J Powell, and D. P Braun, 1996A method for assessing hydrologic alteration within ecosystems. *Conservation Biology* 10 1163 1174

50. B. D Richter, J. V Baumgartner, J Powell, and R Wigington, 1997How much water does a river need? *Freshwater Biology* 37 231 249

51. R. A Rulifson, and C. S. Manooch III. 1990Recruitment of juvenile striped bass in the Roanoke River, North Carolina, as related to reservoir discharge. *North American Journal of Fisheries Management* 10 397 407

52. D. A Rutherford, W. E Kelso, C. F Bryan, and G. C Constant, 1995Influence of physicochemical characteristics on annual growth increments of four fishes from the lower Mississippi River. *Transactions of the American Fisheries Society* 124 687 697

53. P. C Sakaris, 2006Effects of hydrologic variation on dynamics of channel catfish and flathead catfish populations in regulated and unregulated rivers in the southeast USA. Ph.D. Dissertation, Auburn University, Alabama, USA.

54. P. C Sakaris, and E. R Irwin, 2010Tuning stochastic matrix models with hydrologic data to predict the population dynamics of a riverine fish. *Ecological Applications* 20 483 496

55. S. M Sammons, and P. W Bettoli, 2000Population dynamics of a reservoir sport fish community in response to hydrology. *North American Journal of Fisheries Management* 20 791 800

56. R. D Schultz, C. S Guy, and D. A Robinson, Jr. 2002Comparative influences of gizzard shad catch rates and reservoir hydrology on recruitment of white bass in Kansas reservoirs. *North American Journal of Fisheries Management* 22 671 676

57. M. L Scott, G. T Auble, and J. M Friedman, 1997Flood dependency of cottonwood establishment along the Missouri River, Montana, USA. *Ecological Applications* 7 677 690

58. V. H Travnichek, M. B Bain, and M. J Maceina, 1995Recovery of a warmwater fish assemblage after the initiation of a minimum-flow release downstream from a hydroelectric dam. *Transactions of the American Fisheries Society* 124 836 844

59. G. T Watters, 2000Freshwater mussels and water quality: a review of the effects of hydrologic and instream habitat alterations. *Proceedings of the First Freshwater Mollusk Conservation Society Symposium* 1999 261 274

60. R. L Welcomme, 1979Fisheries ecology in floodplain rivers. Longman, New York, New York.

61. M. L Wildhaber, V. M Tabor, J. E Whitaker, A. L Allert, D. W Mulhern, P. J Lamberson, and K. L Powell, 2000Ictalurid populations in relation to the presence of a main-stem reservoir in a Midwestern warmwater stream with emphasis on the threatened Neosho madtom. *Transactions of the American Fisheries Society* 129 1264 1280

62. R. C Wunderlich, B. D Winter, and J. H Meyer, 1994Restoration of the Elwha River ecosystem. *Fisheries* 19 11 19

63. T Yang, Q Zhang, Y. D Chen, X Tao, C Xu, and X Chen, 2008A spatial assessment of hydrologic alteration caused by dam construction in the middle and lower Yellow River, China. *Hydrological Processes* 22 3829 3843

64. P. J Young, B. D Keeland, and R. R Sharitz, 1995Growth response of baldcypress [Taxodium distichum (L.) Rich.] to an altered flow regime. American Midland Naturalist 133 206 212

Chapter 4

UTILIZATION OF THE SUOMI NATIONAL POLAR-ORBITING PARTNERSHIP (NPP) VISIBLE INFRARED IMAGING RADIOMETER SUITE (VIIRS) DAY/NIGHT BAND FOR ARCTIC SHIP TRACKING AND FISHERIES MANAGEMENT

William C. Straka III[1], Curtis J. Seaman[2], Kimberly Baugh[3], Kathleen Cole[4], Eric Stevens[5], and Steven D. Miller[2]

[1]Cooperative Institute for Meteorological Satellite Studies, University of Wisconsin-Madison, Madison, WI 53706, USA

[2]Cooperative Institute for Research in the Atmosphere, Colorado State University, Fort Collins, CO 80523, USA

[3]Cooperative Institute for Research in Environmental Sciences, University of Colorado-Boulder, Boulder, CO 80309, USA

[4]National Weather Service, Alaska Sea Ice Program, Anchorage, AK 99502, USA

[5]Geographic Information Network of Alaska, Fairbanks, AK 99775, USA

ABSTRACT

Maritime ships operating on-board illumination at night appear as point sources of light to highly sensitive low-light imagers on-board environmental satellites. Unlike city lights or lights from offshore gas platforms, whose locations remain stationary from one night to the next, lights from ships typically are ephemeral. Fishing boat lights are most prevalent near coastal cities and along the thermal gradients in the open ocean. Maritime commercial ships also operate lights that can be detected from space. Such observations have been made in a limited way via U.S. Department of Defense satellites since the late 1960s. However, the Suomi National Polar-orbiting Partnership (S-NPP) satellite, which carries a new Day/Night Band (DNB) radiometer, offers a vastly improved ability for users to observe commercial shipping in remote areas such as the Arctic. Owing to S-NPP's polar orbit and the DNB's wide swath (~3040 km), the same location in Polar Regions can be observed for several successive passes via overlapping swaths—offering a limited ability to track ship motion. Here, we demonstrate the DNB's improved ability to monitor ships from space. Imagery from the DNB is compared with the heritage low-light sensor, the Operational Linescan System (OLS) on board the Defense Meteorological Support Program (DMSP) satellites, and is evaluated in the context of tracking individual ships in the Polar Regions under both moonlit and moonless conditions. In a statistical sense, we show how DNB observations of ship lights in the East China Sea can be correlated with seasonal fishing activity, while also revealing compelling structures related to regional fishery agreements established between various nations.

INTRODUCTION

The National Polar-orbiting Partnership (NPP) is the first in a series of next-generation of polar orbiting environmental satellites, serving as risk reduction to the future Joint Polar Satellite System (JPSS) U.S. operational program. It was launched into a sun-synchronous 1330 local time ascending node orbit on 28 October 2011. On 24 January 2012, NPP was formally christened 'Suomi NPP,' (hereafter, S-NPP) in honor of environmental satellite pioneer Verner E. Suomi [1]. S-NPP orbits the Earth at roughly 834 km altitude and completes a single orbit in ~101 min. As such, it provides daily global coverage upon completion of roughly 14 orbits. S-NPP carries five earth-observing sensors: the Visible Infrared Imaging Radiometer Suite (VIIRS), the Cross-track Infrared Sounder (CrIS), the Advanced Technology Microwave Sounder (ATMS), the Ozone Mapping and Profiler Suite (OMPS), and the Clouds and the Earth's Radiant Energy System (CERES). Most of these instruments provide significantly improved observations over their heritage sensor counterparts. S-NPP data are used in NOAA's operational weather forecasts, and provide continuity to NASA's research in climate change, Earth's energy budget, and the global cycling of water and carbon.

VIIRS collects both visible and infrared imagery spanning from 0.4–12 µm, and combines key capabilities of several legacy instruments: Advanced Very High Resolution Radiometer (AVHRR), the Moderate-resolution Imaging Spectroradiometer (MODIS), and the Operational Linescan System (OLS). Perhaps the most unique component of VIIRS is the Day/Night Band (DNB) [2], which has the ability to collect visible/near-infrared (500–900 nm spectral response) imagery during both day and night. The DNB has very high sensitivity to small amounts of light present in its band pass, and is capable of detecting from its orbital altitude the light emitted from a single isolated street lamp [3,4,5]. The DNB offers a wide range of applications at night [6], ranging from fire detection, meteorological phenomena to observations of anthropogenic light sources. It has become apparent that the scope of possible applications far exceeds the basic nocturnal imagery requirement that drove the DNB's design.

While meteorological phenomena observed via natural light sources (e.g., moonlight) are mainstays of DNB's utility to forecasters, this paper examines the DNB's usage to observe anthropogenic light emission sources, with particular emphasis on the activity of maritime vessels. While navigation lighting on maritime vessels, defined by the International Regulations for Preventing Collisions at Sea 1972 (Colregs) [7], notionally is too weak to be detected by the DNB as a point light source, many commercial maritime vessels operate far more powerful lights than those required by the regulations. For example, some fishing boats use strong lights as part of their operations. In particular, squid boats utilize an extreme form of light-luring, operating large arrays of high-powered lamps that emit up to 300 kW per vessel [8]. Certain non-fishing vessels also emit levels of light sufficient for DNB detection. Examples include icebreakers, which operate searchlights and other high intensity forms of lighting for navigating ice at night. Also facilitating detection is the fact that most maritime vessels operate away from other natural and anthropogenic light sources, such as cities, making their singular sources easier to discern in DNB imagery.

Whereas the literature is replete with OLS-based studies on various fishing activities [9,10,11], attributes of the higher spatial resolution of the DNB over the OLS (750 m at nadir for the DNB as compared to 5 km for the OLS) which may allow for refined comparisons, e.g., of the complex fishery agreements in various regions of the world, have yet to be examined. This paper explores three illustrative case studies highlighting the improved ability of the DNB to observe ship lights. We compare the capabilities of the DNB and the OLS for ship light detection, demonstrate the utility of this improved detection in the arena of operational support, and examine the research potential of these measurements in the context of gaining a statistical perspective on fishing activity over a given region, as well as adherence to international fishing agreements. Through these examples, we aim to spur a reinvigoration of research in these areas based on the wealth of new information content present in the DNB observations.

TRACKING SHIP MOVEMENTS

A more expedient and economical intercontinental transport of goods between the mid-latitude nations of the Northern hemisphere has long stood as motivation for a northern passage, via the Arctic, as an alternative to the Suez Canal in Egypt. While the original idea of such a route dated back to the 1500s, it was not until 1878 that the first passage of the Northern Sea Route (NSR) was completed [12]. Some segments of the NSR are free of ice for only two months per year and require the use of icebreakers to escort convoys of commercial boats through the ice packed portions of the route. The development of nuclear-powered icebreakers by the former Soviet Union first allowed for the regular transit of the NSR. These ships are among the largest and most powerful icebreakers in the world. The *Taimyr* class ships are shallow draft ships intended for coastal icebreaking operations [13], equipped to handle ice up to 2 m thick. As with most large vessels, sufficient on-deck lighting permits their detection by the DNB.

Figure 1 and Figure 2 demonstrate the DNB's unique ability to monitor ship transits in the Polar Regions. DNB imagery from two successive passes of Suomi NPP on 18 November 2013 are shown in Figure 1a (1554 UTC) and Figure 1b (1735 UTC). Here, the light emitted by the *Taimyr* at roughly 72N, 158E (circled in red to highlight the ship light and shown as an inset in Figure 1a,b). Positions of the *Taimyr* from 0600UTC on 18 November 2013 (71.3°N, 164°E) and 0000 UTC on 19 November 2013 (72.9°N 153.9°E) were available from the World Meteorological Organization's Voluntary Observing Ship (VOS) scheme [14]. While these data do not match the exact times of the satellite overpasses, they provide approximate locations of ships and offer a notional ground truth for ship identification. The VOS positions for the *Taimyr* overlaid inFigure 1 were obtained from sailwx.info (http://www.sailwx.info/ shiptrack/shipdump.phtml?call=UEMM), who allowed us to utilize their database (Personal Communication with Hal Mueller, 2014).

Multiple satellite observations separated in time can be used to estimate vectors of ship travel. An estimation of the speed of the *Taimyr* was derived from the two successive VIIRS passes. In the 15:54:43 UTC pass, the ship light is located at 72.16221°N 158.4802°E. In the 17:35:44 UTC image, the light from the ship is located at 72.3142°N 157.5319°E. Thus, in roughly 1 h and 41 min, the *Taimyr* traveled roughly 36.38 km, suggesting an average velocity of 21.61 km/h (~11.67 knots). Using the VOS positions, a distance of roughly 387.7 km was traversed in 18 h, with an average speed of 21.5 km/h (~11.61 knots). In light of assumptions on constant motion between the two points and a single vector of travel, the satellite and VOS-derived vectors are in reasonable agreement. The estimates are also consistent with the conditions. For reference, the open water speed of the *Taimyr* is 18.5 knots (about 60% faster than what was observed) and is 3 knots when traveling through 2.2 m thick ice. With additional data from the operators of the *Taimyr* (Atomflot and the Russian government), an empirical relationship between observed speed and mean ice thickness over the path traversed may be inferable from the ship light data.

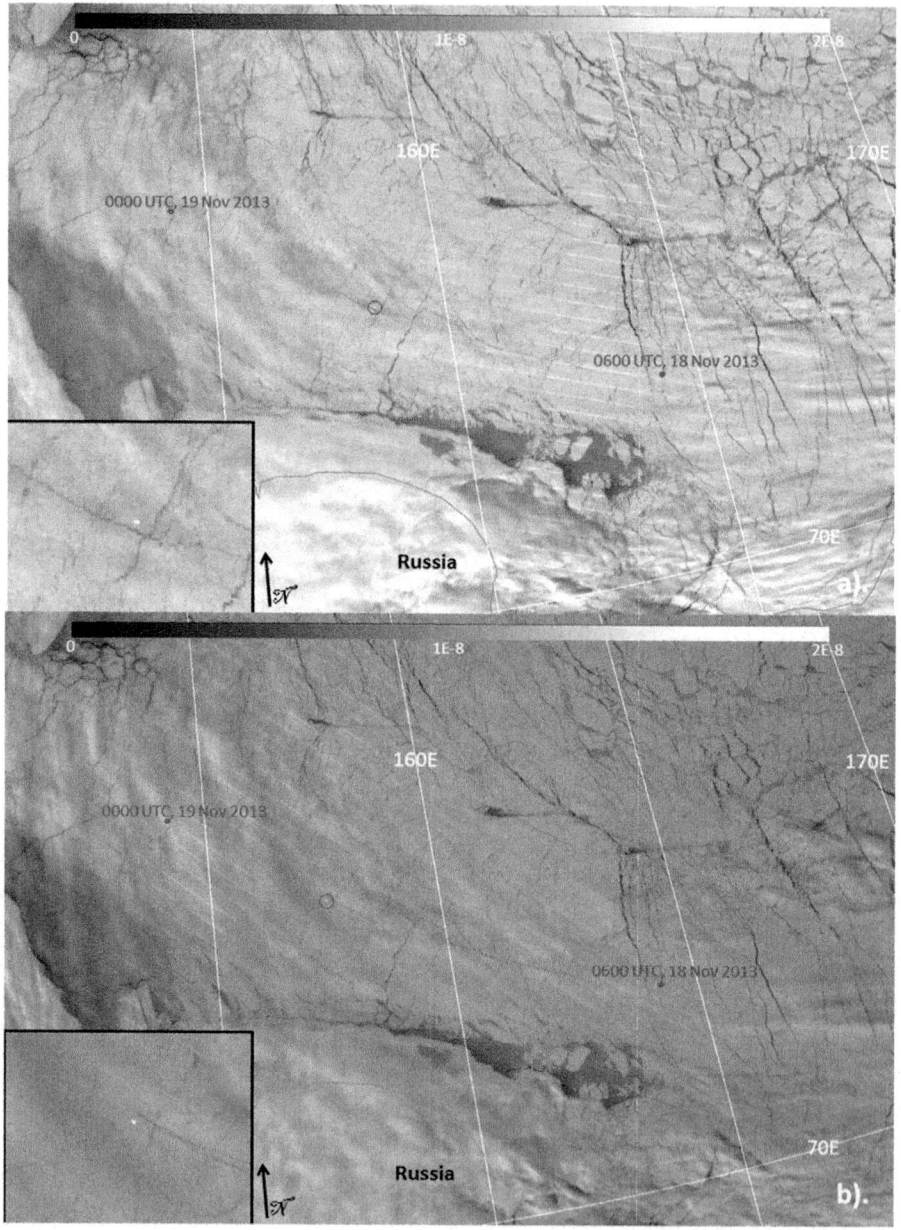

Figure 1: DNB imagery from 1554UTC on 18 November 2013 (a) and 1735UTC on 18 November 2013 (b). The VOS locations of Taimyr from sailwx.info (http://www.

sailwx.info/shiptrack/shipdump.phtml?call=UEMM) are also shown as red dots on the image. The location of the Taimyr is circled in red. Inset shows zoomed image of ship light. Values are W·cm^{-2}·sr^{-1}.

While the transit of the *Taimyr* illustrates a unique way in which the DNB can assist in the tracking of various ships as they transit the NSR, a particularly compelling additional capability emerged during close inspection of this case study. By virtue of moonlight, details of the highly reflective sea ice, including fissures and leads where darker (low albedo) ocean waters provide strong reflective contrast, were easily discerned in the DNB imagery. Careful inspection of the multi-pass series of satellite imagery revealed the subsequent breakup of the sea ice along the path of the *Taimyr* as it transited the Arctic Sea. The emerging ice lead, oriented along the *Taimyr*'s vector of travel is highlighted by arrows in Figure 2a–c. The transit of the *Taimyr* in the 1553 UTC and 1734 UTC passes on 18 November 2013 is marked by the red circle in Figure 2a,b. By the next S-NPP overpass at 1543 UTC pass on 19 November 2013, Figure 2c, the *Taimyr* was well to the west of the region shown, as indicated by the lack of the light from the ship. However, a fresh sea ice lead formed by the transit of the *Taimyr* is indicated by the set of blue arrows in all three images. Many other naturally occurring fissures are also present across this ice sheet. The ability to observe these detailed spatial changes in the nocturnal sea ice from satellite-based low-light visible imaging is unprecedented.

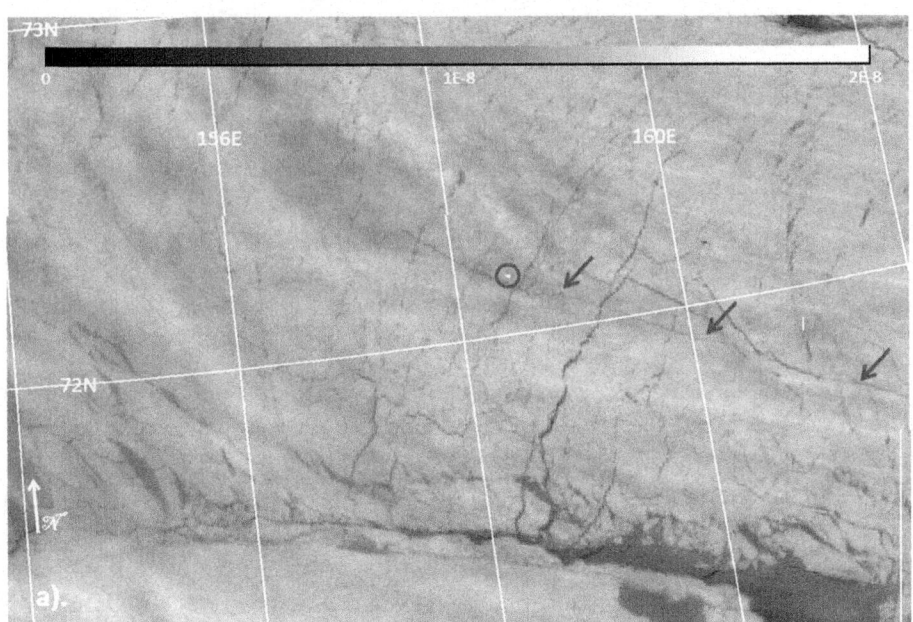

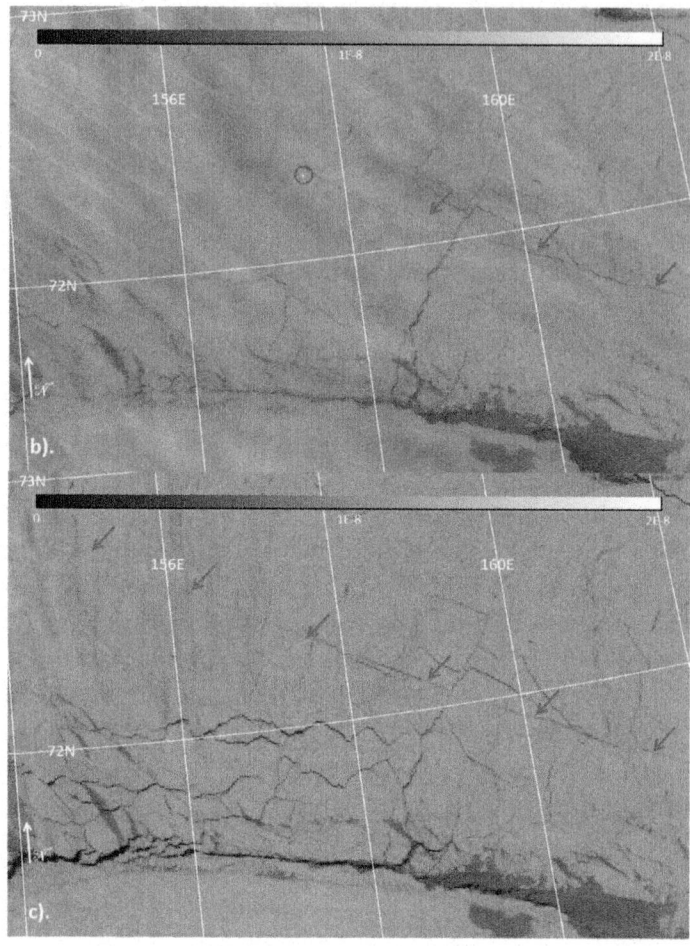

Figure 2. DNB imagery from 1553UTC on 18 November 2013 (a), 1734UTC on 18 November 2013 (b), and 1534UTC on 19 November 2013 (c). The location of the light of the Taimyr is circled in red. The blue arrow shows the crack formed by the transit of the Taimyr. Values are $W \cdot cm^{-2} \cdot sr^{-1}$.

The DNB's low-light visible capability bears particular relevance to increased maritime operations in the Arctic during the winter months, when very short days and copious cloud cover serve to confuse and obscure sea ice details in the thermal channels [6]. So long as ships are far removed from any other light sources (*i.e.*, aurora, other ships, cities, gas flares, *etc.*, that might make it difficult to identify an isolated ship) and are not completely obscured by clouds, such as in this example, their progress can be tracked. It is worth pointing out that even with thin clouds, the bright lights from the icebreakers, though diffused, can be seen.

To qualify the advances of the DNB over heritage sensors in the capacity of ship tracking, we consider the OLS from the Defense Meteorological Satellite Program (DMSP). The DMSP satellites also operate in a low earth orbit. One of the current operational DMSP satellites, designated F18, which flies a sun-synchronous orbit with ~0800 local time, descending node, crossed the region the *Taimyr* was located (per VOS data) four times on 18 November 2013 at between 0503UTC and 1009 UTC. Given the previously demonstrated ability for the OLS to pick out fishing activities [9,10,11], it might be possible for it to resolve the powerful lights from the *Taimyr*. One of the OLS passes, from 0827 UTC on 18 November 2013, is shown inFigure 3.

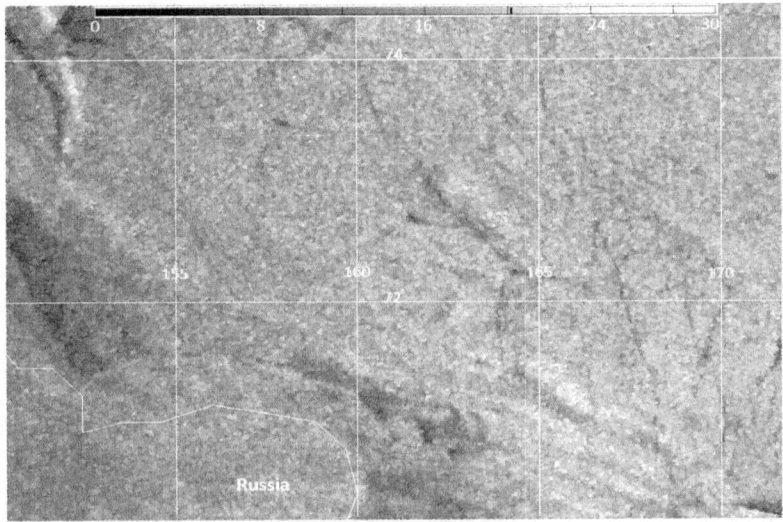

Figure 3: Operational Linescan System (OLS) imagery from 18 November 2013 at 0827 UTC. Values are in OLS count space.

Because the *Taimyr* was moving at roughly 12 knots, the ship is present somewhere in the region spanned by the OLS imagery. However, the F18 OLS imagery (Figure 3) gives no obvious indication of the same isolated ship light feature seen in Figure 2 by the DNB. Under aggressive scaling, other features such as ice leads are marginally observable, but the ship is not discernible. The cause for this missed detection is most likely tied to spatial resolution limitations. The OLS offers significantly lower spatial resolution than the DNB (45–88 times coarser, depending on position in scan; Miller *et al.*, 2013), meaning that small features such as a ship light would be far more difficult to detect uniquely in OLS imagery due to spatial averaging. The challenges are augmented when the surrounding background (in this case, widespread sea ice) reflects moonlight strongly. As inferred from the sea ice details, during the period of these observations there was a waning gibbous moon, with significant moonlight present.

SUPPORTING OPERATIONAL NAVIGATION

We demonstrated in the previous section that even under moonlit conditions, the DNB has sufficient sensitivity to identify and notionally track ship lights. We also demonstrated that under the moonlit conditions the DNB has a unique ability to provide high-resolution detail of sea ice structure. These capabilities can in principle be leveraged to the benefit of maritime operations, and we explore here one noteworthy example of this concept put into practice under the most extreme environmental conditions and challenging levels of illumination.

Alaskan king crab fishing has an occupational fatality rate that is 36 times the rate of all other U.S. occupations [15,16], and for this it earns the distinction of being among the world's most dangerous professions. The season is conducted during the fall and winter months in the waters off the coast of Alaska and the Aleutian Islands. The boats often operate near the ice pack in order to avoid freezing ocean spray, which would freeze upon contact and coat the upper portions of the vessel, making it unstable and prone to capsizing under the force of accompanying strong winds. However, operating near ice pack introduces its own array of hazards.

During the second week of February 2013, the *F/V Kiska Sea*, a fishing vessel that has been featured as part of the *Deadliest Catch* reality television series on the Discovery Channel, was operating as the northernmost vessel of the Bering Sea crab fleet, close to the sea ice edge. Recent passage of a polar low-pressure system brought strong northerly winds across the central Bering Sea, causing a rapid advance of sea ice that threatened to encroach on the region where the fleet had placed over 150 crab pots. On 10 February 2013 the *Kiska Sea* encountered significant sea ice near the location where they had placed these pots, prompting the ship's captain to request assistance from the Ice Desk at the National Weather Service, based in Anchorage, AK. Ice Desk analysts examined the previous night's DNB imagery, provided by Geographic Information Network of Alaska's (GINA) direct broadcast antenna located in Fairbanks, AK, along with the navigation information provided by the captain of the *Kiska Sea*, and confirmed that the ice was indeed encroaching on the ship's location. The *Kiska Sea* and the Ice Desk continued to exchange information regarding the situation as the ship navigated the increasing ice field.

However, on 13 February 2013 the conditions deteriorated rapidly, and the *Kiska Sea* found herself surrounded by sea ice, some in excess of 3 feet thick. This was a potentially dangerous situation for the *Kiska Sea* as the ship is not an ice class vessel, meaning that it cannot break through thick ice without running the risk of sinking. Lacking solar or even lunar illumination (the new moon occurred on 10 February 2013) to assist in visual navigation, it was imperative that the *F/V Kiska Sea* navigate out of the sea ice quickly.

Figure 4 shows the DNB imagery at 05:26 AM Alaska Standard Time (UTC-9:00) on 13 February 2013. Similar to previous examples of *Taimyr*, the *Kiska Sea* can be seen here as a discrete point of light (red circle) close to the sea ice edge, with the main Bering Sea crab fishing fleet deployed to the south, near St. Paul Island (57.19N, −170.26W). We can be certain of this as the *Kiska Sea* communicated their intentions to move away from the main crab fishing fleet to retrieve their crab pots, along with communicating their coordinates to the NWS Ice Desk when they requested assistance. The sea ice edge, along with other meteorological features such as low level stratus clouds, can also be seen in the DNB imagery, despite the fact that these data were collected during a moonless night. The detection is due to illumination from atmospheric nightglow, which offers nocturnal low-light visible imagery even on moonless nights [17]. Although some of these features are also evident in the VIIRS I05 (11 μm) infrared brightness temperature (BT) imagery, shown in Figure 5, critical details of the sea ice edge near the *Kiska Sea*'s location are confused or obscured completely by low clouds, which appear as linear streets aligned parallel to strong northwesterly winds. In particular, the DNB nightglow imagery provides specific detail on sea ice edge extent in the immediate vicinity of the ship under duress.

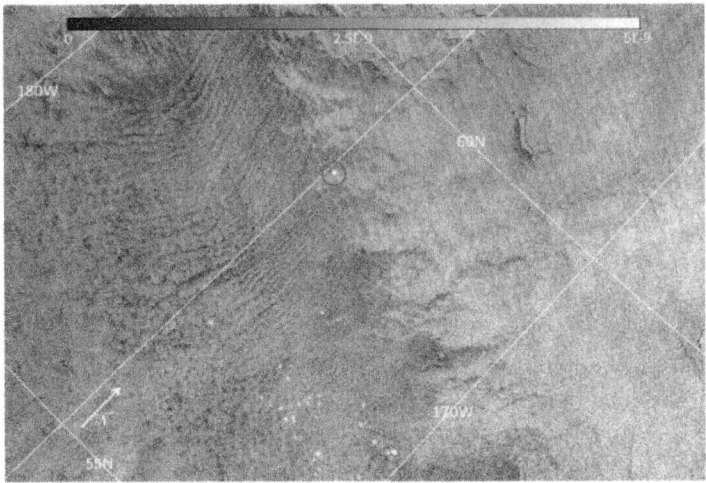

Figure 4: Day Night Band imagery of the Bering Sea at 1426 UTC on 13 February 2013. The red circle indicates the position of the *Kiska Sea*. Ship lights from the main

crab fishing fleet can also be seen near St. Paul Island (near the bottom of the image). The moon is in waxing crescent and well below the horizon at the time of overpass; the illumination for clouds and sea ice in this case is atmospheric nightglow. Values are $W \cdot cm^{-2} \cdot sr^{-1}$.

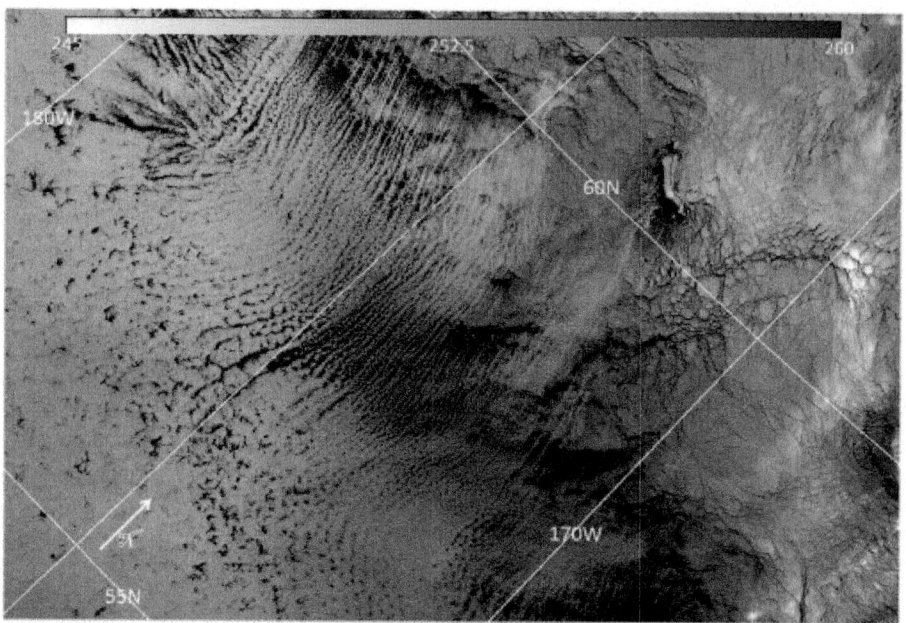

Figure 5: VIIRS 11μm BT from of the Bering Sea at 1426 UTC on 13 February 2013. Red circle indicates Kiska Sea. Units are Kelvins (K).

The NWS Ice Desk was able to use the timely direct-broadcast DNB imagery from S-NPP to chart a track safely out of the growing and shifting ice pack. The guidance helped *Kiska Sea* avoid areas of thicker and higher concentration sea ice, and a possible tragedy at sea.

As with the *Taimyr* case study, we considered the DNB imagery in the context of heritage sensor capabilities as a way of assessing the advanced utility. Here again, we enlisted OLS from the F18 DMSP satellite, which overflew the area multiple times during the *Kiska Sea* incident. Imagery from an OLS pass collected on 13 February 2013 at 0556 UTC is shown in Figure 6. There are virtually no anthropogenic lights present in the scene. The *Kiska Sea* does not appear as a unique light within the noisy imagery. In addition, the meteorological clouds and sea ice and edge, features which were easy to distinguish in the DNB imagery in Figure 4, and noted in Figure 6 by "A", are not visible in OLS imagery—owing to the lower sensitivity and spectral shift away from the nightglow emissions [17].

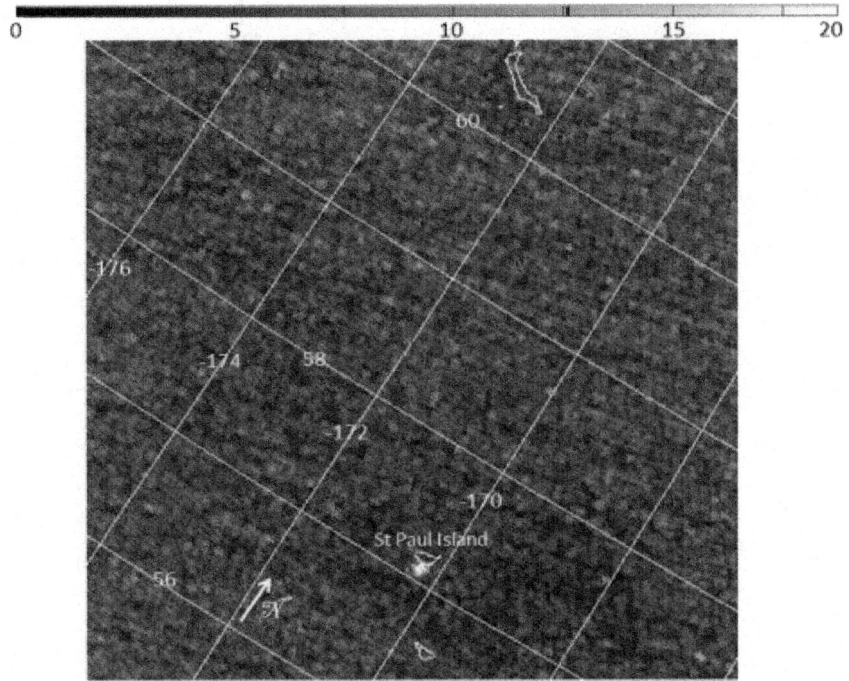

Figure 6: OLS imagery from 13 February 2013 at 0556 UTC. Values are in OLS count space. "A" marks the region where sea ice can be seen in the DNB imagery.

There is one anthropogenic source of light which can be seen in the OLS imagery, located roughly at 58N, 170W. However, this light source does not move from pass to pass and, upon closer inspection, one finds it is associated with Saint Paul Island (noted in Figure 6). This shows that, even with the lack of moonlight as a source of background reflectance contamination, the OLS does not have the spatial and radiometric sensitivity to distinguish individual ship lights at the level of fidelity demonstrated by the next-generation DNB sensor.

FISHERY BOUNDARIES

The world's oceans, comprising 71% of the Earth's surface, provide an important source of global economic activity. In this section, we highlight a specific application of DNB ship detection bearing high importance for the monitoring of fishing activities and adherence to international agreements. The United Nations Convention on the Law of the Sea (UNCLOS) [18] defines regions where a given nation holds special rights over the exploration and use of marine resources, such as fishing or exploitation of mineral resources, which are called Exclusive Economic Zones (EEZs). These EEZs extend out to 200 nautical miles from the coast of a given nation. However, there are certain regions in the world where the EEZs of multiple nations overlap. These are designated by UNCLOS as "semi-enclosed seas".

One notable example is the East China Sea, which is shared by People's Republic of China (PRC), Japan and the Republic of Korea (ROK). In order to resolve the numerous disagreements which have arisen in the past and continue to arise to this day, several bilateral agreements were reached once the UNCLOS came into effect in 1994. A map of the various agreements in the East China Sea as they exist today is shown in Figure 7.

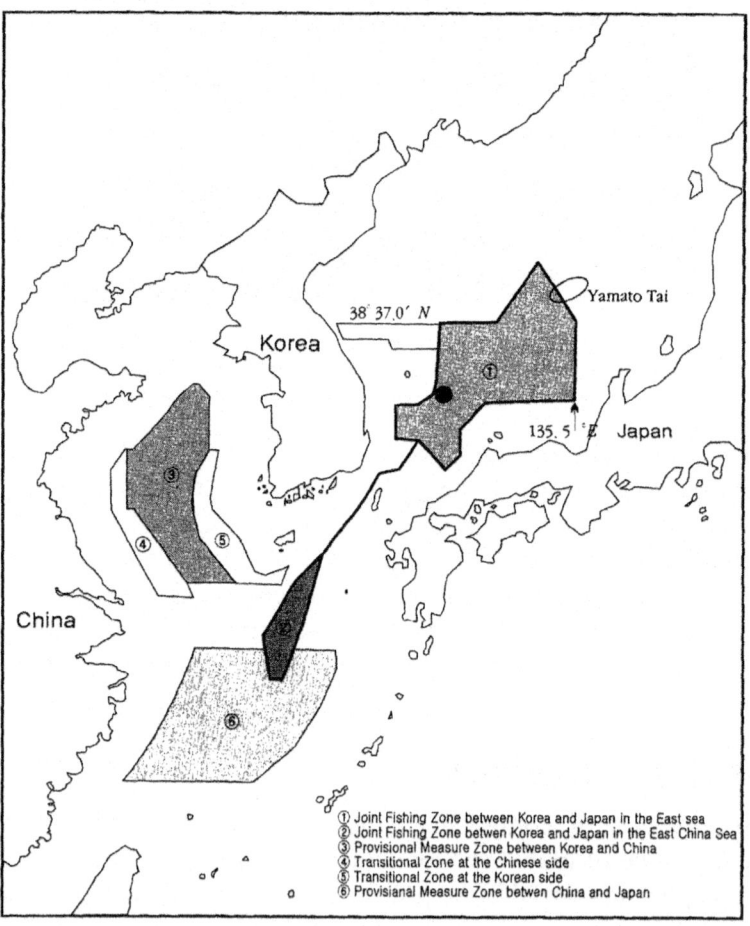

Figure 7: Map of the various fishery agreements, Sun Pyo Kim, Maritime Delimitation and Provisional Arrangements in North East Asia (The Hague/London/Boston: Kluwer Law International, 2004), p. 226 [19]. Reproduced with permission from Sun Pyo Kim; Kim (2004).

One of these agreements, the Sino-Japanese Fishery Agreement of 1997, establishes a Provisional Measures Zone (PMZ), which is a jointly managed fishing zone by the PRC and Japan. Within this PMZ, which extends from roughly 124°E to 127.75°E, and from 27°N to 30.667°N [20,21,22,23](shown as Zone 6, or the southernmost region depicted in Figure 7), ships from both nations may conduct fishing operations. The intent of this jointly managed region is to prevent overfishing (particularly of squid) of these waters. North of 30.667°N are regions where there are multiple fishery agreements between Japan, China and South Korea as well as a region where no fishing permits are required (hereafter referred to as the NFPZ).

Figure 8 shows the DNB imagery from a single pass on 11 November 2012 at 1717 UTC covering the East China Sea region. One identifiable boundary that can be seen is the eastern edge of the PMZ of the Sino-Japanese Fishery Agreement of 1997 south of 30.667°N. In addition, another boundary is oriented north/south at 127.5°E. This appears to be the eastern edge of the region of shared waters where vessels from both nations can fish without any licenses. The NFPZ extends north of 30.667°N and has a western boundary of 124.75°E and an eastern boundary of 127.5°E. However, as seen in Figure 7, there is also a Provisional Measures Zone between South Korea and Japan in this same area.

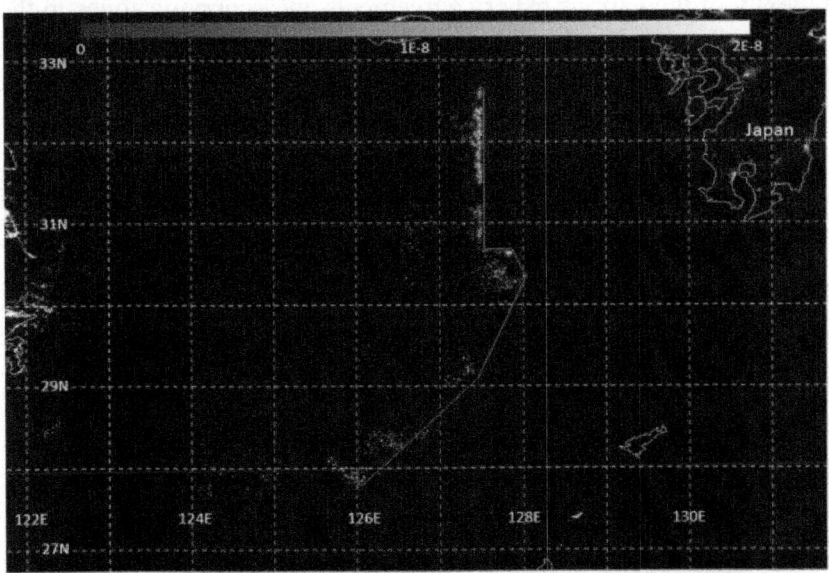

Figure 8: Day Night Band imagery from 11 November 2012. The red line is placed along the approximate eastern boundary of the ship lights. Values are W·cm^{-2}·sr^{-1}.

Temporal composites of nighttime lights have been produced by the Earth Observation Group (EOG) at NOAA's National Geophysical Data Center (NGDC) using the OLS imagery for many years [24]. This group is now starting to produce similar composites using the DNB [25]. For this study, monthly and annual composites from August 2013 to July 2014 were created by EOG using cloud-free data from moonless nights. This time period was selected to avoid data impacted by stray light [3], as the stray light mitigation algorithm went into operation in August 2013.

An annual composite of the region, shown in Figure 9, reveals that there are three distinct boundaries which are denoted by the precise edge of the ship lights. The first is the eastern edge of the Sino-Japanese PMZ, shown in red. The second, denoted by a green dashed line, extends north of 30.667°N and is at 127.5°E. This is the eastern edge of the "no fishing permits" zone (NFPZ) of the Sino-Japanese Fishery Agreement of 1997. For completeness, the western edge of the NFPZ is also shown as a green dashed line at 124.75°E. The final boundary, denoted in orange, is located around the Republic of Korea island of Jeju, and appears to coincide with the northern boundary of the Korea-Japanese PMZ. It is important to note that the country of origin cannot be determined from the ship light properties. However, the orientation of these lights with respect to the PMZ and other boundaries provides insight as to the complexity of the various fishery agreements and economic zones in the East China Sea. Presumably the fact that ships are so closely aligned with respect to the various boundaries is the result of these vessels being equipped with Global Positioning Satellite (GPS) receivers.

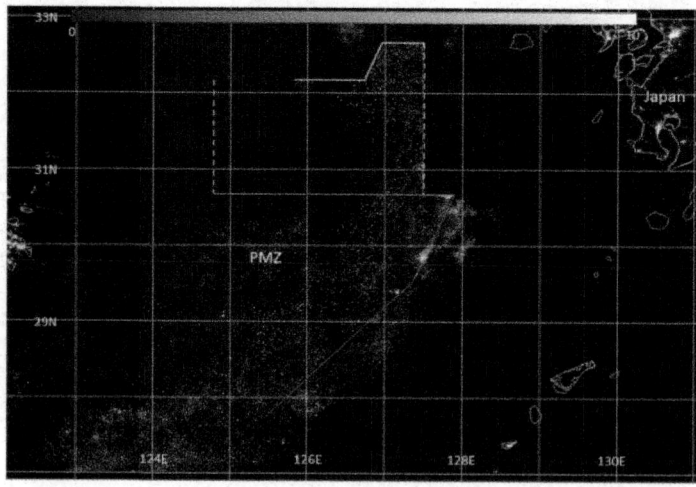

Figure 9: Average Day Night Band radiance for August 2013 to July 2014. Values are nanoWatts·cm^{-2}·sr^{-1}.

Using the monthly composites of the DNB, one can also observe the changes of the fishing patterns throughout the various seasons in each of these zones. Figure 10 shows the average DNB radiance for the month of August 2013.

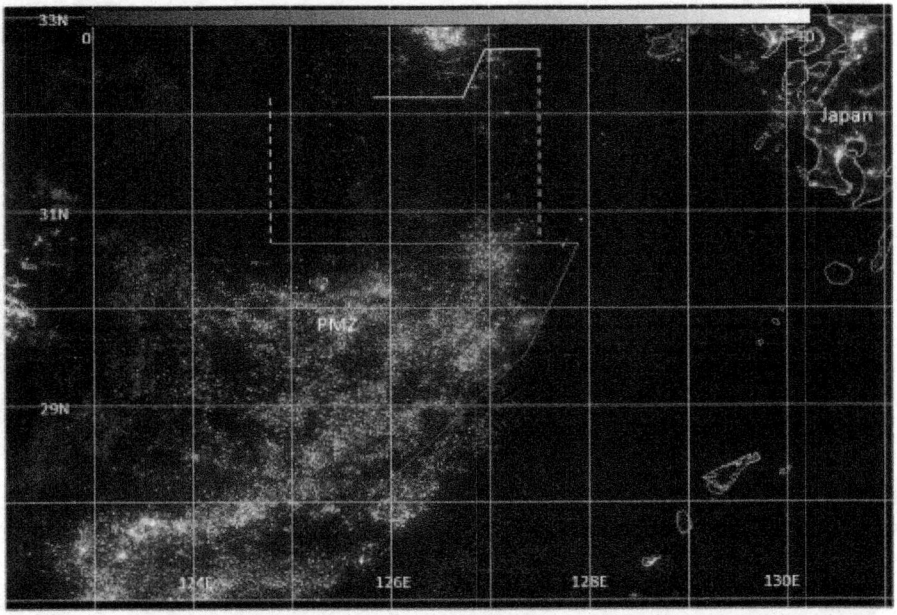

Figure 10: Average Day Night Band radiance for August 2013. Boundaries are the same as in Figure 9. Values are nanoWatts·cm^{-2}·sr^{-1}.

Based on the spatial extent and brightness of this composite, it is evident that fishing activity is prolific during the warm summer months. Contrasting this composite with the corresponding composite from February 2014 (Figure 11), which is during the winter, we observed that there are far fewer ships across the entire East China Sea region, save for a strong localized cluster within the Japanese EEZ.

A seasonal trend is also apparent by comparing DNB composites made using 3 months of data, with virtually no fishing during the late winter/early spring (February through April), fishing picking up from May through the summer months, and falling off through the fall and early winter (January). This can be visualized by comparing the three month composite from March through May 2014 (Figure 12), to other seasonal composites.

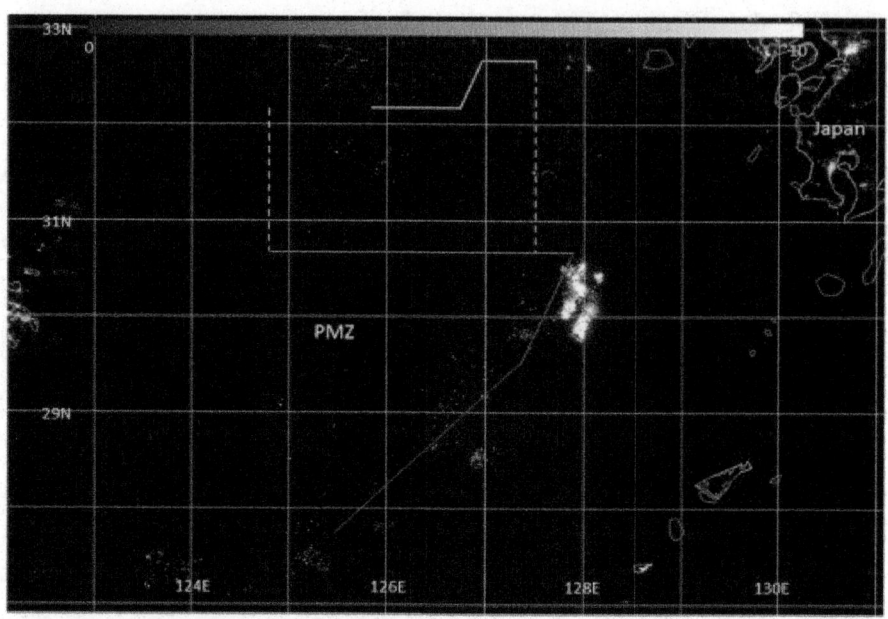

Figure 11: Average Day Night Band radiance for February 2014. Boundaries are the same as in Figure 9. Values are nanoWatts·cm^{-2}·sr^{-1}.

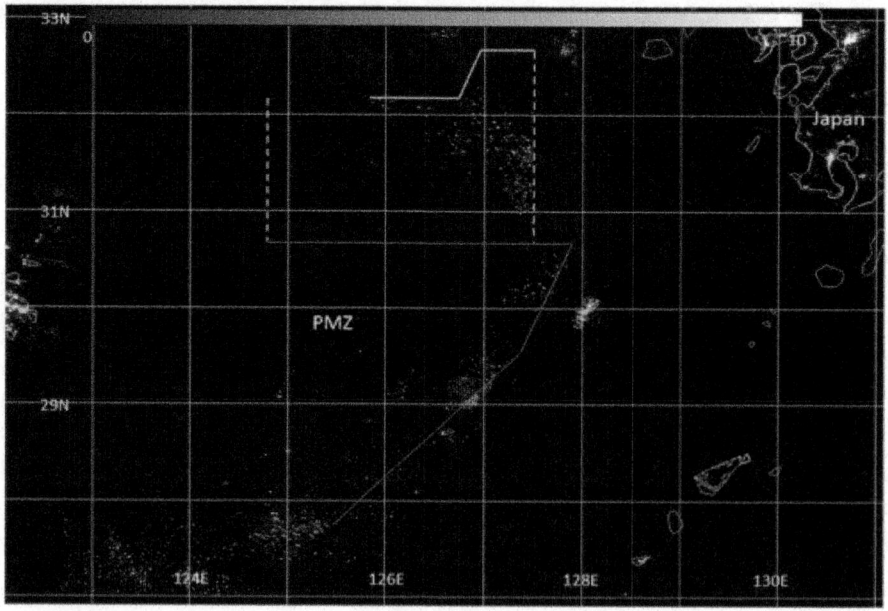

Figure 12: Average Day Night Band radiance composite for March, April and May 2014. Boundaries are the same as in Figure 9. Values are nanoWatts·cm^{-2}·sr^{-1}.

These variations in fishing lights likely correspond to the seasonal variations of nutrients in the East China Sea [26], which drop off significantly as cold water intrudes in the East China Sea during the winter. Another interesting seasonal occurrence, which can be seen in the DNB composites, is a significant decrease in the number of ships in the NFPZ, or "no fishing permit zone" (the east and west boundaries are denoted by green dashed lines) during June through August (shown in Figure 13), as compared to other months during the height of the fishing season.

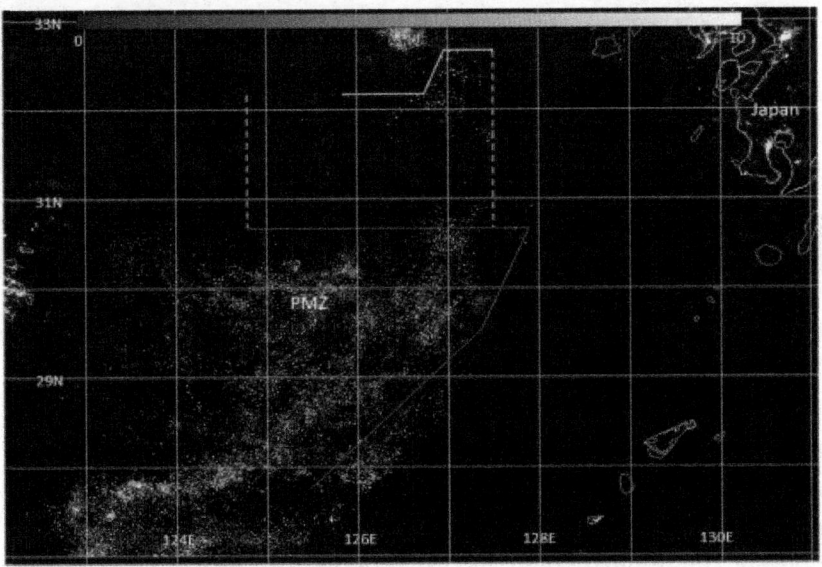

Figure 13: Average Day Night Band radiance composite for June, July and August. Boundaries are the same as in Figure 9. Values are nanoWatts·cm^{-2}·sr^{-1}.

Figure 14 shows the three month composite from September through November 2013. As can be seen, there is a significant increase in the number of ships in the NFPZ, particularly on the eastern edge of the boundary.

The difference in the number of ship lights seen in the NFPZ between the June through August (Figure 13) and the September through November (Figure 14) composites seems to correspond to the "hot" season fishing moratorium that imposed by the Chinese government each year from 1 June through 1 September in the Yellow and East China Seas. This moratorium affects ~120,000 fishing vessels and over a million fishermen, and is aimed at preventing overfishing. Given that the Japan and the Republic of Korea do not impose a similar moratorium suggests that the difference in number of ships in the NFPZ between the two composites is due to the lack of Chinese fishing vessels during the June through August period.

Figure 14: Average Day Night Band radiance composite for September, October and November 2013. Boundaries are the same as in Figure 9. Values are nanoWatts·cm^{-2}·sr^{-1}.

CONCLUSIONS

We have highlighted the improved ability of the VIIRS/DNB over the OLS instruments to observe ship lights under various lunar conditions in the Polar Region and highlighted both the operational and research potential of the DNB in ship tracking in various regions. As was expected given the improved spatial and radiometric resolution (45–88 times higher and 256 times finer, respectively) and increased sensitivity (~100 times higher) of the DNB over the heritage DMSP/OLS, the DNB showed dramatically improved detail and ability to resolve anthropogenic marine light sources. This was true even under conditions when there was no lunar illumination and aggressive scaling was used to bring out features at lowest counts detectable by the OLS. While not unexpected, it was important to demonstrate given the usage of the OLS data in the Polar Regions today on a routine basis.

Both the operational and research potential of the DNB in ship tracking in the three case studies were shown. In the cases of the *Taimyr* and *F/V Kiska Sea*, it was shown that the lights from an individual ship could be monitored over the course of its voyage. In the case of the *F/V Kiska Sea*, this was the first documented time that satellite based visible imagery of the light on a civilian ship was used to distinguish the location of an individual ship along with the pack ice present, and guide it back to safety. The ability to track and monitor ships in those regions, especially in the polar regions where VOS or Satellite-based Automatic Identification System (S-AIS) are delayed, is of particular importance given the retreat of the sea ice as well as the increase in commercial maritime traffic in these regions. While the DNB is able to observe the Polar Regions roughly every 102 minutes, these examples speak to the growing importance of low-light visible imagery and point to the significance of such imagery were it to one day become available at higher temporal refresh. The geostationary platform highly elliptical (Molniya) orbits may one day include such capabilities and a paradigm shift to our ability to watch as ships pass in the night at all latitudes.

While past research has examined how the OLS instrument was used to look at fisheries in various locations in the world, the current research shows how the DNB is able to more accurately show the positions of various fishing fleets, right down to the individual ships along the fishery boundaries of the world. This is of particular importance in order to provide researchers the ability to study human impacts in the ecosystem of the Earth. Monthly and seasonal composites show both the seasonal variation of the fishing season in the East China Sea as well as what appears to be the effect on the fishing fleet during the "hot" season moratorium imposed by the Chinese government. As more data become available over the lifetime of the S-NPP mission and the subsequent Joint Polar Satellite System (JPSS) satellites, which will also carry the DNB on-board, more comprehensive studies on how often fishing occurs in a given region will become possible. Such information can help scientists determine which regions may be at risk for over-fishing worldwide. Government agencies could use monthly and quarterly composited DNB data to determine the effectiveness of their seasonal fishing moratoriums.

ACKNOWLEDGMENTS

We thank the National Weather Service Sea Ice Program at the Anchorage, AK National Weather Service Forecast Office for their assistance in providing information regarding the *Kiska Sea* case. We would also like to thank Sun Pyo Kim for allowing us to use the graphic of the various East China Sea fishery agreements. Finally we would like to acknowledge and thank Hal Mueller of sailwx.info for allowing and providing us the VOS location information on the *Taimyr* (*Таймыр*) nuclear powered icebreakers. Support of the NOAA Joint Polar Satellite System Cal/Val and Algorithm Program is gratefully acknowledged. We also thank our Reviewers and Remote Sensing's Chief Editor for their helpful recommendations toward improving this manuscript.

AUTHOR CONTRIBUTIONS

William Straka III and Steve Miller conceived the idea for this study. Curtis Seaman provided the data for the analysis of the tracking of the *Taimyr* (*Таймыр*). Kim Baugh provided the OLS data as well as the monthly and seasonal composites over the East China Sea. Kathleen Cole, the National Weather Service Sea Ice Program and Eric Stevens provided invaluable information regarding the details of the *Kiska Sea* incident. Steven Miller provided guidance on the capabilities of the DNB and OLS instruments. William Straka III wrote the paper with input from all of the authors, all of the authors reviewed and approved the submitted manuscript and agreed to be listed and accepted the version for publication.

REFERENCES

1. Lewis, J.M.; Martin, D.W.; Rabin, R.M.; Moosmüller, H. Suomi: Pragmatic visionary. *Bull. Am. Meteorol. Soc.* 2010, *91*, 559–577.

2. Lee, T.E.; Miller, S.D.; Turk, F.J.; Schueler, C.; Julian, R.; Deyo, S.; Dills, P.; Wang, S. The NPOESS/VIIRS day/night visible sensor. *Bull. Am. Meteor. Soc.* 2006, *87*, 191–199.

3. Liao, L.B.; Weiss, S.; Mills, S.; Hauss, B. Suomi NPP VIIRS Day and Night Band (DNB) on-orbit performance. *J. Geophys. Res. Atmos.* 2013, *118*, 12705–12718.

4. Mills, S.; Jacobson, E.; Jaron, J.; McCarthy, J.; Ohnuki, T.; Plonski, M.; Searcy, D.; Weiss, S. Calibration of the VIIRS Day/Night Band (DNB). In Proceedings of American Meteorological Society 6th Annual Symposium on Future National Operational Environmental Satellite Systems-NPOESS and GOES-R, Atlanta, GA, USA, 16–21 January 2010.

5. Jacobson, E.; Ibara, A.; Lucas, M.; Menzel, R.; Murphey, H.; Yin, F.; Yokoyama, K. Operation and characterization of the Day/Night Band (DNB) for the NPP Visible/Infrared Imager Radiometer Suite (VIIRS). In Proceedings of The 6th Annual Symposium on Future National Operational Environmental Satellite Systems-NPOESS and GOES-R, Boston, MA, USA, 20 January 2010; p. 349.

6. Miller, S.D.; Straka III, W.; Mills, S.P.; Elvidge, C.D.; Lee, T.F.; Solbrig, J.; Walther, A.; Heidinger, A.K.; Weiss, S.C. Illuminating the Capabilities of the Suomi National Polar-Orbiting Partnership (NPP) Visible Infrared Imaging Radiometer Suite (VIIRS) day/night band. *Remote Sens.* 2013, *5*, 6717–6766.

7. Commandant, U.C.G. *International Regulations for Prevention of Collisions at Sea, 1972 (72 Colregs)*; Commandant Instruction M 16672. US Department of Transportation, US Coast Guard: Washington, DC, USA, 1999.

8. Rodhouse, P.G.; Elvidge, C.D.; Trathan, P.N. Remote sensing of the global light-fishing fleet: An analysis of interactions with oceanography, other fisheries and predators. *Adv. Mar. Biol.* 2001, *39*, 261–303.

9. Kiyofuji, H.; Saitoh, S.I. Use of nighttime visible images to detect Japanese common squid *Todarodes pacificus* fishing areas and potential migration routes in the Sea of Japan. *Mar. Ecol. Prog. Ser.* 2004, *276*, 173–186.

10. Waluda, C.M.; Yamashiro, C.; Elvidge, C.D.; Hobson, V.R.; Rodhouse, P.G. Quantifying light-fishing for *Dosidicus gigas* in the eastern Pacific using satellite remote sensing. *Remote Sens. Environ.* 2004, *91*, 129–133.

11. Waluda, C.M.; Trathan, P.N.; Elvidge, C.D.; Hobson, V.R.; Rodhouse, P.G. Throwing light on straddling stocks of *Illex argentinus*: Assessing fishing intensity with satellite imagery. *Can. J. Fish. Aquat. Sci.* 2002, *59*, 592–596.

12. Nordenskiöld, A.E.; Leslie, A. *The Voyage of the Vega Round Asia and Europe*; Macmillan and Co.: New York City, NY, USA, 1885.

13. Pavlenko, V.I.; Glukhareva, E.K.; Kutsenko, S.Y. Development of the arctic fleet in the Russian Federation. In Proceedings of The Twenty-second International Offshore and Polar Engineering Conference, Rhodes (Rodos), Greece, 17–23 June 2012.

14. Kent, E.; Hall, A.D.; Leader, V.T.T. The Voluntary Observing Ship (VOS) scheme. In Proceedings from the 2010 AGU Ocean Sciences Meeting, Portland, OR, USA, 22–26 February 2010.

15. Lincoln, J.M.; Lucas, D.L.; McKibbin, R.W.; Woodward, C.C.; Bevan, J.E. Reducing commercial fishing deck hazards with engineering solutions for winch design. *J. Saf. Res.* 2008, *39*, 231–235.

16. Thomas, T.K.; Lincoln, J.M.; Husberg, B.J.; Conway, G.A. Is it safe on deck? Fatal and non-fatal workplace injuries among Alaskan commercial fishermen. *Am. J. Ind. Med.* 2001, *40*, 693–702.

17. Miller, S.D.; Mills, S.P.; Elvidge, C.D.; Lindsey, D.T.; Lee, T.F.; Hawkins, J.D. Suomi satellite brings to light a unique frontier of environmental imaging capabilities. *Proc. Nat. Acad. Sci.* 2012, *109*, 15706–15711.

18. United Nations Convention on the Law of the Sea, UN Doc A/CONF.62/122 (1982). Available online: http://legal.un.org/diplomaticconferences/

lawofthesea-1982/docs/vol_XVII/a_conf-62_122_CONVENTION.pdf (accessed on 2 September 2014).

19. Kim, S.P. *Maritime Delimitation and Interim Arrangements in North East Asia*; Martinus Nijhoff Publishers: Leiden, Boston, MA, USA, 2004; Volume 40.

20. Kang, J.S. The United Nation convention on the law of the sea and fishery relations between Korea, Japan and China.*Mar. Policy* 2003, *27*, 111–124.

21. Keyuan, Z. Sino-Japanese joint fishery management in the East China Sea. *Marine Policy* 2003, *27*, 125–142.

22. Kim, S.P. The UN convention on the law of the sea and new fisheries agreements in north East Asia. *Marine Policy*2003, *27*, 97–109.

23. Xue, G. *China and International Fisheries Law and Policy*; Martinus Nijhoff Publishers: Leiden, Boston, MA, USA, 2005; Volumn 50.

24. Elvidge, C.D.; Baugh, K.E.; Kihn, E.A.; Kroehl, H.W.; Davis, E.R. Mapping city lights with nighttime data from the DMSP operational linescan system. *Photogramm. Eng. Remote Sens.* 1997, *63*, 727–734.

25. Baugh, K.; Hsu, F.-C.; Elvidge, C.D.; Zhizhin, M. Nighttime lights compositing using the VIIRS day-night band: Preliminary results. *Proc. Asia-Pacific Adv. Netw.* 2013, *35*, 70–86.

26. Guo, X.; Zhu, X.; Wu, Q.; Huang, D. The Kuroshio nutrient stream and its temporal variation in the East China Sea. *J. Geophys. Res. Oceans (1978–2012)* 2012, *117*.

Chapter 5

FISHPRINT OF COASTAL FISHERIES IN JALISCO, MEXICO

Myrna Leticia Bravo-Olivas[1], Rosa María Chávez-Dagostino[1], Carlos Antonio López-Fletes[2], and Elaine Espino-Barr[3]

[1]Biological Sciences Department, Centro Universitario de la Costa, Universidad de Guadalajara, Puerto Vallarta, Jalisco 48280, Mexico

[2]Master's Program in Human Ecology, Vrije Universiteit Brussel, Brussel 1050, and Belgium

[3]Regional Center of Fisheries Research, National Fisheries Institute, Manzanillo, Colima 28200, Mexico

ABSTRACT

Coastal fisheries contribute to global food security, since fish are an important source of protein for many coastal communities in the world. However, they are constrained by problems, such as weak management of fisheries and overfishing. Local communities perceive that they are fishing less, as in other fisheries in the world. The aim of this study was to evaluate the fisheries sustainability in the Jalisco coast through the fishing footprint, or fishprint (FP), based on the primary productivity required (PPR) and the appropriated surface by the activity (biocapacity). The total catch was 20,448.2 metric tons from 2002–2012, and the average footprint was calculated to be 65,458 gha/year, a figure that quadrupled in a period of 10 years; the biocapacity decreased, and the average trophic level of catches was 3.1, which implies that it has remained at average levels, resulting in a positive balance between biocapacity and ecological footprint. Therefore, under this approach, the fishing activity is sustainable along the coast of Jalisco.

INTRODUCTION

The coast of Jalisco is located on Mexico's Pacific coast, bordered to the north by the state of Nayarit and to the south by the state of Colima, with a length of 351 km, including five municipalities: Puerto Vallarta, Cabo Corrientes, Tomatlán, La Huerta and Cihuatlán [1] (Figure 1). It is an occasional upwelling area [2] with the presence of an immediate current to the coast of Cabo Corrientes, which intensifies on the surface [3]. Fishing activity is only artisanal, but it maintains a good level of labor employment and generates a significant demand for technical and commercial services [4].

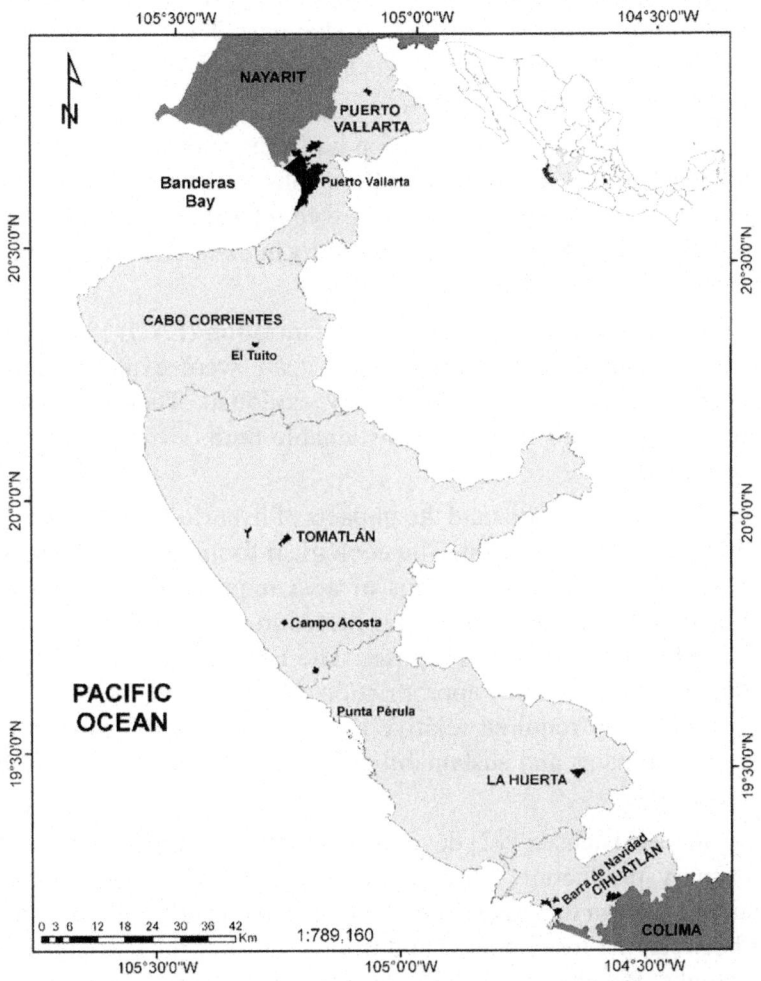

Figure 1: The coast of Jalisco, México, located in the Pacific Central-American Coast Large Marine Ecosystem (www.seaaroundus.org/lme) and in the Pacific, East Central Food and Agriculture Organization (FAO) fishing area [5].

It is estimated that about 14,274 people live on fishing directly [6]. Most of the people depend on fishing in this region not only to feed themselves, but for cultural reasons. If fishing is an important activity here, how sustainable is it?

Fishing is important worldwide: more than a billion people, mostly in poor countries, depend on fish products to meet their need for animal protein in their diet.

The consumption of fishery resources in the world has increased to 130.8 million tons intended for human consumption due to aquaculture, increased fishing and improvement of distribution channels. This situation allowed the growth of the world's supply of fish per capita from 9.9 kg (live weight) in 1960 to 18.8 kg per capita in 2011 [7].

Although world fisheries production has remained stable over the last ten years and reached a peak of 86.4 million tons in 1996, concerns about the sustainability of fisheries have been expressed related to overfishing, depletion of some stocks, ecosystem changes induced by humans and its potential impact on supplies and equity at the local level [8].

In 2011, the Food and Agriculture Organization (FAO) [9] reported that 28.8% of the stocks were overexploited, 9.9% were exploited moderately or underexploited and the rest were fully exploited. Therefore, catches are considered close to their maximum sustainable limits, without the possibility of increasing [10].

Given the need to understand the impacts of fisheries and their status, some indicators have been developed. The ecological footprint (EF) is an indicator that measures human needs in terms of area required for the generation of products and waste absorption during the course of the production process [11] and can be used for many purposes. The EF of fishing, a tool to measure the spatial extent of human appropriation of marine ecosystems based on primary productivity required relative to the catch, can be used to establish the ecological impacts and sustainability of fish production and consumption at different levels.

Pauly and Christensen [12] developed a method to improve the estimation of the primary production required to sustain global fisheries catches and proposed an equation that takes into account the catch efficiency of the energy transfer between the trophic level and trophic level of the species or group of species caught. Results showed in general that 8% of primary productivity was necessary to sustain catch levels, almost four-times more than previously estimated [13]. The requirements were only 2% for the open ocean, but fluctuated between 24% and 35% in fresh water upwelling systems and the continental shelf, which justifies the current concerns about sustainability and biodiversity.

In collaboration with *The Sea around Us* project, the Redefining Progress organization published The Fishprint of Nations [14], which extends the analysis of EF to aquatic resources. The fish print (FP) provides a method to quantify the pressure that the human population has on marine ecosystems at different scales and allows one to distinguish between levels of sustainable or unsustainable fisheries. They established the worldwide FP between 1950 and 2003, including 149 countries, and their results were interpreted as "unsustainable levels of fishing" probably since mid-1970, where 60.4% of countries had a negative ecological balance in 2003. The highest deficit scores were obtained by Japan, Indonesia, China, Philippines, Thailand and Norway. Mexico resulted also in a negative ecological balance in terms of its biocapacity, occupying the 23rd place.

Tyedmers et al. [15] quantified fuel consumption due to sea fishing, using statistical data from more than 250 fisheries around the world, and found that about 50 million liters of fuel were consumed during the capture of about 80 million tons of marine resources. This represents 1.2% of world oil consumption and directly emits more than 130 million tons of carbon dioxide into the atmosphere. From an efficiency perspective, the energy content of the fuel consumed by the world's fisheries is 12.5-times greater than the energy content of protein in the catch [15].

The State of World Fisheries and Aquaculture report 2008, stated that these activities had a small, but significant contribution to the emission of greenhouse gases during fishing operations, transportation, processing and storage of the product captured [10].

Regarding the impact of fishing on marine ecosystem, Galván-Piña [16] constructed a trophic model based on Ecopath (a program used to characterize the trophic structure and function of an ecosystem based on the mass balance [17]) to describe the structure and biomass flows of the ecosystem at the continental shelf along the south coast of the state of Jalisco and the north coast of Colima state. The model includes 38 functional groups, where 22 are fishes; nine are invertebrates and a group of marine mammals, seabirds, zooplankton, phytoplankton, dead fish and detritus. He found that most of the significant negative impacts were on the groups of higher trophic levels, confirming the hypothesis about the negative effect of fishing on the ecosystem. Therefore, he proposed as a strategy for the area to increase the fishing effort by 10% for gillnet fleet, to increase the commercial diving fleet three times and to reduce the shrimp fleet by 10%.

Most fisheries scientists agree that fisheries have declined in the world in the past 20 years. Fishermen along the coast of Jalisco perceive an outflow of resources related to poor catches. This situation is not reflected by the official catch data in the last 20 years in the region, but the fishing effort is greater, as perceived by fishermen. In this global and regional context, the aim of this study was to determine the sustainability of coastal fisheries along the coast of Jalisco, under the approach of the FP, based on primary productivity required (PPR) to catch resources in the period from 2002 to 2012.

RESULTS

The catch totaled 20,448.2 tons between 2002 and 2012. Marked variations are observed; there was a steady increase, reaching the highest records in 2008, and a sharp decline in 2009 (Figure 2). The best represented group in the total catch were fishes (about 84.9% of the biomass); the second were mollusks (12.8%), and the rest were crustaceans. In official catch records, there were 73 groups identified, where 67 corresponded to fish. The most frequent species in catches were snapper (Lutjanidae family), grouper (Serranidae family), octopus (*Octopus hubbsorum*) and Pacific sierra (*Scomberomorus sierra*), which together account for 47.5% of the total catch during the period analyzed (Figure 3).

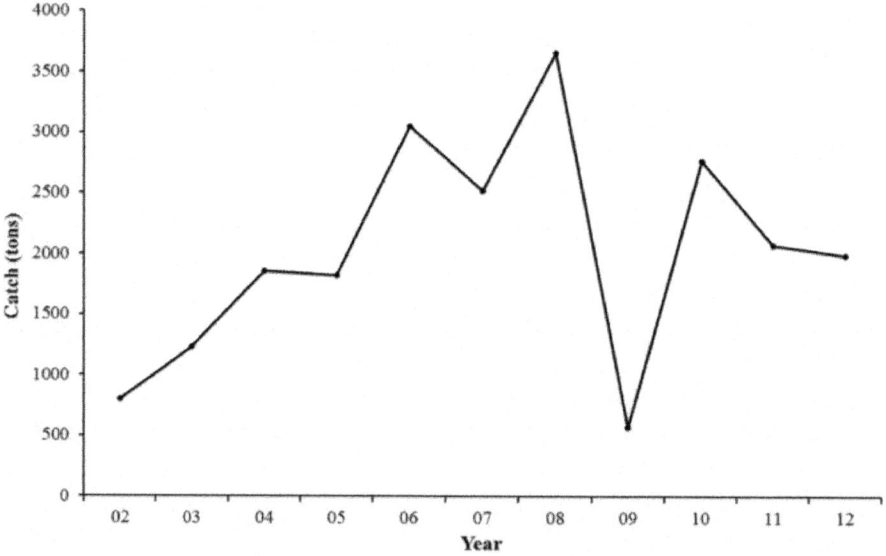

Figure 2: Annual catch trends off the coast of Jalisco in the period 2002–2012. Data obtained from SAGARPA (Secretaría de Agricultura, Ganadería, Desarrollo Rural, Pesca y Alimentación (Secretariat of Agriculture, Livestock, Rural Development, Fisheries and Food)).

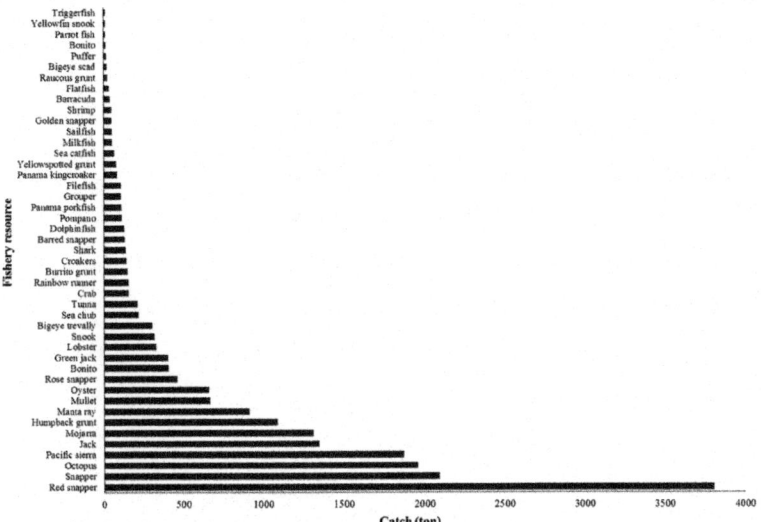

Figure 3: Catches off the coast of Jalisco in the period 2002–2012. Resources with catches higher than 10 tons are shown.

The fish caught by artisanal fishermen were classified into 39 groups, with trophic levels ranging between two (parrot fish, *Scarus compressus*) and 4.5 (roosterfish, *Nematistius pectoralis*) (Table 1).

PPR to support the catch was calculated to be 598,642 tons; the groups that require greater primary productivity were carangids, the Pacific sierra and scombrids (Table 1, Figure 4).

Table 1: Trophic level, trophic category, catch and primary production required (PPR) of fishery production along the coast of Jalisco during the period 2002–2012

Trophic Group	Trophic Level	Trophic Category	Catch (tons)	PPR (tons)
Carangids	3.63	Carnivorous	2336	140,616
Pacific sierra	3.72	Carnivorous	1869	138,411
Scombrids	4.09	Carnivorous	627	108,850
Lutjanus peru adults	3.17	Carnivorous	3803	79,376
Hemulids	3.00	Carnivorous	1451	20,475
Other lutjanids	2.62	Carnivorous	2727	16,042
Octopus	2.69	Carnivorous	1957	13,525
Sharks	3.79	Carnivorous	144	12,529
Sphyrenids	4.28	Carnivorous	42	11,293
Gerreids	2.78	Carnivorous	1306	11,105
Billfishes	4.03	Carnivorous	59	8921
Dolphin fish	3.57	Carnivorous	129	6763
Rays	2.70	Carnivorous	909	6429
Gasterosteids	3.03	Carnivorous	327	4944
Scienids	3.05	Carnivorous	230	3642
Arius	3.57	Carnivorous	67	3513

Kyphosids	2.94	Herbivorous	215	2642
Serranids	3.10	Carnivorous	133	2363
Mollusks	2.10	Herbivorous	658	1169
Mugilids	2.01	Herbivorous	663	957
Filefish	2.76	Carnivorous	107	869
Belonidae	4.46	Carnivorous	2	814
Ten pounder	4	Carnivorous	4	564
Other crustaceans	2	Carnivorous	327	461
Balistids	3.34	Carnivorous	12	370
Belonids	4.50	Carnivorous	1	446
Lobotes pacificus	4.04	Carnivorous	2	309
Other fishes	3.08	Omnivore	12	204
Pleuronectids	2.69	Carnivorous	34	235
Brachyura	2.02	Omnivore	156	231
Tetraodontids	2.91	Omnivore	18	206
Peneids	2.10	Herbivorous	49	87
Chanidae	2.03	Herbivorous	52	79
Sparids	3.52	Carnivorous	2	93
Flying fish	4.01	Carnivorous	0.2	29
Nematistius	4.50	Carnivorous	0.1	45
Scarids	2.00	Herbivorous	13	18
Clupeids	2.89	Omnivore	1	11
Pristigasterids	3.31	Carnivorous	0.2	6

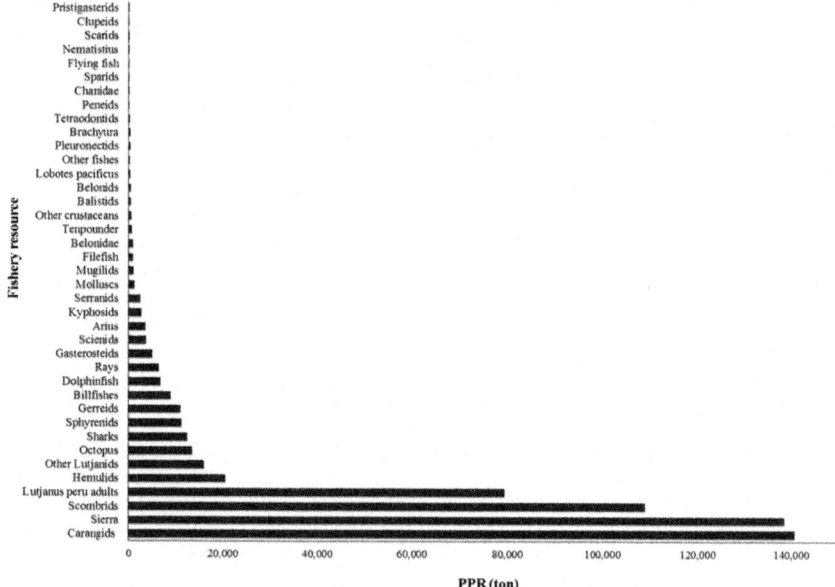

Figure 4: Primary productivity required in marine catches off the coast of Jalisco.

The footprint of coastal catches in Jalisco for all years analyzed did not exceed the biocapacity of the area. The year with the highest mark was 2008, representing 0.05% of the biocapacity of the fishing area. Consequently, coastal catches in Jalisco State had a positive ecological balance every year, measured in global hectare units (gha). That means the average productivity of the entire area of biologically productive land and sea in the world in a given year (Table 2).

Table 2: Mean trophic level, fishprint, biocapacity and ecological balance of catches on the coast of Jalisco from 2002 to 2012. gha, global hectare unit

Year	Catch (tons)	Mean Trophic Level	FP (gha)	Biocapacity (gha)	Ecological Balance (gha)
2002	789	3.1	21,716	35,872,580	35,850,864
2003	1025	3.1	28,152	35,840,538	35,812,386
2004	1524	3.1	40,802	35,661,473	35,620,671
2005	1478	3.1	40,494	35,541,767	35,501,273
2006	2382	3.1	69,916	35,192,021	35,122,105
2007	2289	3.2	76,817	35,364,783	35,287,966
2008	1970	3.2	123,126	35,232,310	35,109,184
2009	546	3.0	21,744	34,969,782	34,948,038
2010	2724	3.2	104,556	34,969,782	34,865,225
2011	2074	3.1	103,673	31,975,955	31,872,282
2012	1986	3.1	89,047	31,658219	31,569,172

DISCUSSION AND CONCLUSIONS

In order to promote sustainable fishing practices, the UN Convention on the Law of the Sea requires countries to maintain and restore fisheries using best management practices by controlling their catches. Such measures have been designed to maintain or restore populations at adequate levels to produce the maximum sustainable yield according to environmental and economic factors.

The fishing footprint tool enables users to quantify, with a non-traditional approach, the impact on marine ecosystems and can be used as a tool for assessing the sustainability of catch levels in countries or regions, taking into account the effects of trophic level and the biocapacity of a specific area of the ocean, where the catch is expressed in terms of PPR, which is largely a function of the trophic level of the species caught.

In terms of catch, it is important to consider the unreliability of the official catch data in Mexico, as many other countries, due to the unreported and illegal catch data by fishermen, which was an estimate of 15% of the total catch for the 2000–2003 period in the Eastern Central Pacific region [18].

That means that the real catch for the coast of Jalisco would have been 2283.9 instead of 1986 tons for the year 2012 (Table 2). This fact would affect the PPR, as well as the FP for the same year. However, the ecological balance would still be positive. Even assuming a number over 20% of illegal or unreported catch for this region, PPR and FP would not be modified significantly.

Overfished zones now cover most of the world's oceans, including ones of low productivity [19], coupled with the evident global decline in catches since the late 1990s [7], indicating that fishing has reached its limit.

If biocapacity for marine fisheries is depleted and excessive PPR occurs in many regions of the world, the option for sustainable fishing is to reduce the PPR, focus fishing on lower trophic levels, reduce the high trophic catch level, establish protected areas and eliminate destructive fishing practices.

However, it should be remembered that capture at the maximum sustainable level in the lower trophic levels can also have large impacts on the ecosystem [20].

The results for the coast of Jalisco show that the FP has quadrupled over a period of 10 years, while the biocapacity has decreased and is related to the local population increase. The average trophic level catch between 2002 and 2012 was 3.1, ranging from 3.0 to 3.2, which implies that it has remained at average levels.

The trophic level decreasing phenomenon in fisheries catch was released in 1998 [21]; since, then researchers have been looking for evidence at regional and local scales.

While average catch trophic level is noted as an indicator of the sustainability of catches, it is important to note that the decline of the trophic level of fish catch can not only be due to ecological problems, but also to changes in the price of fish products, leading to the capture of specific resources, as well as to the natural increase of low level trophic species or the development of new fisheries targeted at low level trophic species.

The existence of species at lower trophic levels in the Jalisco coast can be explained by the fact that in this area, the continental shelf is narrow, and the presence of an oxygen minimum zone represents a physiological barrier to vertical migration [22].

Other environmental characteristic of the area that may have an influence, in days' or months' scale, in the biological behavior of the species providing favorable or unfavorable habitat conditions, is the presence of a shallow thermocline in the area [23].

The productivity of the coastal zone [2] is related to spatial heterogeneity, since there may be significant differences between littoral and pelagic systems and particularly between biotopes: bays, beaches, estuaries and coastal lagoons [24,25]. Moreover, the small variation in the trophic level over time is consistent with multispecies fisheries, low technological levels and ecosystems with high biodiversity [26,27].

In the same way that the world fisheries exceed biocapacity of the ocean [14] at the national level, Mexico presented a negative ecological footprint and biocapacity balance, while the FP at the coast of Jalisco represented less than 1% of the biocapacity of the fishing area, probably due to the type of fishing performed in the area. It is then concluded under this approach that on the coast of Jalisco, a healthy fishery exists, since there is a positive balance between biocapacity and FP in the area.

The average fishprint at the country level was calculated as 1017.17 million gha/year for Mexico in 2003 [15]. The one we calculated for Jalisco's small scale fishing was 65,458 gha/year (just 3% of the country level), but it cannot be compared to another fishprint along the coast of Mexico, because this approach has not been used to assess fisheries.

The low fishprint can be explained on the basis of the characteristics of fisheries along the coast of Jalisco, which are very similar to the artisanal fishing along the rest of the country. The main characteristic is that Jalisco's marine biocapacity is big enough, but little fishing is done. Jalisco State contributes only 1.3% to the national catch volume. The catch volume is related to the ability for fishing, so although there are 44 fishing cooperatives, they have small boats, mostly less than 30 feet in length, that operate largely without a motor, and till 2011, there were only two docks for unloading catch [6]. Under these conditions and the physical ones already mentioned, the catch is minimal compared nationally.

In addition, fishermen need less fuel for boats and are aware of their dependence on marine resources [26]. The catch is done using fishing gear, like gillnets, hand lines, cast nets, long lines, seines, crab rings and diving equipment, in a few cases. In general, it is more selective, up to 20-times more than industrial fishing [6]. All of these characteristics contribute to sustainability.

Although our results indicate sustainable fishing in the region it can not only be an outcome of FP assessment, it should also include the results of other socioeconomic and environmental indicators, but it is an important tool for fisheries management.

METHOD

Data Collection

Official reported fish catch data were obtained for the coast of Jalisco between 2002 and 2012. These were clustered by taxonomic groups, and the total catch wet weight in tons (biomass) per year for each group was obtained. Then, each one was assigned to a trophic level category, as proposed by Froese and Pauly [28], Galvan-Piña and Arreguín-Sánchez [29] (Table 3).

Table 3: Source for trophic level assignation

Trophic Groups	Author
Lutjanus peru adults	Galvan-Piña and Arreguín-Sánchez (2008)
Arius	Froese and Pauly (2010)
Balistids	Froese and Pauly (2010)
Belonids	Froese and Pauly (2010)
Filefish	Froese and Pauly (2010)
Brachyura	Galvan-Piña and Arreguín-Sánchez (2008)
Carangids	Galvan-Piña and Arreguín-Sánchez (2008)
Chanidae	Froese and Pauly (2010)
Ten pounder	Galvan-Piña and Arreguín-Sánchez (2008)
Clupeids	Froese and Pauly (2010)
Dolphin fish	Galvan-Piña and Arreguín-Sánchez (2008)
Scienids	Galvan-Piña and Arreguín-Sánchez (2008)
Scombrids	Galvan-Piña and Arreguín-Sánchez (2008)
Gasterosteids	Galvan-Piña and Arreguín-Sánchez (2008)
Gerreids	Galvan-Piña and Arreguín-Sánchez (2008)
Hemulids	Galvan-Piña and Arreguín-Sánchez (2008)
Kyphosids	Froese and Pauly (2010)
Lobotes pacificus	Froese and Pauly (2010)
Mollusks	Galvan-Piña and Arreguín-Sánchez (2008)
Mugilids	Froese and Pauly (2010)
Nematistius	Froese and Pauly (2010)
Other crustaceans	Galvan-Piña and Arreguín-Sánchez (2008)
Other lutjanids	Galvan-Piña and Arreguín-Sánchez (2008)
Other fishes	Galvan-Piña and Arreguín-Sánchez (2008)
Peneids	Galvan-Piña and Arreguín-Sánchez (2008)
Billfishes	Galvan-Piña and Arreguín-Sánchez (2008)
Pleuronectids	Galvan-Piña and Arreguín-Sánchez (2008)
Pristigasterids	Froese and Pauly (2010)
Octopus	Galvan-Piña and Arreguín-Sánchez (2008)
Rays	Galvan-Piña and Arreguín-Sánchez (2008)
Belonidae	Froese and Pauly (2010)
Scaridae	Froese and Pauly (2010)
Serranids	Galvan-Piña and Arreguín-Sánchez (2008)
Sierra	Galvan-Piña and Arreguín-Sánchez (2008)
Sparids	Froese and Pauly (2010)
Sphyrenids	Froese and Pauly (2010)
Tetraodontids	Galvan-Piña and Arreguín-Sánchez (2008)
Sharks	Galvan-Piña and Arreguín-Sánchez (2008)
Flying fish	Froese and Pauly (2010)

Primary Productivity Required to Sustain Fisheries

In order to calculate the PPR for capturing each group of species or to sustain fishing in a determined trophic level, the following equation [30] was used:

$$PPR = CC \times DR \times \left(\frac{1}{TE}\right)^{TL-1}$$

(1)

CC is the carbon content of the total catches, *DR* is the discarded rate by catch, *TE* is the transfer efficiency of biomass between trophic levels and *TL* is the trophic level of the group or species. A ratio of 9:1 was used to convert units of wet weight (ton) to grams of carbon. An overall average value of 1.27 for all species is assigned for *DR*, which means that for each ton of fish obtained, 0.27 tons constitute the catch [12]. This rate is used due to the lack of local by-catch data, as a constant factor in the corresponding equation, assuming that the by-catch trophic level is the same as that of the species caught. The *TE* value is also constant and equal to 0.1 for all groups, meaning that 10% of the biomass is transferred to successive trophic levels [12].

The corresponding trophic level for each group for the coast of Jalisco was taken from that proposed by Galvan-Piña and Arreguín-Sánchez [29]; the missing data in this study were obtained from the Fishbase [28] for the central Mexican Pacific area.

Fishing Footprint and Biocapacity

The footprint of a given area *A* in a year is:

$$FP_A = \frac{TPP_{Ay}}{YFPP_{Gy}} \times EQF_{NPP}$$

(2)

where TPP_{Ay} is the total primary productivity required in the area, $YFPP_{Gy}$ is the overall yield factor, calculated dividing TPP_{Ay} by the global biocapacity for marine fisheries (*BCg*), and EQF_{NPP} an equivalence factor for marine fisheries (1.66). This factor convert a specific land type into a global hectare, a universal unit of biologically productive area.

In turn, the global biocapacity of marine fisheries was calculated by multiplying the open ocean area, including its equivalency factor by the surface of the exclusive economic zone (EEZ) with its equivalence factor. Table 4 provides the calculations and indicates a marine biocapacity of 33.94 billion global hectares.

Table 4: Global marine fisheries biocapacity [14]

Region	Area (billion ha)	Equivalence Factor	Biocapacity (billion ha)
Exclusive economic zones	13.88	1.66	23.18
Open oceans	22.41	0.48	10.76
Total	36.29	0.94	33.94

The fishing area biocapacity (BC) for the coast of Jalisco State was obtained through the following formula:

$$BC = \frac{BC_G}{POP_G} \times POP_A$$

(3)

BC_G is the global biocapacity of marine fisheries (equivalent to 23.18 billion hectares), which is obtained by multiplying the area of EEZ (13.88 billion hectares) and equivalence factor EEZ (1.66); POP_G is the world population (taken from Population Reference Bureau [31]), and POP_A is the population of the area of Jalisco State (taken from National Population Council (CONAPO) [32]).

The calculation takes into account the fact that the fishing fleet in a given country can catch fish in different parts of the world. Some countries have international or bilateral agreements allowing them to capture fish in a much larger area than their own EEZ. Another advantage of this approach is the implication that landlocked countries also have biocapacity, which by default would be that its fishing footprint exceeds its biocapacity.

ACKNOWLEDGMENTS

The authors are deeply thankful to the Consejo Nacional de Ciencia y Tecnología (CONACYT), to Secretaría de Agricultura, Ganadería, Desarrollo Rural, Pesca y Alimentación (SAGARPA) for sharing the historical catch data and to the University of Guadalajara for providing research facilities during the research. The authors thank Claudia de Jesús Avendaño for elaborating the map of the study area.

AUTHOR CONTRIBUTIONS

Bravo-Olivas and Chávez-Dagostino designed the research, collected data and wrote the paper. López-Fletes collaborated in the literature review, performed research, checked the statistical results and extensively updated the paper. Espino-Barr co-designed the research and edited the paper. All authors read and approved the final manuscript, analyzed the data and took part in the discussion conjointly.

REFERENCES

1. National Institute Of Statistics and Geography (INEGI). *Los Municipios de Jalisco. Colección: Enciclopedia de los Municipios de México*; Secretaría de Gobernación: Ciudad de México, Mexico, 1995; p. 264. (In Spanish)

2. Lara-Lara, J.R.; Arenas, F.V.; Bazán, G.C.; Díaz, C.V.; Escobar, B.E.; García-Abad, M.C.; Gaxiola, C.G.; Robles, J.G.; Sosa, A.R.; Soto,

G.L.A.; *et al.* Los ecosistemas marinos. In *Capital Natural de México. Volume I: Conocimiento Actual de la Biodiversidad*; Comisión Nacional para el Conocimiento y Uso de la Biodiversidad: Ciudad de Mexico, Mexico, 2008; pp. 135–159. (In Spanish)

3.   Roden, G. Termohaline structure and baroclinic flow across the Gulf of California entrance and in the Revillagigedo Island region. *J. Phys. Oceanogr.* 1972, *2*, 177–183.

4.   Gallegos, A.; Rodríguez, R.; Márquez, E.; Lecuanda, R.; Zavala, J. Una climatología de la temperatura de la superficie del mar de las aguas oceánicas adyacentes a las costas de Jalisco, Colima y Michoacán, México: 1996–2003. In *Los Recursos Pesqueros y Acuícolas de Jalisco, Colima y Michoacán*; Jiménez-Quiroz, M.C., Espino-Barr, E., Eds.; Secretaría de Agricultura, Ganadería, Desarrollo Rural, Pesca y Alimentación: Manzanillo, México, 2006; pp. 17–28. (In Spanish)

5.   Food and Agriculture Organization (FAO). CWP Handbook of Fishery Statistical Standards. Available online: http://www.fao.org/fishery/cwp/handbook/h/en (accessed on 29 October 2014).

6.   Martínez-González, P.; Corgos, A. Pesca artesanal en la costa de Jalisco. Conflictos en torno a la conservación biocultural. *Obs. Desarro.* 2013, *7*, 38–46. (In Spanish).

7.   Food and Agriculture Organization (FAO). *The State of World Fisheries and Aquaculture 2012*; Fisheries and Aquaculture Department: Rome, Italy, 2012; p. 209.

8.   Food and Agriculture Organization (FAO). *Indicators for Sustainable Development of Marine Capture Fisheries*; Fisheries and Aquaculture Department: Rome, Italy, 1999; p. 68.

9.   Food and Agriculture Organization (FAO). *The State of World Fisheries and Aquaculture 2014*; Fisheries and Aquaculture Department: Rome, Italy, 2014; p. 223.

10.   Food and Agriculture Organization (FAO). *The State of World Fisheries and Aquaculture 2008*; Fisheries and Aquaculture Department: Rome, Italy, 2009; p. 196.

11.   Wackernagel, M.; Rees, W. *Our Ecological Footprint. Reducing Human Impact on the Earth*; New Society Publishers: Philadelphia, PA, USA, 1996; p. 160.

12.   Pauly, D.; Christensen, V. Primary production required to sustain global fisheries. *Nature* 1995, *374*, 255–257.

13.   Christensen, V.; Pauly, D. The ECOPATH II—A software for

balancing steady-state ecosystem models and calculating network characteristics. *Ecol. Model.* 1992, *61*, 169–185.

14. Talberth, J.; Venetoulis, J.; Wolowicz, K. Recasting Marine Ecological Fishprint Accounts. Available online: http://rprogress. org/publications/2006/Fishprint%20Technical%20Supplement. pdf (accessed on 10 September 2014).

15. Tyedmers, P.; Watson, R.; Pauly, D. Fueling global fishing fleets. *AMBIO* 2005, *34*, 635–638.

16. Galván-Piña, V. Impacto de la Pesca en la Estructura, Función y Productividad del Ecosistema de la Plataforma Continental de las Costas de Jalisco y Colima, México. Ph.D. Thesis, Centro Interdisciplinario de Ciencias Marinas-Instituto Politécnico Nacional, La Paz, Mexico, 2005; p. 106. (In Spanish).

17. Ulanowicz, R.E. *Growth and Development: Ecosystems Phenomenology*; Springer-Verlag: New York, NY, USA, 1986; p. 203.

18. Agnew, D.D.; Pearce, J.; Pramod, G.; Peatman, T.; Watson, R.; Beddington, J.R.; Pitcher, T.J. Estimating the worldwide extent of illegal fishing. *PLoS One* 2009, *4*, 1–8.

19. Swartz, W.; Sala, E.; Tracey, S.; Watson, R.; Pauly, D. The spatial expansion and ecological footprint of fisheries (1950 to present). *PLoS One* 2010, *5*, 1–6.

20. Smith, A.D.M.; Brown, C.J.; Bulman, C.J.; Fulton, E.A.; Johnson, P.; Kaplan, I.C.; Lozano-Montes, H.; Mackinson, S.; Marzloff, M.; Shannon, L.J.; *et al.* Impacts of fishing low-trophic level species on marine ecosystems. *Science* 2011, *333*, 1147–1150.

21. Pauly, D.; Christensen, V.; Dalsgaard, J.; Froese, R.; Torres, F. Fishing down marine food webs. *Science* 1998, *272*, 860–863.

22. Hendrickx, M.E.; Serrano, D. Impacto de la zona de mínimo de oxígeno sobre los corredores pesqueros en el Pacífico Mexicano. *Interciencia* 2010, *35*, 12–18. (In Spanish).

23. Filonov, A.E.; Monzón, C.; Tereshchenko, I. Acerca de las condiciones de generación de las ondas internas de marea en la costa occidental de México. *Cienc. Mar.* 1996, *22*, 255–272. (In Spanish).

24. Caddy, J.F.; Sharp, G.D. *An Ecological Framework for Marine Fishery Investigations*; FAO: Rome, Italy, 1986; p. 152.

25. Longhurst, A.; Pauly, D. *Ecology of Tropical Oceans*; Academic Press: San Diego, CA, USA, 1987; p. 407.

26. Bravo-Olivas, M.L. Huella Ecológica de las Pesquerías Ribereñas en

la Costa de Jalisco. Ph.D. Thesis, Universidad de Guadalajara, Puerto Vallarta, Mexico, 2014; p. 167. (In Spanish).

27. Reyes-Bonilla, H.; Calderón-Aguilera, L.E.; Aburto-Oropeza, O.; Díaz-Uribe, J.G.; Pérez-España, H.; del Monte-Luna, P.; Lluch-Cota, S.E.; López-Lemus, L.G. La disminución en el nivel trófico de las capturas pesqueras en México.*Ciencia* 2009, *60*, 1–9.

28. Froese, R.; Pauly, D. *FishBase 2010: Conceptos, Estructura y Fuentes de Datos*; Froese, R., Pauly, D., Eds.; ICLARM: Manila, Philippines, 2010; p. 322. (In Spanish)

29. Galván-Piña, V.H.; Arreguín-Sánchez, F. Interacting industrial and artisanal fisheries and their impact on the ecosystem of the continental shelf on the Central Pacific coasts of Mexico. In Proceedings of the 4th World Fisheries Congress, Vancouver, BC, Canada, 2–6 May 2004; pp. 587–600.

30. Pérez-España, H.; Abarca-Arenas, L.G.; Jiménez Badillo, M.L. Is fishing-down trophic web a generalized phenomenon? The case of Mexican fisheries. *Fish. Res.* 2006, *79*, 349–352.

31. Haub, C.; Kaneda, T. *2012 World Population Data Sheet*; Population Reference Bureau: Washington, DC, USA, 2012; p. 20.

32. National Popularion Council (CONAPO). Dinámica demográfica 1990–2010 y Proyecciones de Población 2010–2030. Available online: http://www.omi.gob.mx/es/CONAPO/Proyecciones_Analisis (accessed on 10 September 2014). (In Spanish).

Chapter 6

MIGRATION AND FISHERIES OF NORTH EAST ATLANTIC MACKEREL (SCOMBER SCOMBRUS) IN AUTUMN AND WINTER

Teunis Jansen[1], Andrew Campbell[2], Ciarán Kelly[2], Hjálmar Hátún[3], and Mark R. Payne[1, 4]

[1]DTU AQUA - National Institute of Aquatic Resources, Technical University of Denmark, Charlottenlund, Denmark

[2]Fisheries Ecosystems Advisory Services, Marine Institute, Galway, Ireland

[3]Faroe Marine Research Institute, To´rshavn, Faroe Islands

[4]ETHZ Swiss Federal Institute of Technology, Zurich, Switzerland

ABSTRACT

It has been suggested that observed spatial variation in mackerel fisheries, extending over several hundreds of kilometers, is reflective of climate-driven changes in mackerel migration patterns. Previous studies have been unable to clearly demonstrate this link. In this paper we demonstrate correlation between temperature and mackerel migration/distribution as proxied by mackerel catch data from both scientific bottom trawl surveys and commercial fisheries. We show that mackerel aggregate and migrate distances of up to 500 km along the continental shelf edge from mid-November to early March. The path of this migration coincides with the location of the relatively warm shelf edge current and, as a consequence of this affinity, mackerel are guided towards the main spawning area in the south. Using a simulated time series of temperature of the shelf edge current we show that variations in the timing of the migration are significantly correlated to temperature fluctuations within the current. The proposed proxies for mackerel distribution were found to be significantly correlated. However, the correlations were weak and only significant during periods without substantial legislative or technical developments. Substantial caution should therefore be exercised when using such data as proxies for mackerel distribution. Our results include a new temperature record for the shelf edge current obtained by embedding the available hydrographic observations within a statistical model needed to understand the migration through large parts of the life of adult mackerel and for the management of this major international fishery.

INTRODUCTION

Changes in global climate and the aspiration for sustainable fisheries management have highlighted the requirement for improved understanding of the effects of the marine climate on the behaviour of important fish species [1]. Mackerel (*Scomber scombrus*) is an abundant migratory pelagic fish in the north-east Atlantic, where it plays an important ecological role by feeding on zooplankton and on the pelagic larval and juvenile stages of a number of commercially important fish stocks [2], [3]. Furthermore, mackerel is itself targeted by whales, fish and a large pelagic fishing fleet with annual landings of between 500 000 and 1 000 000 tonnes [2], [4]. The largest mackerel fishery targets and follows mackerel aggregations throughout autumn and winter. Marked historical changes in the timing and spatial distribution of this fishery have been observed, but remain unexplained [4]–[7]. The fishing fleet is composed of modern pelagic trawlers and seiners that use sonar to locate schools of adult mackerel and are highly mobile, regularly steaming hundreds of kilometres from port. As a result of this adaptive behaviour, it is feasible that the observed changes in the timing and spatial distribution of commercial landings are representative of the spatiotemporal dynamics of the mackerel population. It has been hypothesized that temperature is an important modulator of the autumn/winter spawning migration. An acoustic and oceanographic survey in December 1995 demonstrated a relationship between the location of mackerel in the Northern North Sea prior to the onset of migration and the local temperature field [8]. It has also been noted that mackerel behaviour appeared to be related to temperature while the mackerel stayed to the north and west of the Shetland [9], [10]. If the distribution of the fishery reflects the distribution of the mackerel and the mackerel distribution is related to the water temperature, then we would expect the temperature field to be reflected in the spatiotemporal distribution of the fishery. However, previous studies have not revealed any simple correlation between these variables [5]–[7].

Using fisheries independent data from scientific bottom trawl surveys and commercial landings statistics we investigate the mackerel migration from October to March and test

- whether data from commercial fisheries and scientific bottom trawl surveys can form the basis for useful proxies of the distribution of adult mackerel

- whether changes in the temperature of the shelf edge current are related to the significant temporal and spatial variation observed in these proxies

We consider our results in the light of other factors that influence the fishing fleet behaviour such as fisheries development, legislation and distance to home port. Finally, we discuss our findings within a larger oceanographic context of circulation patterns and global warming, review possibilities for hindcasts and forecasts, and implications for fisheries management.

MATERIALS AND METHODS

Fisheries Data

Quarterly landings in the autumn-winter fishery were used as reported to the International Council for Exploration of the Sea (ICES). Due to the fact that the autumn-winter fishery overlaps two calendar years, first quarter landings were treated as being a '5th' quarter of the previous year. Thus, Q4 landings are those reported in October–December and Q5 corresponds to January–March of the following year. The study area encompasses the northern limit of the reported catches and includes the majority of the total reported catch (83% in Q4 and 56% in Q5) (Figure 1).

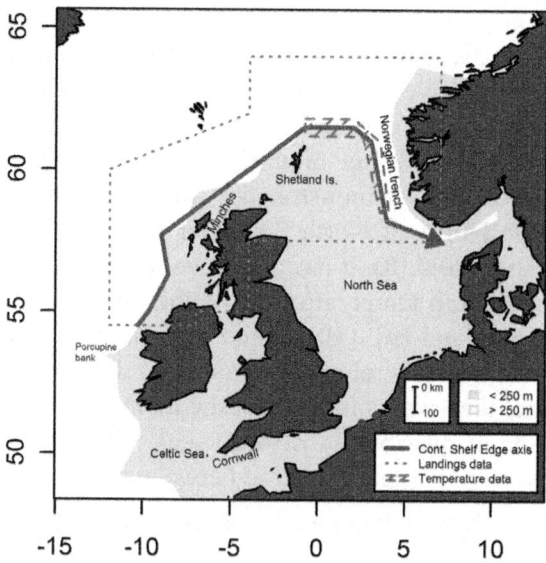

Figure 1: Map of study area and place names referred in the text.

Continental shelf marked in grey (bottom depth <250 m). Blue polygon indicates the study area. Blue bold arrow shows the Continental shelf edge axis. Red shaded area marks the area of temperature profiles.

doi:10.1371/journal.pone.0051541.g001

Commercial landings data were reported to ICES as quarterly totals per ICES statistical rectangle (1° latitude by 0.5° longitude). The position and time of the catch was assumed to be at the center of the reported rectangle and midway through the quarter. The landings consisted primarily (>95%) of adult fish [4].

To investigate the spatial variations in the behavior of the fleet the reported landings were projected onto a curvilinear 'Continental Shelf Edge' (CSE) axis in the style of [11], from 54.5 N 10.5 W in the south, and following the 200 m isobath, passing north of the Shetland Islands before turning south and following the Norwegian Trench into the North Sea (Figure 1). The total length of the CSE axis is approximately 1700 km. Each reported landing was projected onto the CSE axis by selecting the closest of 1000 equally spaced positions along the CSE axis. Distances were calculated based on great circle (WGS84 ellipsoid) distances. Both the position projected onto CSE axis and the distance of the reported landing from the axis were calculated and stored for further analysis.

The quarterly CSE axis distributions were then represented by a single metric for further comparison with temperature. Two alternative metrics were explored;

- the center of gravity of landings (CoG)
- the position of 50% cumulative landings (Po50%CL)

CoG was calculated by year and quarter as the weighted average of distances. The weighting factor was the mass (in kg) of each projected landing record. Po50%CL was calculated as the position along the CSE where the cumulative landings represented 50% of the total landings by year and quarter.

A literature survey and an interview with the skipper of a vessel that fished throughout the study period were carried out in order to identify periods where changes in the behavior of the commercial fishery were driven by factors other than mackerel behavior.

Bottom Trawl Survey Data

Data from international bottom trawl surveys (IBTS) carried out in quarter 1 (January–March) between 1985 and 2011 on the shelf out to 500 m were downloaded from the ICES repository (http://datras.ices.dk). The study area was limited to the area described for the commercial landings. Relatively few mackerel were caught outside the study area, e.g. in Kattegat/Skagerrak [12] and over 90% were from surveys in March. Further south, in the Bay of Biscay, mackerel arrive at the spawning grounds around the time of this survey [13]:

the present dataset therefore covers the northern part of the NEA mackerel population. Catch per Unit Effort (CPUE) of adult mackerel was calculated as catch in numbers per trawl hour, where adult mackerel were defined as being longer than 27 cm (most mackerel first spawn at the age of 2 (58%) and the mean length at age 2 in Q1 west of Scotland is 27 cm [4]). For ease of comparison with the commercial landings dataset, first quarter surveys were treated as being a '5th' quarter of the previous year. Hauls were projected onto the CSE axis as described for commercial landings and the CoG and Po50%CL of CPUEs calculated.

Temperature Data and Modelling

In the present study, we investigate links between water temperature and mackerel distribution that could support the hypothesis of a temperature-driven migration. The continental shelf edge current which flows along the shelf edge to the northwest of Scotland, north and then east of the Shetland Islands, along the western edge of the Norwegian trench and into the northern North Sea, is warmer than both the surrounding coastal waters and the oceanic waters off the shelf during winter (Figure 2) [8], [9]. It is the temperature of this water mass that is of interest in this study. Unfortunately, relevant observations are not available for the entire study period. A relevant temperature record was therefore obtained by embedding the available hydrographic observations within a statistical model. The modelled area is shown in Figure 1 and was selected because it is the coldest area of the warm core of the current (Figure 2) and therefore the area where cold avoidance by mackerel would be most pronounced. Also, there are a significant number of observations available for this area.

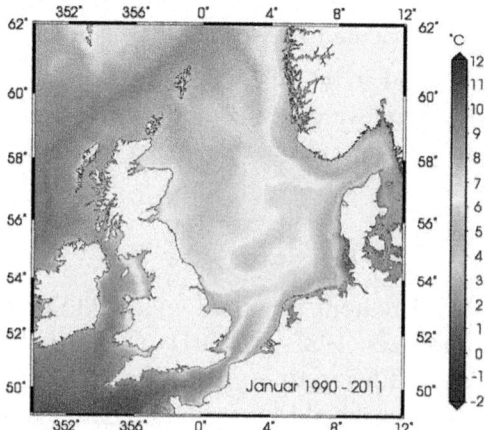

Figure 2: Map of average sea surface temperature in January 1990–2011 showing the relatively warm high-saline eastern Atlantic water flowing north-eastwards on and along the continental shelf edge, flanked by cooler water masses.

Temperature measurement measured by satellite and mapped with permission from Bundesamt für Seeschifffahrt und Hydrographie, Germany (www.bsh.de).

doi:10.1371/journal.pone.0051541.g002

It is within this core of relatively warm water in the northern North Sea that acoustic surveys found mackerel to aggregate in 50–220 m depth in early winter [8]–[10], [14]. Due to the fact that water is cooled throughout the winter, both downstream (along) and away from the CSE, temperature was modeled with year, day of year, distance parallel (CSE) and perpendicular (dCSE) to the CSE axis as explanatory variables i.e:

$$Temperature = \beta_4(Year) + \beta_3 S_3(Day) + \beta_2 S_2(dCSE)$$

$$+ \beta_1 S_1(CSE) + \beta_0 + \varepsilon,$$

where *CSE* is the distance along the CSE axis from the start of the axis (in the south) to the projected sample position, *dCSE* is the distance from the sample site to the projected position, *day* is the number of days elapsed in the year, from 1st of February (day 32) to 31st of January (Day 386). *Year* is the year of the observation and, *S()* is the penalized cubic regression spline smoothing function implemented in the "mgcv"-R-package as cardinal spline [15]. *Day*, *CSE* and *dCSE* were thus modeled as smoothed predictor variables with smoothing parameters (k=number of "knots") set to 3, in order to allow for a non-linear temperature development through the season and along the CSE whilst avoiding overfitting, whilst *Year* is treated as a categorical factor (i.e. one parameter per year). 1056 temperature profiles from CTD stations and bottle sampling between November and January were downloaded from the ICES hydrographic database [16] and used to fit the model using the "mgcv" package in R [15]. Model building was done by sequentially removing non-significant parameters (*i.e.* those with p>0.05). The final model was then used to predict a time series of temperatures in early winter (15th of December), at the center of the area (1326 km along CSE axis from starting point) where mackerel were known to be present [8].

For validation purposes, we compared the GAM temperature time series with

- a similarly modeled time series further upstream (west of Scotland, 35 km from CSE in the area 55–65°N 10°W-5°E) in February–March, and

- a coarser modeled and validated dataset of sea surface temperatures (SST) obtained from the Hadley Centre SST data set (HadSST2) [17], by averaging over a larger geographical box covering the North Sea-SE Norwegian Sea area (55°N- 65°N, 0–5°E) and including the months from November to January.

Finally, correlation analysis of the mackerel distribution metrics described above and modelled temperature field were performed. All correlation analyses were adjusted for autocorrelation if this exceeded the 95% confidence limits of white noise ($\pm 2\sqrt{N^{-1}}$, where N is sample size)[18]. Adjustments were done by substituting the degrees of freedom with the effective number of degrees of freedom [19].

RESULTS

The final temperature model identified *Year, Day of Year* and *CSE* as significant explanatory variables. In line with expectations, temperature decreased through the winter (Figure 3, p<0.001) and downstream along the CSE axis (Figure 4, p<0.001). The modeled temperature time series shows an overall increase throughout much of the study period with a decrease in the most recent years (Figure 5). The model explained 81% of the variance in the data (adj. R^2=0.81). As a rough validation for the overall development of the temperature time series, we found it to be significantly positively correlated to a modeled temperature time series in the area west of Scotland in February–March 1985–2010 (P=0.005, R^2=0.36, same GAM model structure as the primary temperature series), and also to the Hadley time series of sea surface temperature in November–January 1948–2010 (P<0.001, R^2=0.48).

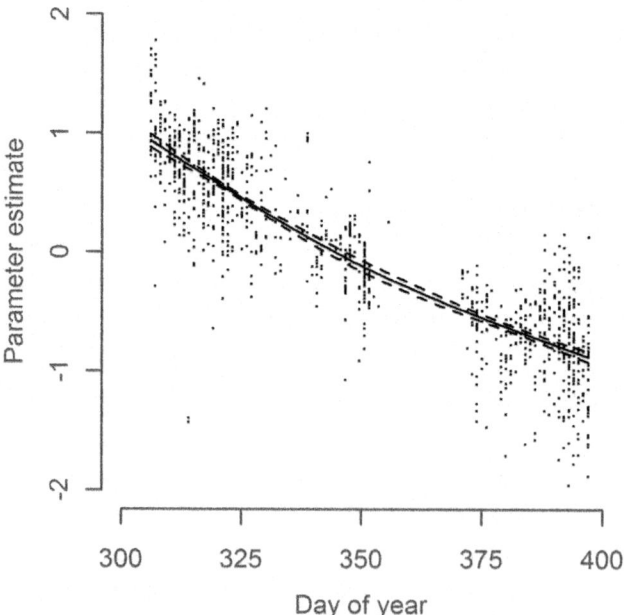

Figure 3: *Day of Year* **parameter in the temperature model.**

Parameter estimate (solid line) with 95% confidence interval (dashed lines) and partial residuals (dots) relative to mean predicted value.

doi:10.1371/journal.pone.0051541.g003

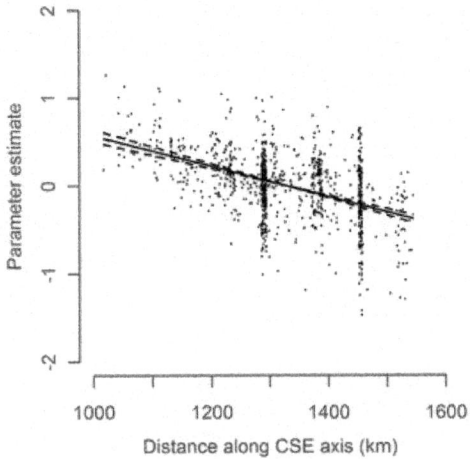

Figure 4: *CSE* **parameter in the temperature model.**

Parameter estimate (solid line) with 95% confidence interval (dashed lines) and partial residuals (dots) relative to mean predicted value.

doi:10.1371/journal.pone.0051541.g004

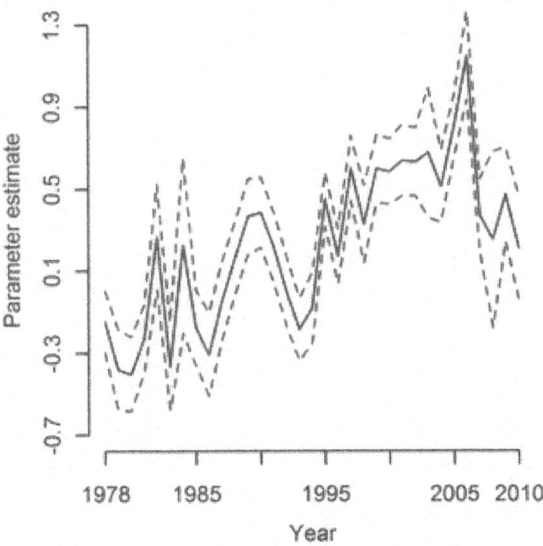

Figure 5: *Year* **parameter in the temperature model.**

Parameter estimate (solid line) with 95% confidence interval (dashed lines) relative to mean predicted value.

doi:10.1371/journal.pone.0051541.g005

There was a strong tendency for commercial and bottom trawl catches to be associated with the area along the CSE axis, with 74% of the commercial landings in Q4, 92% in Q5 and 87% of the survey catches were taken within a 75 km distance of the CSE axis (Figure 6). We therefore chose to reduce the complexity of the spatial distributions by disregarding the across-axis information, *i.e.* considering the catches projected onto the CSE axis. Visual inspection of Center of Gravity (CoG) and Position of 50% Cumulative Landings (Po50%CL) overlaid on the distributions (Figure 7) indicates that both metrics are appropriate representations of the commercial landings and survey catches.

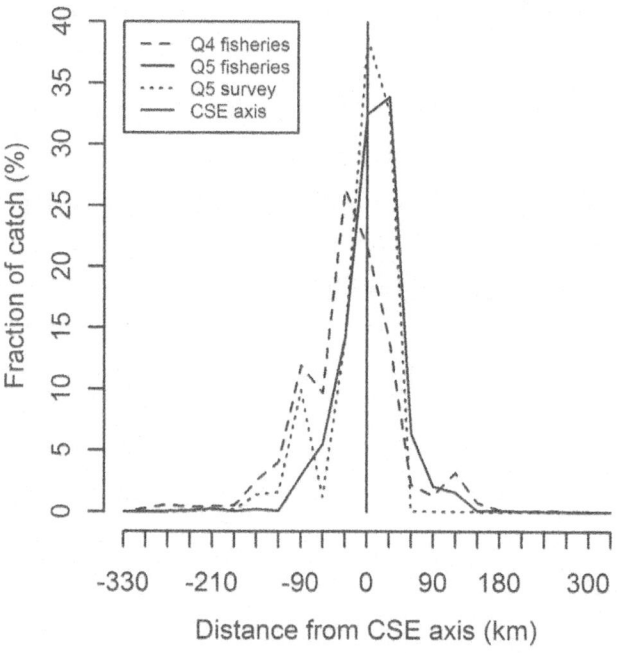

Figure 6: Distance from catch position to continental shelf edge (CSE) axis.

Positive values are off the shelf. In the North Sea positive values are northeast of the axis.

doi:10.1371/journal.pone.0051541.g006

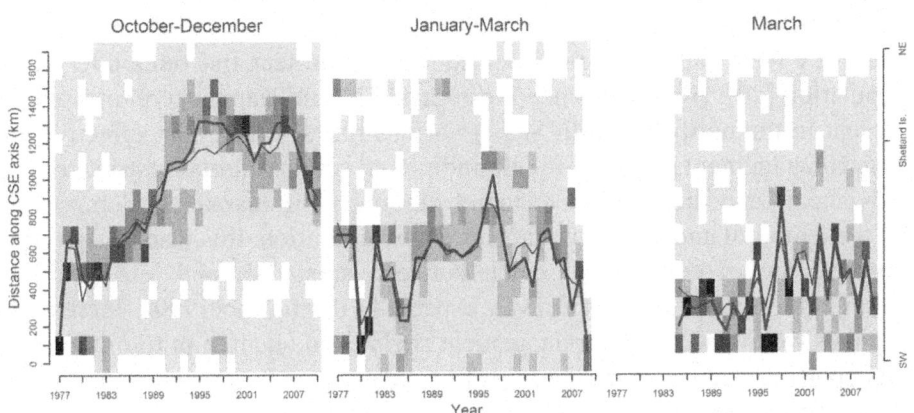

Figure 7: Hovmüller plot of mackerel distributions proxies from commercial landings in October to December (left), January to March (mid) and bottom trawl surveys in March (right).

The spatial aspect have been reduced to one dimention by projecting the catch location onto the CSE axis. Greyscale in cells range in 10%-steps from 0–10% (white), to 90–100% (black). Thick line represents position of 50% cumulative landings (left, mid) or CPUE (right) and thin line shows the center of gravity of the distances.

doi:10.1371/journal.pone.0051541.g007

Landings in Q4 followed a consistent spatial pattern with generally small variance within and between years (Figure 7, left). Landings in Q5 and especially bottom trawl survey catches show greater variance (Figure 7, mid-right).

A progressive southwesterly shift along the CSE axis is evident in the commercial landings data from quarter 4 to 5 (Figure 7, left-mid) and also in the survey catches in late Q5 (Figure 7, right). The average shift of the CoG was found to be 360 km from Q4 to Q5, and 140 km from landings in Q5 to the survey in late Q5.

On a decadal scale, commercial landings (Figure 7, left-mid) show spatial shifts of the commercial fisheries over several hundreds of kilometers, consistent with that reported in the literature [4].

A literature review and an interview with an experienced fishing skipper with first-hand experience of the mackerel fishery during the study period (Tables 1, 2), suggests that factors other than the distribution of mackerel could have influenced the behavior of the fishing fleet, particularly for the Q4 fishery between 1990–1995 and also prior to 2000 for Q5 (see Tables 1,2).

After the collapse of the North Sea Mackerel stock in the 1970s, management measures were put in place in an attempt to protect the remainder of the population [20]. However, since Western and North Sea mackerel mix and are present in the northern North Sea at various times of the year, effective area based management proved difficult. Individual country quotas restricted vessel movements and their ability to target the migrating mackerel. Compounded by the temporal and spatial variability in the migration, this lead to significant misreporting of commercial catch between areas IVa and VIa (and to a lesser extent between IIa and IVa), especially during the 1990s. Incremental changes were made to the management regimes in an attempt to mitigate this misreporting, including partial relaxation of the area-based quotas, modifying area closures, and increased monitoring of the fishery.

Table 1: Factors affecting spatiotemporal distribution of the commercial fishery in Q3–4

Years	Q3–4
1977–1983	Landings data reflected the traditional Q3 Norwegian fishery in the Northern North Sea, and the development of Q3 fisheries more coastal to Eastern Scotland and in the Minches.
1984–1995	The Q3 landings reflect a putative temporal and spatial change in fish availability. Main landings were caught progressively later (ending up in Q4) and north-eastwards from 1983 to 1997 [7]. The large north-eastwards shift from the mid-1980s to mid-1990s occurred in times when fisheries were developing and legislation were changing. However, fisherman observations confirm the spatial development of the fishery was, at least in the beginning, a response to changes mackerel migration patterns as they encountered the mackerel progressively further north-east (Pers. Com. Capt. Alex Wiseman, July 2011). This statement seems reliable, because if the mackerel had been available further north-east in the late 1970s and early 1980s, it would have been economically beneficial to fish on those schools rather than steaming all the way to the Minches from the pelagic ports in north-east Scotland. Later, this fishery (now a Q4 fishery) fluctuates between the coast of Norway and the Shetlands, but remains predominantly east of 4°W.
1996–2010	From about 1996 onwards the fishery was well established in Q4, and its movements through this period was not known to be affected by other large changes than movements of the mackerel stock.

doi:10.1371/journal.pone.0051541.t001

doi:10.1371/journal.pone.0051541.t001

Table 2: Factors affecting spatiotemporal distribution of the commercial fishery in Q5

Years	Q5
1977–1983	Fishery was predominantly in the Cornwall area. However, in this period a new fishery was developing to the north-west of Ireland and west of Scotland
1984	The area around Cornwall was then closed in 1984 to protect the juveniles in this nursery area
1985–1990	The bulk of the landings were from the north of Ireland and west of Scotland moving progressively northwards. The fishery were mainly targeting adult mackerel when they were resident in an area or migrating slowly. However, during this period, development of the pair-trawling technique facilitated the fishery on fast migrating mackerel. Movement of landings in this period may therefore represent a development of the fishery as well as a movement of the stock.
1991–1999	Landings are clustered west of 4 W. This may reflect area misreporting from further east, as the northern North Sea was closed from 31st December.
2000–2010	From 1999 legislation were changed to allow fishing in the northern North Sea up to the 15th of February, and even though this should have ended area misreporting (as the fish were available in the northern North Sea at this time) there appears to have been a "habit" of misreporting to a series of rectangles on the 4 W line which persisted [35].

doi:10.1371/journal.pone.0051541.t002

doi:10.1371/journal.pone.0051541.t002

Further data analysis was restricted to periods where the influence of management measures on the fleet behavior was expected to be minimal. This restricted the landings data from Q5 to only 10 observations (2000–2009), and is therefore why we draw our main conclusions based on the correlation analysis of landings in Q4 and scientific surveys.

The spatial development of the fishery (Figure 7) during these periods, shows i) a southwestern distribution in Q4 in 1977–1989, ii) a steady northeastern distribution in 2000–2007 (Q4+Q5), followed by iii) a movement toward southwest in 2008–2010 (Q4+Q5). Detailed maps of relative distributions of commercial landings and CPUE from bottom trawl survey in these three periods confirm this pattern (Figure 8). Annual maps of relative distributions as well as annual and periodic maps of actual catches are given in Figure.

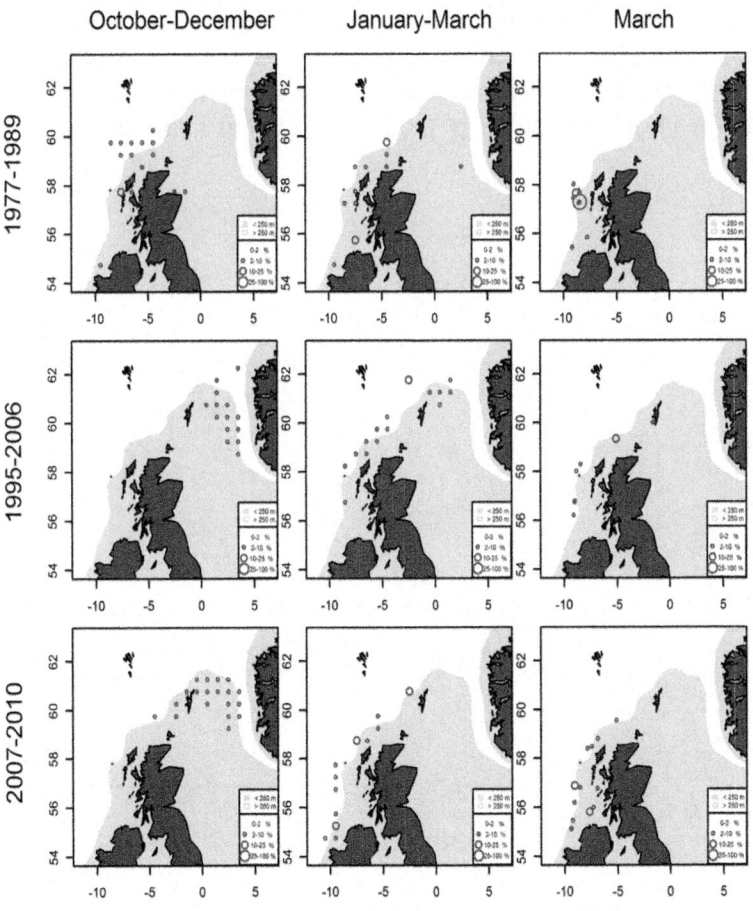

Figure 8: Relative distribution of mackerel landings from the commercial fisheries and mackerel catches from fisheries independent bottom trawl surveys.

Data from January–March are shifted back one year to match data in the same season from October–December.

doi:10.1371/journal.pone.0051541.g008

An examination of the consistency between the three Po50%CL proxies for spatial distribution showed significant positive correlations between the quarter 4 fisheries and the quarter 5 trawl survey (1985–2010 ex.1990–1995, p=0.031, R^2=0.23). This was also the case when the quarter 4 and quarter 5 fisheries were analysed (2000–2009, p=0.040, R^2=0.43). However, no significant correlation was found between the short time series of commercial landings in Q5 and the trawl survey (2000–2009, p>0.05).

Comparisons of the modelled temperature time series with the Po50%CL proxies for mackerel distribution (Figure 9) reveal a significant positive correlation with fisheries-independent surveys (1985–2010, p=0.007, R^2=0.27), and with commercial landings in Q4 from 1977–2010 (ex. 1990–1995) (p<0.001, R^2=0.59), but not with the short time series of commercial landings in Q5 (2000–2009, p>0.05). Correlation analyses are summarized in table 3.

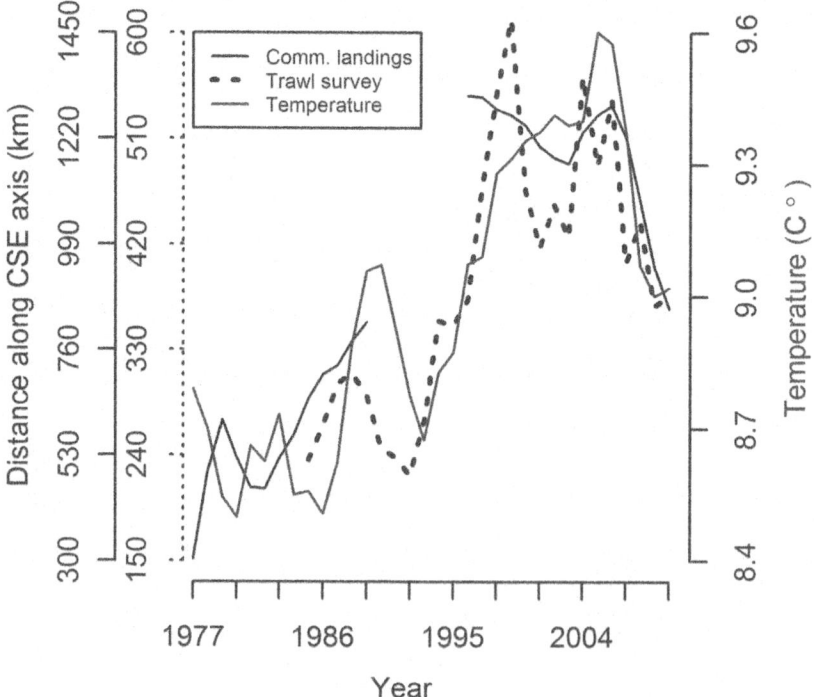

Figure 9: Mackerel distribution and temperature trends (3y rm).

Position of 50% cumulative landings in Q4 and in survey CPUE along the continental shelf edge axis (blue) and temperature around the shelf edge in the Northern North Sea from off Shetland Is. to the western edge of the Norwegian trench in November–January (red).

doi:10.1371/journal.pone.0051541.g009

Table 3: Correlation analyses between proxies for spatial patterns of the mackerel represented by Position of 50% Cumulative Landings (Po50%CL) and modeled temperature in the shelf edge current

	Landings Q5	Trawl survey	Temperature
Landings Q4	$p = 0.040$, $R^2 = 0.43$	$p = 0.020$, $R^2 = 0.25$	$p < 0.001$, $R^2 = 0.58$
Landings Q5		$p > 0.05$	$p > 0.05$
Trawl survey			$p = 0.007$, $R^2 = 0.27$

doi:10.1371/journal.pone.0051541.t003

doi:10.1371/journal.pone.0051541.t003

Discussion

Our analyses demonstrate that when the NEA mackerel return in late summer from the feeding areas on the European shelf and in the Nordic Seas [4], they aggregate through autumn and early winter along the continental shelf edge, where they are targeted by commercial trawlers and purse seiners. Later in winter the commercial fleets and the fisheries independent bottom trawl survey find the mackerel further towards the southwest. The path of the migration, as suggested by the location of commercial and survey catches coincides with the location of the relatively warm high-saline eastern Atlantic water flowing north-eastwards on and along the continental shelf edge, flanked by cooler water masses. We present a modelled new time series of temperature in this current and find it to be significantly correlated with two proxies for spatiotemporal mackerel distribution. The proxies are derived from data over a significant period of time and a large proportion of the European shelf and encapsulate large scale changes in distribution. Our results indicate that

- the mackerel population is found further upstream in warmer waters as the current cools through winter
- this process is associated via climatic variability, with large impacts on the mackerel migration and fisheries, and suggest a mechanism where

- this affinity for warm water leads the mackerel towards the main spawning areas.

These results are in accordance with earlier studies of mackerel during autumn and winter[5]–[10].

The present work illustrates the limitations associated with the available data and underlines that caution should be exercised when utilising catch data as a proxy for distribution. The relatively low trawling speed and small scale trawls employed by standardized scientific surveys are unsuited for catching a fast pelagic species like mackerel. Furthermore, changes in vertical distribution and schooling behaviour reduce the signal-to-noise ratio in the trawl survey data and contributes to the low levels of explained variance (R^2) in correlations that include this variable. In contrast, commercial fishing employs much more efficient methods. Commercial landings data are, however, only appropriate for inferring changes in stock movements over time when other factors remain relatively constant. This was not the case for the Q4 fishery between 1990 to 1995, when the management regime restricted the ability of vessels to target fish migrating through areas IVa and VIa and fisheries technology and techniques changed the behaviour and increased the efficiency of the fleet (Table 1). An approach to circumvent this problem has been used in a previous study, where high resolution catch data from a validated subset of the fleet showed that the observed change from late 1970s to late 1990s leveled out from 1989 to 1994 [5]. This is consistent with our conclusions, as this was the period where fisheries and temperature deviated (Figure 9).

Other major changes in mackerel fisheries have occurred through the period 1977–2010, such as the summer fishery in Icelandic waters that commenced in recent years [4]. While this fishery is outside the main scope of this study, it is related to the westward expansion of the summer distribution [21]. Changes in the summer distribution could lead to a change in the path taken during the return migration in late summer and early autumn, which could potentially affect the autumn-winter distribution. Further investigation of this effect is therefore warranted.

The results presented are in accord with recent investigations that link climatic variability and spatiotemporal dynamics of mackerel spawning [12], [22], [23], [33]. Mackerel differ from most other exothermal organisms by being i) purely pelagic through all life stages, and ii) relatively fast and constantly swimming [24], able to react to the environment by migrating over long distances. This dynamic spatial behavior enables the mackerel to avoid poor temperature conditions during its migration in search of optimal areas for reproduction and feeding. This seems to be most evident during the cold season when other constraints such as feeding and reproduction are reduced or absent. The effect of temperature on the spatial shifts of the mackerel distribution is suggested to be on a scale of hundreds of kilometers during winter (Figure 9), much larger than in spring where spawning has been moving only 40 km north per °C [23] and in summer where polar water merely forms an outer boundary of the extremely large area occupied by mackerel [4], [25]. It is understood that the primary activity during winter is the maturation of eggs and sperm. It may be that the specific temperature conditions selected by the mackerel are an adaptation to optimize development of reproductive products. The present findings facilitate testing of this hypothesis and exploration of further importance for spawning.

The physical environment within the shelf edge current is related to large scale oceanographic circulation patterns. Conditions in the Bay of Biscay and the European shelf seas, to the east of the continental shelf edge current, are related to the Northern Hemisphere Temperature trend[26]. This differs from the oceanic region west of the shelf edge current, which to a greater extent is regulated by the dynamics of the subpolar gyre [27], [28]. The physical environment within the shelf edge current is related to the northern hemisphere temperature type of variability, but may also be influenced by the oceanic domain during periods when the subpolar gyre circulation is particularly strong, such as during the period 1990–1995 [27]. The shelf edge waters are furthermore modulated by smaller sub-decadal oscillations, caused by pulses of eastern water from the Bay of Biscay [29]. Once warm and saline anomalies have passed the Porcupine Bank, the geographic divide between the subtropical and the subpolar gyres, they are destined to continue northward as baroclinic Rossby waves [30], [31], with an advection time of one-two years, to the entrance of the Nordic Seas [27], [32]. This oceanic inertia holds promise for making projections one-two years into the future. Shorter-term predictions may be possible based on measurements of the temperature further "upstream": such predictions could be of value for the fishing industry as it may reduce the time spend on searching for mackerel. However, detailed forecasting of mackerel behavior outside the observed temperature range is not possible before any additional causal effects and their interactions are sufficiently clarified.

The results presented have implications for the management, fishery and monitoring of mackerel. Recent changes in mackerel distribution have resulted in political disputes over zonal attachments and led to a break-down of the international management agreements since 2008. Furthermore, in 2009 fishermen were taken by surprise when the mackerel had departed the northern North Sea east of 4° (which separates management areas IVa and VIa) by October [34], significantly earlier than in previous years. As a consequence, quotas worth over 100 M € could not be utilized in that year by the Norwegian and Danish industries [35] whilst, at the same time, Scottish seiners had little difficulty in catching the mackerel further west. We have demonstrated that cooling of the continental shelf edge current, possibly triggered this early migration. In a climate change scenario where temperatures increase further, our results suggest that mackerel distribution is likely to be affected with subsequent effects for the fishery and mackerel prey.

ACKNOWLEDGMENTS

The authors wish to thank members of the ICES mackerel assessment working group from 1978–2011 for providing catch data. BSH (Bundesamt für Seeschiffahrt und Hydrographie) for kindly providing figure 2. Alex Wiseman and other members of the pelagic fishing industry who provided background information. Finally we would like to thank John Molloy whose book "The Irish mackerel fishery and the making of an industry", provided much useful historical information particularly on the ICES working groups in the 1970 & 1980s.

AUTHOR CONTRIBUTIONS

Analyzed the data: TJ. Wrote the paper: TJ AC CK HH MRP.

REFERENCES

1. Graham CT, Harrod C (2009) Implications of climate change for the fishes of the British Isles. J Fish Biol 74: 1143–1205. doi: 10.1111/j.1095-8649.2009.02180.x

2. Trenkel VM, Huse G, MacKenzie B, Alvarez P, Arizzabalaga H, et al.. (2012) Comparative ecology of widely-distributed pelagic fish species in the North Atlantic: implications for modelling climate and fisheries impacts. Prog Oceanogr. In press.

3. Payne MR, Egan A, Fassler SMM, Hatun H, Holst JC, et al. (2012) The rise and fall of the NE Atlantic blue whiting (Micromesistius poutassou). Marine Biology Research 8: 475–487. doi: 10.1080/17451000.2011.639778

4. ICES (2011) Report of the Working Group on Widely Distributed Stocks (WGWIDE). ICES CM 2011/ACOM: 15.

5. Reid DG, Eltink A, Kelly CJ (2003) Inferences on the changes in pattern in the prespawning migration of the western mackerel (Scomber scombrus) from commercial vessel data. ICES CM 2003/Q: 19.

6.   Reid DG, Eltink A, Kelly CJ, Clark M (2006) Long-Term Changes in the Pattern of the Prespawning Migration of the Western Mackerel (Scomber scombrus) Since 1975, using Commercial Vessel Data. ICES CM 2006/B: 14.

7.   Walsh M, Martin JHA (1986) Recent changes in the distribution and migration of the western mackerel stock in relation to hydrographic changes. ICES CM 1986/H: 17.

8.   Reid DG, Walsh M, Turrel WR (2001) Hydrography and mackerel distribution on the shelf edge west of the Norwegian deeps. Fish Res 50: 141–150. doi: 10.1016/s0165-7836(00)00247-2

9.   Reid DG, Turrell WR, Walsh M, Corten A (1997) Cross-shelf processes north of Scotland in relation to the southerly migration of western mackerel. ICES J Mar Sci 54: 168–178. doi: 10.1006/jmsc.1996.0202

10.   Walsh M, Reid DG, Turrell WR (1995) Understanding Mackerel Migration Off Scotland - Tracking with Echosounders and Commercial Data, and Including Environmental Correlates and Behavior. ICES J Mar Sci 52: 925–939. doi: 10.1006/jmsc.1995.0089

11.   Hátún H, Payne M, Jacobsen JA (2009) The North Atlantic subpolar gyre regulates the spawning distribution of blue whiting (Micromesistius poutassou). Can J Fish Aquat Sci 66: 759–770. doi: 10.1139/f09-037

12.   Jansen T, Gislason H (2011) Temperature affects the timing of spawning and migration of North Sea mackerel. Cont Shelf Res 31: 64–72. doi: 10.1016/j.csr.2010.11.003

13.   Punzon A, Villamor B (2009) Does the timing of the spawning migration change for the southern component of the Northeast Atlantic Mackerel(Scomber scombrus, L.1758)? An approximation using fishery analyses. Cont Shelf Res 29: 1195–1204. doi: 10.1016/j.csr.2008.12.024

14.   ICES (2005) Report of the Working Group on the Assessment of Mackerel, Horse Mackerel, Sardine and Anchovy (WGMHSA). ICES CM 2005/ACFM: 08.

15.   Wood SN (2006) Generalized Additive Models: An Introduction with R. Boca Raton, FL: Chapman and Hall/CRC Press.

16.   ICES (2012) ICES hydrographic database. Available: http://www.ices.dk. Accessed 2012 Nov 13.

17.   Rayner N, Brohan P, Parker D, Folland C, Kennedy J, et al. (2006) Improved analyses of changes and uncertainties in sea surface temperature measured in situ sice the mid-nineteenth century: The HadSST2 dataset. J Clim 19: 446–469. doi: 10.1175/jcli3637.1

18. Madsen H (1998) Tidsrækkeanalyse, 3rd edition. Informatik Og Matematisk Modelling, Danish Technical University : 145–151.

19. Pyper BJ, Peterman RM (1998) Comparison of methods to account for autocorrelation in correlation analyses of fish data. Can J Fish Aquat Sci 55: 2127–2140. doi: 10.1139/f98-104

20. ICES (1995) Report of the Working Group on the Assessment of Mackerel, Horse Mackerel, Sardine and Anchovy (WGMHSA). ICES CM 1995/Assess: 2.

21. ICES (2012) Ad hoc Group on Distribution and Migration of Mackerel (AGDMM). ICES CM 2012/ACOM: Xx.

22. Jansen T (2012) North Sea Mackerel or Mackerel in the North (Sea)? PhD thesis. DTU Aqua, Technical University of Denmark. In press.

23. Hughes K, Johnson M, Dransfeld L (2012) Changes in the spatial distribution of spawning activity by northeast Atlantic mackerel in warming seas: 1977–2010. ICES J Mar Sci. In press.

24. Lockwood SJ (1988) The Mackerel–Its biology, assessment and the management of a fishery. London: Fishing News Books Ltd. 181 p.

25. Utne KR, Huse G, Ottersen G, Holst JC, Zabavnikov V, et al. (2011) Horizontal distribution and overlap of planktivorous fish stocks in the Norwegian Sea during summers 1995–2006. Mar Biol Res 8: 420–441. doi: 10.1080/17451000.2011.640937

26. Beaugrand G, Reid PC, Ibanez F, Lindley JA, Edwards M (2002) Reorganization of North Atlantic marine copepod biodiversity and climate. Science 296: 1692–1694. doi: 10.1126/science.1071329

27. Hátún H, Sando AB, Drange H, Hansen B, Valdimarsson H (2005) Influence of the Atlantic subpolar gyre on the thermohaline circulation. Science 309: 1841–1844. doi: 10.1126/science.1114777

28. Hátún H, Payne M, Beaugrand G, Reid PC, Sandø AB, et al. (2009) Large bio-geographical shifts in the north-eastern Atlantic Ocean: From the subpolar gyre, via plankton, to blue whiting and pilot whales. Prog Oceanogr 80: 149–162. doi: 10.1016/j.pocean.2009.03.001

29. Larsen KMH, Hátún H, Hansen B, Kristiansen R (2012) Atlantic water in the Faroe area: sources and variability. ICES J Mar Sci 69.

30. Eden C, Willebrand J (2001) Mechanism of interannual to decadal variability of the North Atlantic circulation. J Clim 14: 2266–2280. doi: 10.1175/1520-0442(2001)014<2266:moitdv>2.0.co;2

31. Kauker F, Gerdes R, Karcher M, Koberle C (2005) Impact of north Atlantic current changes on the Nordic Seas and the Arctic Ocean. Journal of

Geophysical Research-Oceans 110: C12002. doi: 10.1029/2004jc002624

32. Orvik KA, Skagseth O (2003) The impact of the wind stress curl in the North Atlantic on the Atlantic inflow to the Norwegian Sea toward the Arctic. Geophys Res Letters 30: 1884–1887. doi: 10.1029/2003gl017932

33. Jansen T, Kristensen K, Payne M, Edwards M, Schrum C, et al.. (2012) Long-term Retrospective Analysis of Mackerel Spawning in the North Sea: A New Time Series and Modeling Approach to CPR Data. PLoS ONE 7.

34. Norges Sildesalgslag (2009) Blanke skjermer på makrellen - Finvær i Nordsjøen hjalp ikke for makrellfiske i norsk sone. Available: https://www.sildelaget.no/. Accessed 2012 Nov 13.

35. ICES (2010) Report of the Working Group on Widely Distributed Stocks (WGWIDE). ICES CM 2010/ACOM: 15.

Chapter 7

LOCAL ECOLOGICAL KNOWLEDGE AND SCIENTIFIC DATA REVEAL OVEREXPLOITATION BY MULTIGEAR ARTISANAL FISHERIES IN THE SOUTHWESTERN ATLANTIC

Mariana G. Bender[1], Gustavo R. Machado[2], Paulo José de Azevedo Silva[3], Sergio R. Floeter[1], Cassiano Monteiro-Netto[2], Osmar J. Luiz[4], Carlos E. L. Ferreira[2]

[1]Departamento de Ecologia e Zoologia, Universidade Federal de Santa Catarina, Floriano´polis, SC, Brazil

[2]Departamento de Biologia Marinha, Universidade Federal Fluminense, Niteroi, RJ, Brazil

[3]Fundac,a~o Instituto de Pesca, Arraial do Cabo, RJ, Brazil

[4]Department of Biological Sciences, Macquarie University, Sydney, NSW, Australia

ABSTRACT

In the last decades, a number of studies based on historical records revealed the diversity loss in the oceans and human-induced changes to marine ecosystems. These studies have improved our understanding of the human impacts in the oceans. They also drew attention to the shifting baseline syndrome and the importance of assessing appropriate sources of data in order to build the most reliable environmental baseline. Here we amassed information from artisanal fishermen's local ecological knowledge, fisheries landing data and underwater visual census to assess the decline of fish species in Southeastern Brazil. Interviews with 214 fishermen from line, beach seine and spearfishing revealed a sharp decline in abundance of the bluefish *Pomatomus saltatrix*, the groupers *Epinephelus marginatus, Mycteroperca acutirostris, M. bonaci* and *M. microlepis*, and large parrotfishes in the past six decades. Fisheries landing data from a 16-year period support the decline of bluefish as pointed by fishermen's local knowledge, while underwater visual census campaigns show reductions in groupers' abundance and a sharp population decline of the Brazilian endemic parrotfish *Scarus trispinosus*. Despite the marked decline of these fisheries, younger and less experienced fishermen recognized fewer species as overexploited and fishing sites as depleted than older and more experienced fishermen, indicating the occurrence of the shifting baseline syndrome. Here we show both the decline of multigear fisheries catches – combining anecdotal and scientific data – as well as changes in environmental perceptions over generations of fishermen. Managing ocean resources requires looking into the past, and into traditional knowledge, bringing historical baselines to the present and improving public awareness.

INTRODUCTION

Fishing is the most ancient form of exploitation of coastal resources, preceding all other human disturbances to marine ecosystems, such as pollution, eutrophication, habitat loss, disease outbreaks, human induced climate change and species invasions [1]–[4]. Fishing has caused the worldwide depletion of large predatory fishes, including shark populations [5], [6], and driven several species across different ecosystems to ecological extinction [1]. Moreover, fishing is reported to impact fish populations since prehistoric times [2], [7]–[10], affecting species diversity and size [11], [12]. Today, fishing remains as the major source of impact upon marine and coastal environments, contributing to global biodiversity loss [13]–[15]. The continuity of unsustainable historical fishing has left few truly pristine ecosystems in the marine realm [1], [5]. In the Southwestern Atlantic, patterns of marine resources' overexploitation are similar to what have been reported worldwide [5], [14], [16], with widespread population declines and collapsed stocks [17]–[19]. It has been estimated that 23% of all Brazilian marine fish stocks are fully exploited and 33% are overexploited, including species from low trophic levels [20]–[22]. Overfishing has been changing the density and the size structure of reef fish top predators occurring in Brazil [17], [23], consequently elevating their threatened status [24], [25].

Environmental changes and lack of baselines for pristine marine ecosystems have profound implications in our perceptions of what is a natural environment [4], [26]. The compromised ability of people in perceiving environmental modifications and past ecological conditions is called the shifting baseline syndrome (SBS) [27]. This phenomenon was first noted among fisheries scientists, who perceived as a 'natural' baseline for stock size and composition the condition they observed at the start of their careers, while failing to incorporate past and historical data [27]–[29]. As generations change, environmental baselines become increasingly shifted, misinforming fisheries management. In order to adjust shifted environmental perceptions, old travel diaries, naturalists› observations, historical data and fishermen› anecdotes are important sources of information [28]–[32], as well archaeological and paleontological data [1], [4]. Baselines constructed from historical data are critical to better gauge and interpret long-term changes and to set appropriate targets for management and restoration [1], [4], [33].

The Local Ecological Knowledge (LEK) held by traditional fishing communities is considered an important tool for the assessment and management of tropical fisheries [29], [30], [37]. Local Ecological Knowledge is a set of perceptions and experiences of traditional communities regarding its surrounding natural environment [34], [35], this knowledge being handed-down through generations by cultural transmission [36]. This knowledge includes perceptions of fishing resources – fisheries› composition and abundance patterns, as well as fish species biological and ecological aspects [38], [39] – all of which are important to conservation and management strategies [40]. Because humans have modified natural systems, local perceptions of the status of species and ecosystem resources are unlikely to remain constant over time [41]. Thus, the SBS could influence the validity of LEK derived data, whose use for species and ecosystems assessments is becoming more common [42], [43]. However, specific conditions are required for SBS to occur, which include a combination of environmental events and observer perceptions [44]. Here we combine multiple sources of information to understand the status of local fisheries of Arraial do Cabo, Rio de Janeiro state, Southeastern Brazil (Fig. 1). Specifically, we aim to (i) investigate the occurrence of SBS among fishermen communities of Arraial do Cabo; (ii) utilize LEK to obtain past estimates of abundance and population trends for exploited fish species; and (iii) demonstrate the importance of anecdotal data to adjust perceptions regarding marine resources in the subtropical coast of Brazil.

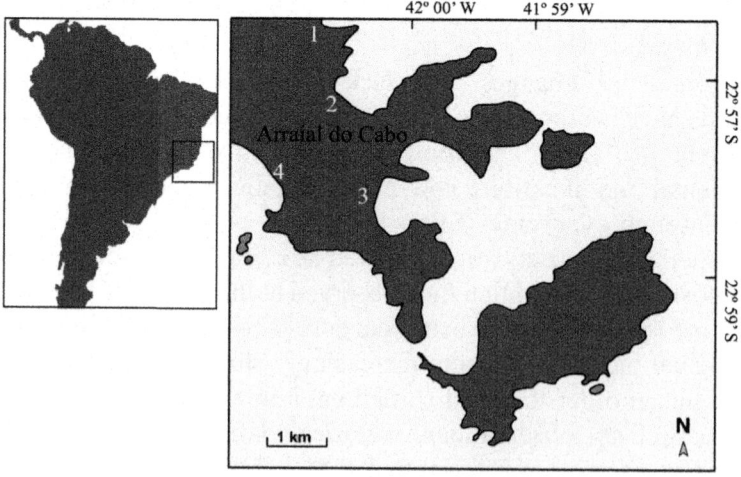

Figure 1: Locations of interview sites at Arraial do Cabo.

METHODS

Study Site

Arraial do Cabo (Rio de Janeiro state, Brazil) constitutes a traditional fishing village with approximately 20,000 residents and 1,340 active fishermen. The area was declared a Marine Extractive Reserve (MER) in 1997 – the first of its kind in Brazil – a collaboratively managed marine protected area where only local fishermen are allowed to exploit resources. In Arraial do Cabo, while there are some fishing regulations, enforcement is poor or non-existent, resulting in a typical overexploitation scenario [17]. Among fishing gears, hook and line, gillnet, beach seine and spearfishing are the most used. The beach seine fishery is the most traditional practice, using large canoes and seine nets that encircle passing schools of bluefish (*Pomatomus saltatrix*), mullets (Mugilidae), jacks and trevallies (Carangidae), and tunas (Scombridae). Spearfishing commenced *c.* 50 years ago in Brazil, being today widespread along the coast. The region is formed by an isthmus and two islands dominated by rocky shores and sand beaches. Coastal morphology associated to prevailing winds trigger small-scale upwelling events, favoring a rich marine environment (Fig. 1).

Data Collection

Ethical Considerations.

The Marine Biology Graduate Program at Universidade Federal Fluminense (PPGBM-UFF), Niterói, Rio de Janeiro state, Brazil, approved the data collection of our study that comprised interviews with local fishermen communities. Prior to each interview, fishermen were informed on research purpose and we obtained verbal consent from participants. This methodology was chosen because some interviewees would lack reading and writing skills. Once the participant agreed on being interviewed, we recorded participant consent by writing the participant's name and interview date on questionnaire sheets. This procedure was considered adequate by the Graduate Program review board. We also informed participants that all information provided in interviews would be anonymized.

Interviews.

To assess fishermen perceptions regarding the status of overexploited fish species, we conducted individual interviews with 214 fishermen – from August 2007 to August 2008 – in Arraial do Cabo, Southeastern Brazilian coast (Fig.1). Interviews with fishermen included three different gears: hook and line, beach seine and spearfishing. As line and beach seine are traditional fishing gears in the region and have been practiced for centuries, this enabled a large number of experienced fishermen to be surveyed. Making possible the unveiling of a rich ecological knowledge related to these gears and a robust historical baseline for local fisheries. Respondents were categorized into three groups according to fishing experience. For line fishing and beach seine fishing the categories were: beginners (<16 years of practice, n=45), intermediate (16–30 years of practice, n=50) and experienced (>30 years of practice, n=43). However, since spearfishing is a recent fishing practice when compared to beach seine and line fishing, experience categories for spearfishermen were adapted to a smaller period: beginners (<8 years of practice, n=25), intermediate (8–15 years of practice, n=28) and experienced (>15 years of practice, n=23).

The survey started a few weeks following arrival to the fishing village. Fishermen were randomly approached at the beach – as they were encountered either going fishing or making repairs on nets and fishing gears – and inquired on the availability to participate on the research. Following this approach, fishermen in different groups (experience category per fishing gear) were selected through snowball sampling [37], [45], where one or more peers indicate the respondent. This method was applied so that we could achieve the target sample size in each experience category and fishing gear (line, beach seine and spearfishing). This method also led us to interview some elder and retired informants at their homes. Respondents were always interviewed separately in order to avoid interfering from others. Our questionnaire constitued of structured and open-ended questions [29].

The bluefish, *Pomatomus saltatrix* was chosen among various species based on its importance as a resource in the region [45], [47]. We asked fishermen about the best bluefish catch they ever remembered landing and the year in which this catch was made [47]. For the spearfishing practice interviews, we selected four Epinephelidae (groupers) species [48] that are important targets in the region (*Mycteroperca acutirostris, M. bonaci, M. microlepis* and*Epinephelus marginatus*) [17]. To explore the possible fishing impacts at lower trophic levels, we included in questionnaires four endemic large herbivorous-detritivorous species of the subfamily Scarinae (parrotfishes), namely, *Scarus trispinosus, Sparisoma axillare, Sparisoma frondosum* and *Sparisoma amplum*. These fishes became targets for spearfishing in the last 20 years [17], [20]. For each species we asked divers about the best catch they remembered spearing (number of fishes/individuals), the largest individual they ever caught (in kilograms), and the year in which these catches were made [29]. When fishermen were asked about the largest fishes they had ever caught, the majority of respondents provided information relative to species weight (in kilograms). However, when fishermen referred to fish body length, species length–weight relationships were used to convert to weight [49]. Finally, we assessed shifts in environmental perceptions among generations of fishermen. We inquired each respondent to cite the number of overexploited species and sites in the region, as well as the number of species that were formerly discarded but are now fishing targets. Given the richness of common names for Brazilian reef fish [46], we used photographs to clarify species identifications during the interviews [37].

Bluefish Landing Data.

We have included fisheries landing data in our analysis in order to improve the overall understanding of the status of fish stocks in the region. Bluefish (*Pomatomus saltatrix*) catch data (total catch – kg and effort – fishing hours) was recorded daily and pooled over the year during 16 years (1992 to 2008) at the local landing point (Marina dos Pescadores, Porto do Forno) in Arraial do Cabo.

Reef Fish Abundance.

The abundance of groupers (Epinephelidae) and parrotfishes (Labridae, Scarinae) was assessed through an underwater visual census (UVC) program performed in Arraial do Cabo region across different years (1992, 2012). The program is led by one of the authors (C.E.L.F.), and includes UVC and other sampling techniques applied to studies/monitoring of the reef community at Arraial do Cabo. Fish density was estimated using 20×2 m strip transects [50]. In each transect, all fishes within the sample area were identified, counted and categorized into five size classes. Data across all sampled sites were pooled together to show temporal changes for target species in Arraial do Cabo region.

Data Analysis

Interviews.

From the data obtained through interviews with local fishermen we measured the decline in fish species catches, as well as the differences in the set of environmental perceptions across fishermen generations. With the purpose of measuring the decline in fish species abundance and body-size patterns across the years, we applied regressions and plotted (i) the best day catch reported by each respondent, versus the year in which the catch was made; and (ii) the larger individual ever caught by the respondent (kg) versus the year in which a given species was caught. We tested distinct regression types (exponential, polynomial and linear regressions) and r^2 values were compared to assess the quality of the regression fit to data distribution. Fish catches measured for time intervals are presented in the results section as mean ± standard deviation.

To investigate the potential differences in respondents perceptions, we tested differences in the mean number of overexploited species, overexploited sites, and the mean number of target species (formerly discarded) mentioned by fishermen in different categories (beginner, intermediate, experienced) through one-way ANOVA (α=0.05; Student-Newman-Keuls post-hoc test) [51]. All data was tested for normality (Kolmogorov-Smirnov test) and transformed when necessary (square root or log transformed) before conducting analysis.

Bluefish Landing Data.

In order to measure *P. saltatrix* catch trends across the years (1992 to 2008) from landing data we have applied regression analysis, plotting the catch per unit effort (tons/hour) throughout the years. Again, we tested for distinct regression types (exponential, polynomial and linear regressions) and r^2 values were compared to assess the quality of the regression fit to data distribution.

Reef Fish Abundance.

In order to examine the population decline trends within targeted reef fish families (groupers and parrotfishes) in Arraial do Cabo we tested the Spearman's rank correlation between the mean density (of groupers and parrotfishes) per transect and year of sampling. Only the most abundant and large sized parrotfish species (*Scarus trispinosus, Sparisoma axillare, S. frondosum, S.amplum*) were used for analysis [52].

RESULTS

Multiple Sources Reveal Fish Species Declines

The pattern for bluefish catches across the years reveals significant reductions in *P. saltatrix*stocks in Arraial do Cabo region over time (Fig. 2a). Captures (mean ±standard deviation) of 1,200 kg (±186) were considered a good catch during the 1960s among line fishermen. This number decreased to an average of 970 kg (±260) in the 1980s. Today, 370 kg (±220) of bluefish is considered a great catch (Fig. 2a). Results reveal that in fifty years (1960–2010) the amount of bluefish caught on a good day catch was reduced by approximately 70%.

Experienced beach seine fishermen also reported great bluefish catches in Arraial do Cabo decades ago, and approximately 80% decline in the past six decades, a pattern similar to that identified by line fishermen (Fig. 2b). During the 1960s, large catches were in the order of 18,000 kg (±3,517) on average. In the 1980s, this number has fallen to13,570 kg (±3,753), and today *P. saltatrix* catches greater than 3,700 kg (±3,340) are rare. Moreover, landing data from*P. saltatrix* catches in the region reveal the same pattern found by accessing the knowledge of local fishermen (Fig. 2c), with a declining trend in bluefish fisheries along 16 years (1992–2008). CPUE varied between 14.2 tons/hour in 1996 and 0.3 tons/hour in 2004, with an annual average of 4 tons hour (Fig. 2c).

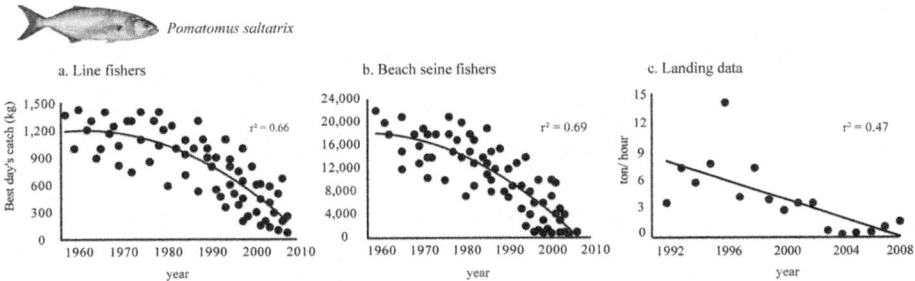

Figure 2: Bluefish (Pomatomus saltatrix) catches according to line fishermen (a) and beach seine fishermen (b), both are second order polynomial regressions; and catch per unit effort (t×h−1) (c) from 1992 to 2008 in Arraial do Cabo, Brazil.

Fish species exploited by spearfishing also exhibit declining trends. There was a decrease in the abundance of the assessed Epinephelidae species (Fig. 3). The relationships considering the largest catch and year again indicate that large individuals of *E. marginatus, M. bonaci, M. acutirostris* and *M. microlepis* have declined in the region, and that the big ones were frequently caught not so far back, in the 1970s (Fig. 3). For the endemic parrotfishes*Sparisoma axillare, S. frondosum* and *S. amplum* there were no significant differences neither in the abundance nor in the size of individuals caught along the years. However, for *Scarus trispinosus*, we have identified reductions on the greatest individuals captured (kg) and on the best catches made (Fig. 3).

Data from visual census campaigns conducted in the region from 1992 to 2012 – revealed a declining trend in the local densities of *Scarus trispinosus*, in fact, individuals of the species are not recorded on transects conducted in the region since 1995 (Fig. 4). The decline in *S. trispinosus* is followed by an overall decline of Scarinae (Labridae) species. For Epinephelidae species, the situation is similar, with decreasing densities in local reef fish communities from 1997 onwards (Fig. 4).

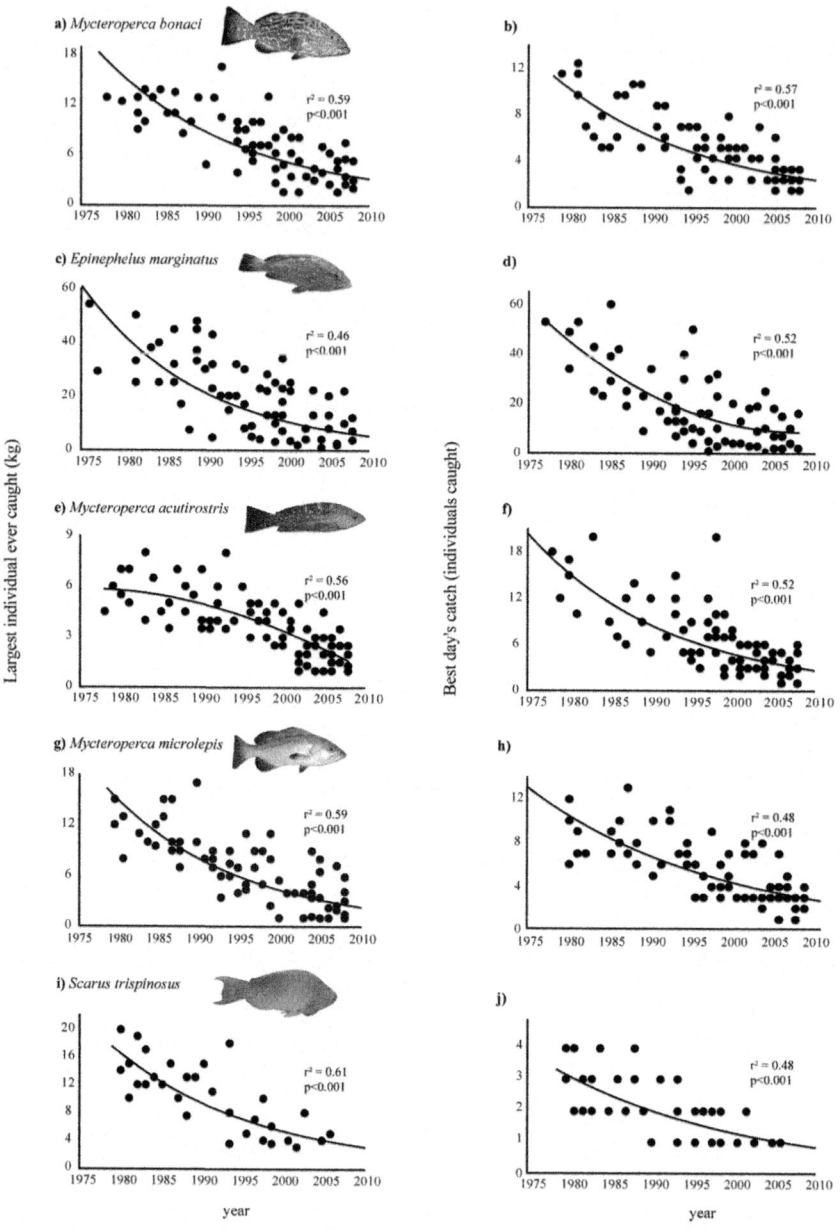

Figure 3: Largest individual (in kilograms) and greatest catch (number of individuals caught) that spearfishermen remembered landing. a) and b) *Mycteroperca bonaci*, second order polynomial regression; c) and d)*Epinephelus marginatus,* exponential regression; e) and f) *Mycteroperca acutirostris*, second order polynomial regression; g) and h) *Mycteroperca microlepis*, second order polynomial regression; i) and j) *Scarus trispinosus*, exponential regression.

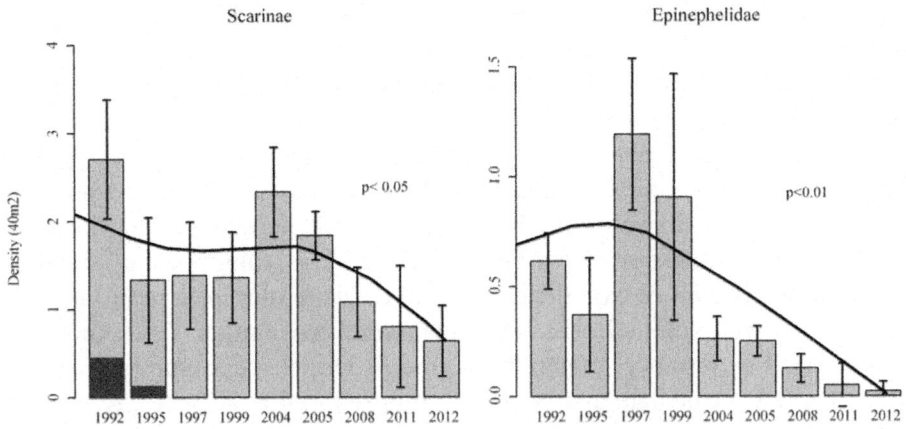

Figure 4: Mean densities per 40 m² **of** Scarinae and Epinephelidae species in Arraial do Cabo from underwater visual census. In the Scarinae plot, black portion of bars correspond to the contribution of *Scarus trispinosus* to densities.

Shifting Baselines Across Multigear Fisheries

Out of 214 interviewed fishermen, ~83% (n=177) recognized reductions in the abundance of exploited fish species in the region of Arraial do Cabo. Less experienced fishermen are less aware of overexploitation, indicating the occurrence of the SBS among local fishermen. Significant age-related differences were evident in assessed fishing communities in terms of respondent perception regarding fisheries declines. Experienced line, beach seine and spearfishermen mentioned a greater number of overexploited species when compared to less experienced ones (Table 1).

Table 1: Mean number of overexploited fish species, new target species and depleted sites mentioned by fishermen from different fishing gears and according to experience categories

Fishing modality	Fishers experience category (n)	Overexploited species (mean ±SD)	ANOVA*	New target species (mean ±SD)	ANOVA*	Depleted sites (mean ±SD)	ANOVA*
Line	Beginners (22)	2.0±1.0	f=139.3	1.6±1.1	f=35.7	2.0±1.0	f< 135
	Intermediate (25)	5.5±1.8	p<0.001	3.1±1.5	p<0.001	4.5±1.9	p<0.001
	Experienced (21)	10.5±1.7		6.0±1.0		11.3±2.3	
Beach seine	Beginners (23)	2.3±1.5	f=76.1	1.4±1.5	f= 60.8	2.4±2.1	f=63.6
	Intermediate (26)	6.0±2.2	p<0.001	3.4±1.1	p<0.001	5.5±2.0	p<0.001
	Experienced (21)	9.0±1.5		5.5 ± 1.9		10.5±3.1	
Spearfishing	Beginners (25)	1.4±0.9	f=187.7	1.8±0.7	f=107.2	1.8±1.4	f=133.5
	Intermediate (28)	5.8±1.5	p<0.001	5.1±1.7	p<0.001	6.3±1.7	p<0.001
	Experienced (23)	8.9±1.7		7.4±1.9		11.4±2.9	

*f and p-values of each ANOVA (one-way) test.
All values associated to different age categories were significantly different according to ANOVA (one-way) and Student-Newman-Keuls comparison test (p<0.05).
doi:10.1371/journal.pone.0110332.t001

We detected significant experience-related differences in terms of respondent perception of shifts in target species (Table 1). Not only for line and beach seine fishermen (line fishing: experienced =6.1, intermediate =3.1 and beginners =1.6; beach seine: experienced =5.5, intermediate =3.4 and beginners =1.4), but also for spearfishing there were differences between the mean number of species cited by each category (spearfishing: experienced =7.4, intermediate =5.1 and beginners =1.8) (see Table 1). For instance, surgeonfishes (e.g.*Acanthurus* spp.) were identified as targets exclusively by beginner spearfishermen. The mean number of sites mentioned as depleted by different fishermen categories also suggests the occurrence of the SBS. The more experienced and older fishermen referred to a greater number of overexploited sites when compared to younger, less experienced fishermen. For instance, experienced line fishermen cited approximately 11 sites, while intermediate and beginners referred to 4 and 2 sites, respectively (see Table 1).

DISCUSSION

From local to Global: the Impacts of Changes in Fishermen's Perceptions

Since Daniel Pauly reported the global phenomenon that he termed 'the shifting baseline syndrome' (SBS) [27], much supporting evidence have been reported across marine [29], [53], freshwater [41], [54] and terrestrial ecosystems [43], [55]. Fishermen gather their environmental references based on personal experiences, which may lead to perception differences among generations. In Arraial do Cabo, experienced line and beach seine fishermen often mentioned the old days when top predators such as the sharks *Carcharhinus milberti*, *Squatina argentina*, *Isurus oxyrinchus* and the shovelnose ray *Rhinobatos percellens*were abundant in their catches, being witnesses to the decline of those species. Even though most of the older beach seine fishermen (73%) recognized the shark *Carcharhinus maculipinnis* as overexploited, beginners believe that the species is naturally rare in the region. Also, contrary to older fishermen, young spearfishermen did not mention the snappers *Lutjanus analis, Lutjanus cyanopterus*, and the groupers *Mycteroperca microlepis* and *Epinephelus morio* as depleted. The decline of these fish species forced fishermen to change their target species. Those species that were not important to fisheries in the past – or were only utilized as bait – turned out to be today's main targets. This replacement process is occurring slowly and gradually over the years, making it more difficult for future generations of fishermen to perceive these changes. Experienced and older fishermen have seen the effects of overexploitation upon marine ecosystems for many more years, whereas young fishermen begun fishing when stocks were either reduced or on the path for depletion. This led young fisherman to assume that sharks and large groupers were naturally rare in Arraial do Cabo, and that overexploited fishing grounds were still very productive. Thus perceptions are subjected to the effects of time and possibly compromised knowledge transmission between community members [30].

Baseline shifts regarding the past conditions of the world›s oceans might lead to a collective amnesia, making people more tolerant to population declines and biodiversity losses [56]. Despite the fact that fishermen are usually in contact with nature on a daily-basis, exhibiting greater awareness of environmental changes, baselines shift rapidly even among them [29],[41]. In Arraial do Cabo, sixty years of fisheries exploitation and impacts upon marine ecosystems have affected the environmental baselines of three consecutive generations. Such changes in the perceptions of fishermen regarding the status of marine ecosystems are of special concern given the importance of LEK to adjust and improve our understanding of the oceans› past conditions.

A possible short-term solution to slow the reference loss process is to recreate old conditions of marine life through historical data [4] such as films, photos [57], booklets and reports [31],[50] of early explorers. Such kind of data can provide important detail on former species size and catch composition [30], potentially improving our knowledge on the effects of overfishing. Recreating past environmental baselines is a unique opportunity to assess the historical changes that ecosystems have undergone, as well as for improving social awareness on the status of biodiversity. Recent studies have pointed that underestimating empirical data is a mistake [27], [29], [45], [56], [58], [59], especially in places like Arraial do Cabo, where written records on species› former size and abundance are rare.

When assessing fishing impacts, few studies make such parallel use of LEK and species population status, mostly because LEK is considered anecdotal, non-scientifically sampled, and of limited application among scientists and managers [40], [60]. Our work is among the first to combine landing, underwater visual census and LEK data to assess the temporal patterns of exploited fish populations [58], [61]. Combining data from different sources renders more robust and reliable assessments, which are especially important in times where decision makers require scientific certainty. More importantly, local management and conservation efforts will certainly gain more commitment from fishermen when their own experience is taken into account on the establishment of management strategies.

Fishing Collapse in the Southwestern Atlantic

The combined methodologies utilized to assess multigear fisheries reveal the decline of different fish species in the region of Arraial do Cabo. The decline of the bluefish (*P. saltatrix*) fishery revealed through information provided by fishermen was supported by evidence from landing data. The declining trend for Epinephelidae species (*E. marginatus, Mycteroperca bonaci, M. acutirostris* and *M. microlepis*) in the subtropical reefs of Arraial do Cabo has been demonstrated through information provided by fishermen, and through data from underwater visual census campaigns conducted in the region. Another study for the region underscores recent declines in the abundance of groupers using the underwater visual census technique[17]. Not only the abundance of fish is diminishing, but also the larger individuals are consistently more difficult to capture and sight in the field [17]. Overfishing is affecting the size of fish individuals, removing the largest ones at an unsustainable rate, and preventing the smallest ones from reaching large sizes. This can seriously compromise fish population rebound since larger and older females produce greater quantity and better quality offspring when compared to younger fish [62]. Most of these grouper species are important fishing targets elsewhere in the Brazilian coast, and the overfished state of populations has been demonstrated by different studies [17], [20]. Moreover, they are highly vulnerable to overexploitation given its critical combination of biological attributes [25].

As populations of top predators collapse, other large-bodied species at lower trophic level become new targets, as is the case of the endemic greenbeak parrotfish, *Scarus trispinosus*. Being the largest of all Brazilian parrotfishes, the greenbeak was the first herbivorous-detritivorous reef fish targeted by spearfishermen, but other large-bodied parrotfishes of the genus *Sparisoma* also became targeted in the last few years. Different species of parrotfishes perform specific functional roles, some reported as critical in maintaining the resistance and resilience of reef ecosystems [63], [64]. Along the Brazilian coast, populations of *S. trispinosus* have strongly declined due to spearfishing [17], and the species is considered ecologically extinct in Arraial do Cabo [50]. The species was recently listed by the IUCN [65] as Endangered (EN) yet there is no regulation or national listing to guarantee its protection[25].

Here we have shown the decline of reef fisheries in the Southwestern Atlantic from multiple sources of information: local ecological knowledge from multigear fisheries (line, beach seine and spearfishing), fisheries landing data and even underwater visual census data. The pattern identified for six different target species follows a similar pattern of global deterioration of marine ecosystems reported elsewhere [1], [66]. In the Southwestern Atlantic, as elsewhere in the world, fishing has moved from upper to lower trophic levels: groupers used to be fisheries› major target, but today, parrotfishes and even surgeonfishes are targeted. If going back to the early days of the 1960›s tell us a very different story on fish abundance and size in the region of Arraial do Cabo, one could only imagine what a 150 or 200-year-old baseline would inform us on those fish species. We emphasize the importance of combining LEK with scientific data in order to improve our understanding of the status of marine habitats and its associated biodiversity, but the occurrence of changes in environmental baselines must be taken into account. Only through a clear picture of the status of natural resources in the past can we help people improve their awareness of the importance of preserving natural resources and ecosystem services.

AUTHOR CONTRIBUTIONS

Conceived and designed the experiments: CELF CMN GRM PJAS. Performed the experiments: CELF CMN GRM PJAS. Analyzed the data: CELF CMN GRM PJAS MGB OJL SRF. Contributed reagents/materials/analysis tools: CELF CMN GRM PJAS MGB OJL SRF. Wrote the paper: CELF CMN GRM PJAS MGB OJL SRF.

REFERENCES

1. Jackson JBC, Kirby MX, Berger WH, Bjorndal KA, Botsford LW, et al. (2001) Historical Overfishing and the Recent Collapse of Coastal Ecosystems. Science 293: 629–637. doi: 10.1126/science.1059199

2. Erlandson JM, Rick TC, Vellanoweth RL (2004) Human impacts on ancient environments: a case study from California's Northern Channel Islands. Voyages of discovery: The archaeology of islands: 51–83.

3. Jackson JB (1997) Reefs since Columbus. Coral reefs 16: S23–S32. doi: 10.1007/s003380050238

4. Lotze HK, Worm B (2009) Historical baselines for large marine animals. Trends in ecology & evolution 24: 254–262. doi: 10.1016/j.tree.2008.12.004

5. Myers RA, Worm B (2003) Rapid worldwide depletion of predatory fish communities. Nature 423: 280–283. doi: 10.1038/nature01610

6. Worm B, Davis B, Kettemer L, Ward-Paige CA, Chapman D, et al. (2013) Global catches, exploitation rates, and rebuilding options for sharks. Marine Policy 40: 194–204. doi: 10.1016/j.marpol.2012.12.034

7. Barrett JH, Locker AM, Roberts CM (2004) The origins of intensive marine fishing in medieval Europe: the English evidence. Proceedings of

the Royal Society of London Series B: Biological Sciences 271: 2417–2421. doi: 10.1098/rspb.2004.2885

8. Barrett JH, Nicholson RA, Cerón-Carrasco R (1999) Archaeo-ichthyological evidence for long-term socioeconomic trends in northern Scotland: 3500 BC to AD 1500. Journal of Archaeological Science 26: 353–388. doi: 10.1006/jasc.1998.0336

9. Erlandson JM, Rick TC (2010) Archaeology Meets Marine Ecology: The Antiquity of Maritime Cultures and Human Impacts on Marine Fisheries and Ecosystems. Annual Review of Marine Science 2: 231–251. doi: 10.1146/annurev.marine.010908.163749

10. Smith I (2005) Retreat and resilience: fur seals and human settlement in New Zealand. The exploitation and cultural importance of sea mammals: 6–18.

11. Morales A, Rosello E, Canas J (1994) Cueva de Nerja (prov. Malaga): a close look at a twelve thousand year ichthyofaunal sequence from southern Spain [Paleolithic, Neolithic, Chalcolithic].

12. Wing S, Wing E (2001) Prehistoric fisheries in the Caribbean. Coral reefs 20: 1–8. doi: 10.1007/s003380100142

13. Dulvy N, Polunin NV, Mill A, Graham NA (2004) Size structural change in lightly exploited coral reef fish communities: evidence for weak indirect effects. Canadian Journal of Fisheries and Aquatic Sciences 61: 466–475. doi: 10.1139/f03-169

14. Jackson JB (2008) Ecological extinction and evolution in the brave new ocean. Proceedings of the National Academy of Sciences 105: 11458–11465. doi: 10.1073/pnas.0802812105

15. Roberts C, Hawkins JP, Campaign WES (2000) Fully-protected marine reserves: a guide: WWF Endangered seas campaign Washington, DC.

16. Halpern BS, Walbridge S, Selkoe KA, Kappel CV, Micheli F, et al. (2008) A global map of human impact on marine ecosystems. Science 319: 948–952. doi: 10.1126/science.1149345

17. Floeter S, Halpern BS, Ferreira C (2006) Effects of fishing and protection on Brazilian reef fishes. Biological Conservation 128: 391–402. doi: 10.1016/j.biocon.2005.10.005

18. Frédou T (2004) The fishing activity on coral reefs and adjacent ecosystems: a case study of the Northeast of Brazil. Cybium 28: 274.

19. Rezende SdM, Ferreira BP (2004) Age, growth and mortality of dog snapper Lutjanus jocu (Bloch & Schneider, 1801) in the northeast coast

of Brazil. Brazilian Journal of Oceanography 52: 107–121. doi: 10.1590/s1679-87592004000200003

20. Ferreira C, Gonçalves J (1999) The unique Abrolhos reef formation (Brazil): need for specific management strategies. Coral reefs 18: 352–352. doi: 10.1007/s003380050211

21. Freire KM, Pauly D (2010) Fishing down Brazilian marine food webs, with emphasis on the east Brazil large marine ecosystem. Fisheries Research 105: 57–62. doi: 10.1016/j.fishres.2010.02.008

22. Pauly D, Christensen V, Dalsgaard J, Froese R, Torres F (1998) Fishing down marine food webs. Science 279: 860–863. doi: 10.1126/science.279.5352.860

23. Luiz OJ, Edwards AJ (2011) Extinction of a shark population in the Archipelago of Saint Paul's Rocks (equatorial Atlantic) inferred from the historical record. Biological Conservation 144: 2873–2881. doi: 10.1016/j.biocon.2011.08.004

24. Bender M, Floeter S, Hanazaki N (2013) Do traditional fishers recognise reef fish species declines? Shifting environmental baselines in Eastern Brazil. Fisheries Management and Ecology 20: 58–67. doi: 10.1111/fme.12006

25. Bender M, Floeter S, Mayer F, Vila-Nova D, Longo G, et al. (2013) Biological attributes and major threats as predictors of the vulnerability of species: a case study with Brazilian reef fishes. Oryx 47: 259–265. doi: 10.1017/s0030605311000144x

26. Knowlton N, Jackson JB (2008) Shifting baselines, local impacts, and global change on coral reefs. PLoS biology 6: e54. doi: 10.1371/journal.pbio.0060054

27. Pauly D (1995) Anecdotes and the shifting baseline syndrome of fisheries. Trends in ecology & evolution 10: 430. doi: 10.1016/s0169-5347(00)89171-5

28. Pinnegar JK, Engelhard GH (2008) The 'shifting baseline'phenomenon: a global perspective. Reviews in Fish Biology and Fisheries 18: 1–16. doi: 10.1007/s11160-007-9058-6

29. Sáenz-Arroyo A, Roberts C, Torre J, Cariño-Olvera M, Enríquez-Andrade R (2005) Rapidly shifting environmental baselines among fishers of the Gulf of California. Proceedings of the Royal Society B: Biological Sciences 272: 1957–1962. doi: 10.1098/rspb.2005.3175

30. Sáenz–Arroyo A, Roberts CM, Torre J, Cariño-Olvera M (2005) Using fishers' anecdotes, naturalists' observations and grey literature to reassess

marine species at risk: the case of the Gulf grouper in the Gulf of California, Mexico. Fish and Fisheries 6: 121–133. doi: 10.1111/j.1467-2979.2005.00185.x

31. Fortibuoni T, Libralato S, Raicevich S, Giovanardi O, Solidoro C (2010) Coding early naturalists' accounts into long-term fish community changes in the Adriatic Sea (1800–2000). PloS one 5: e15502. doi: 10.1371/journal.pone.0015502

32. Sáenz-Arroyo A, Roberts CM, Torre J, Cariño-Olvera M, Hawkins JP (2006) The value of evidence about past abundance: marine fauna of the Gulf of California through the eyes of 16th to 19th century travellers. Fish and Fisheries 7: 128–146. doi: 10.1111/j.1467-2979.2006.00214.x

33. Roberts CM (2003) Our shifting perspectives on the oceans. Oryx 37: 166–177. doi: 10.1017/s0030605303000358

34. Drew JA (2005) Use of traditional ecological knowledge in marine conservation. Conservation Biology 19: 1286–1293. doi: 10.1111/j.1523-1739.2005.00158.x

35. Huntington HP (2000) Using traditional ecological knowledge in science: methods and applications. Ecological applications 10: 1270–1274. doi: 10.1890/1051-0761(2000)010[1270:utekis]2.0.co;2

36. Berkes F, Colding J, Folke C (2000) Rediscovery of traditional ecological knowledge as adaptive management. Ecological applications 10: 1251–1262. doi: 10.1890/1051-0761(2000)010[1251:roteka]2.0.co;2

37. Silvano RA, MacCord PF, Lima RV, Begossi A (2006) When does this fish spawn? Fishermen's local knowledge of migration and reproduction of Brazilian coastal fishes. Environmental Biology of Fishes 76: 371–386. doi: 10.1007/s10641-006-9043-2

38. Johannes RE (1981) Working with fishermen to improve coastal tropical fisheries and resource management. Bulletin of Marine Science 31: 673–680.

39. Johannes RE, Freeman MM, Hamilton RJ (2000) Ignore fishers' knowledge and miss the boat. Fish and Fisheries 1: 257–271. doi: 10.1046/j.1467-2979.2000.00019.x

40. Shackeroff JM, Campbell LM (2007) Traditional ecological knowledge in conservation research: problems and prospects for their constructive engagement. Conservation and Society 5: 343.

41. Turvey ST, Barrett LA, Yujiang H, Lei Z, Xinqiao Z, et al. (2010) Rapidly shifting baselines in Yangtze fishing communities and local memory of extinct species. Conservation Biology 24: 778–787. doi: 10.1111/j.1523-1739.2009.01395.x

42. Jones JP, Andriamarovololona MM, Hockley N, Gibbons JM, Milner-Gulland E (2008) Testing the use of interviews as a tool for monitoring trends in the harvesting of wild species. Journal of Applied Ecology 45: 1205–1212. doi: 10.1111/j.1365-2664.2008.01487.x

43. van der Hoeven CA, de Boer WF, Prins HH (2004) Pooling local expert opinions for estimating mammal densities in tropical rainforests. Journal for nature conservation 12: 193–204. doi: 10.1016/j.jnc.2004.06.003

44. Papworth SK, Rist J, Coad L, Milner-Gulland EJ (2009) Evidence for shifting baseline syndrome in conservation. Conservation Letters.

45. Silvano RAM, Begossi A (2010) What can be learned from fishers? An integrated survey of fishers' local ecological knowledge and bluefish (*Pomatomus saltatrix*) biology on the Brazilian coast. Hydrobiologia 637: 3–18. doi: 10.1007/s10750-009-9979-2

46. Freire KMF, Carvalho-Filho A (2009) Richness of common names of Brazilian reef fishes. Pan-American Journal of Aquatic Sciences 4: 96–145.

47. Brito R (1999) Modernidade e tradição. Construção da identidade social dos pescadores de Arraial do Cabo, RJ Niterói: EDUFF.

48. Craig MT, Hastings PA (2007) A molecular phylogeny of the groupers of the subfamily Epinephelinae (Serranidae) with a revised classification of the Epinephelini. Ichthyological Research 54: 1–17. doi: 10.1007/s10228-006-0367-x

49. Bohnsack JA, Harper DE, Center SF (1988) Length-weight relationships of selected marine reef fishes from the southeastern United States and the Caribbean: National Oceanic and Atmospheric Administration, National Marine Fisheries Service, Southeast Fisheries Center.

50. Floeter SR, Krohling W, Gasparini JL, Ferreira CE, Zalmon IR (2007) Reef fish community structure on coastal islands of the southeastern Brazil: the influence of exposure and benthic cover. Environmental Biology of Fishes 78: 147–160. doi: 10.1007/s10641-006-9084-6

51. Underwood AJ (1997) Experiments in ecology: their logical design and interpretation using analysis of variance: Cambridge University Press.

52. Ferreira C, Gonçalves J (2006) Community structure and diet of roving herbivorous reef fishes in the Abrolhos Archipelago, south-western Atlantic. Journal of Fish Biology 69: 1533–1551. doi: 10.1111/j.1095-8649.2006.01220.x

53. Bunce M, Rodwell LD, Gibb R, Mee L (2008) Shifting baselines in fishers' perceptions of island reef fishery degradation. Ocean & Coastal

Management 51: 285–302. doi: 10.1016/j.ocecoaman.2007.09.006

54. Humphries P, Winemiller KO (2009) Historical impacts on river fauna, shifting baselines, and challenges for restoration. BioScience 59: 673–684. doi: 10.1525/bio.2009.59.8.9

55. Rittenhouse CD, Pidgeon AM, Albright TP, Culbert PD, Clayton MK, et al. (2010) Conservation of forest birds: evidence of a shifting baseline in community structure. PloS one 5: e11938. doi: 10.1371/journal. pone.0011938

56. Ainsworth CH, Pitcher TJ, Rotinsulu C (2008) Evidence of fishery depletions and shifting cognitive baselines in Eastern Indonesia. Biological Conservation 141: 848–859. doi: 10.1016/j.biocon.2008.01.006

57. McClenachan L (2009) Documenting loss of large trophy fish from the Florida Keys with historical photographs. Conserv Biol 23: 636–643. doi: 10.1111/j.1523-1739.2008.01152.x

58. O'Donnell K, Pajaro M, Vincent A (2010) How does the accuracy of fisher knowledge affect seahorse conservation status? Animal Conservation 13: 526–533. doi: 10.1111/j.1469-1795.2010.00377.x

59. Paterson B (2010) Integrating fisher knowledge and scientific assessments. Animal Conservation 13: 536–537. doi: 10.1111/j.1469-1795.2010.00419.x

60. Brook RK, McLachlan SM (2008) Trends and prospects for local knowledge in ecological and conservation research and monitoring. Biodiversity and Conservation 17: 3501–3512. doi: 10.1007/s10531-008-9445-x

61. Boudreau SA, Worm B (2010) Top-down control of lobster in the Gulf of Maine: insights from local ecological knowledge and research surveys. Marine Ecology Progress Series 403: 181–191. doi: 10.3354/meps08473

62. Palumbi SR (2003) Population genetics, demographic connectivity, and the design of marine reserves. Ecological applications 13: 146–158. doi: 10.1890/1051-0761(2003)013[0146:pgdcat]2.0.co;2

63. Hughes TP, Rodrigues MJ, Bellwood DR, Ceccarelli D, Hoegh-Guldberg O, et al. (2007) Phase shifts, herbivory, and the resilience of coral reefs to climate change. Curr Biol 17: 360–365. doi: 10.1016/j.cub.2006.12.049

64. Mumby PJ, Dahlgren CP, Harborne AR, Kappel CV, Micheli F, et al. (2006) Fishing, trophic cascades, and the process of grazing on coral reefs. Science 311: 98–101. doi: 10.1126/science.1121129

65. IUCN (2013) IUCN Red List of Threatened Species. Version 2013.2.

66. Worm B, Barbier EB, Beaumont N, Duffy JE, Folke C, et al. (2006) Impacts of biodiversity loss on ocean ecosystem services. Science 314: 787–790. doi: 10.1126/science.1132294

Chapter 8

A NOVEL TOOL TO MITIGATE BY-CATCH MORTALITY OF BALTIC SEALS IN COASTAL FYKE NET FISHERY

Sari M. Oksanen[1], Markus P. Ahola[2], Jyrki Oikarinen[3], and Mervi Kunnasranta[1]

[1]Department of Biology, University of Eastern Finland, Joensuu, Finland

[2]Natural Resources Institute Finland, Turku, Finland

[3]Perämeren Kalatalousyhteisöjen Liitto ry, Oulu, Finland

ABSTRACT

Developing methods to reduce the incidental catch of non-target species is important, as by-catch mortality poses threats especially to large aquatic predators. We examined the effectiveness of a novel device, a "seal sock", in mitigating the by-catch mortality of seals in coastal fyke net fisheries in the Baltic Sea. The seal sock developed and tested in this study was a cylindrical net attached to the fyke net, allowing the seals access to the surface to breathe while trapped inside fishing gear. The number of dead and live seals caught in fyke nets without a seal sock (years 2008–2010) and with a sock (years 2011–2013) was recorded. The seals caught in fyke nets were mainly juveniles. Of ringed seals (*Phoca hispida botnica*) both sexes were equally represented, while of grey seals (*Halichoerus grypus*) the ratio was biased (71%) towards males. All the by-caught seals were dead in the fyke nets without a seal sock, whereas 70% of ringed seals and 11% of grey seals survived when the seal sock was used. The seal sock proved to be effective in reducing the by-catch mortality of ringed seals, but did not perform as well with grey seals.

INTRODUCTION

Fisheries worldwide capture several non-target species as incidental by-catch. Populations of many large aquatic predators, such as marine mammals and sharks are especially vulnerable to by-catch mortality in consequence of their slow reproductive rates [1]. Mitigation of by-catch mortality has therefore become increasingly important in both fisheries management [2] and wildlife conservation [3]. However, the interactions between aquatic predators and fisheries can be complex, and in particular, many seal species cause losses to fish catches and damage to fishing gear [4,5]. Developing fishing methods can provide means to mitigate both aspects of the seal-fisheries conflict. For example, physical barriers mounted on the fish traps, such as wire grids, prevent seals from entering the trap [6–8]. Alternatively, exclusion devices that facilitate the seals' exit from the fishing gear [9] and techniques that trap seals alive inside the gear have also been developed [10].

The Baltic grey seal (*Halichoerus grypus*) and ringed seal (*Phoca hispida botnica*) have had significant interactions with human activities. Both populations declined drastically to only about 5000 seals in the 1970s as a result of excessive hunting and reproductive disorders caused by environmental pollution. The seal populations have been recovering, but the present number of grey seals (census size: 30 000 seals) is still less than half and ringed seals (13 000 seals) less than one tenth of the estimated historical abundance at the beginning of the 20th century [11–13]. The recovery of these populations has led to an increase in seal-induced losses to coastal fisheries, which has had negative effects on attitudes towards the conservation of seal populations [14]. The Baltic grey seal in general causes more losses to fisheries than the ringed seal [6,15]. However, increasing numbers of ringed seals are assumed to cause substantial catch losses in the Bothnian Bay, the northernmost part of the Baltic Sea (Fig 1) [16]. The hunting of seals has become a key action for the mitigation of damage in the northern Baltic Sea [14,17]. In addition to hunting, by-catch mortality may be another significant component of the anthropogenic mortality of seal populations. Estimated annual by-catch of the Baltic grey seal is in the order of 2000 animals or more [18] but for ringed seals the magnitude is unknown. An information gap regarding by-catch and its mitigation methods in coastal and small scale fisheries has been recognized globally [19]. Also in the Baltic Sea broader knowledge of by-catch is of great importance for sustainable fishery and population management [20].

Considerable efforts have been put into research on interaction between Baltic grey seal and fisheries [10,14,17,18,21,22] and on the development of seal-proof fishing gear. For example, trap nets with modified fish chambers, i.e. pontoon traps, are reported to reduce grey seal-induced losses to fisheries and simultaneously to reduce by-catch [8,23]. However, relatively little attention has been paid to estimating and reducing the by-catch mortality of seals in other stationary gears. In addition, there is little information on the interaction between ringed seals and fisheries, not only in the Baltic Sea [24], but also globally. Methods for reducing interactions of Baltic ringed seals and coastal fishery are of great importance, as climate change can be expected to further threaten the population growth of this ice-breeding species, especially in the southern breeding areas, where the stocks have recovered slowly, if at all [12]. In this study we tested the effectiveness of a new type of pinniped by-catch reduction device for fyke nets, called a seal sock. We also examined the demographic properties, such as age and sex distribution, of the by-caught seals.

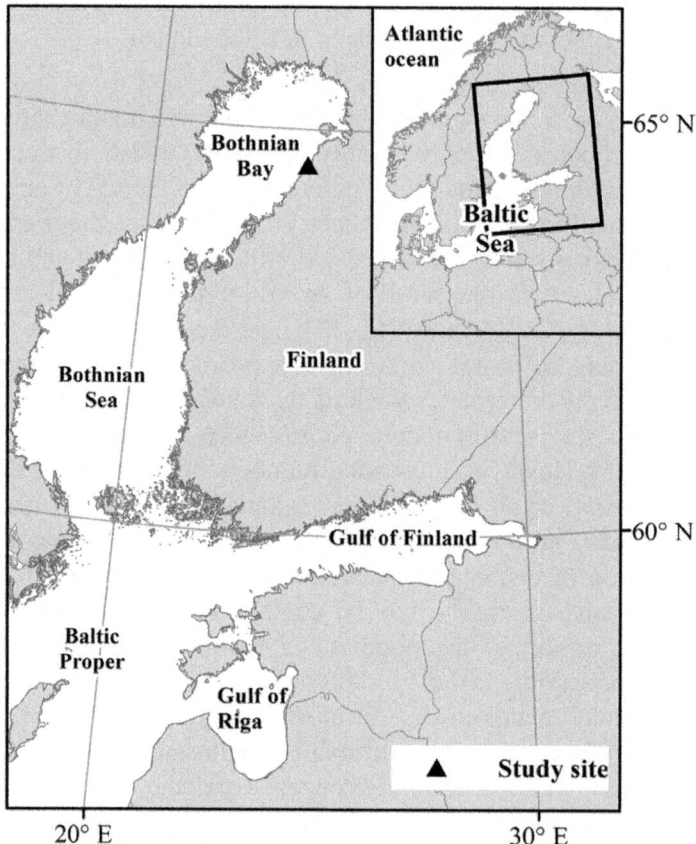

Figure 1: The Baltic Sea.

doi:10.1371/journal.pone.0127510.g001

METHODS

Around 70% of the Baltic ringed seal population is found in the Bothnian Bay (Fig 1), and grey seal is also encountered in this region. Coastal fyke net fishery in the area targets whitefish (*Coregonus lavaretus*), vendace (*C. albula*), Atlantic salmon (*Salmo salar*) and brown trout (*S.trutta*) in particular. Various modifications of fyke nets are used, but in general they are large, bottom-anchored fishing gear that consists of a leader net and wings guiding the fish through the middle chamber into the fish chamber. Seals entering the chambers may not find their way out and drown, as the chambers are usually submerged (approximately 2 m from the surface). The seal sock is a by-catch reduction device that is designed to enable seals to have access to the surface to breathe while trapped inside the fyke net. The sock was initially developed and constructed by a Finnish fish trap manufacturer (Ab Scandi Net Oy).

We developed a modification of the seal sock, comprising a cylindrical net (diameter 0.7 m, length 2.5–4.0 m) made of strong seal-proof netting (Dyneema, mesh size 30 mm). It was attached to the roof of the fish chamber in a fyke net (Fig 2A). The sock was fitted with one or two hoops, to keep its shape, and a small float. To examine the effectiveness of the seal sock in reducing by-catch, we compared the number of seals caught alive in a set of fyke nets between years when the socks were not used (2008–2010) and when they were used (2011–2013). Annually, between May and October–November, a commercial fisherman fished with 4–6 fyke nets within the same fishing area in the Bothnian Bay (64°32N, 24°19E, Fig 1). The fyke nets comprised a leader net (mesh size 200–300 mm), wings (60–100 mm) and two cylindrical chambers (30–35 mm, Dyneema, Fig 2B). Fishing effort was determined based on the fisherman's logbooks and was relatively same between both periods (2106 days for years 2008–2010 and 1767 days for years 2011–2013).

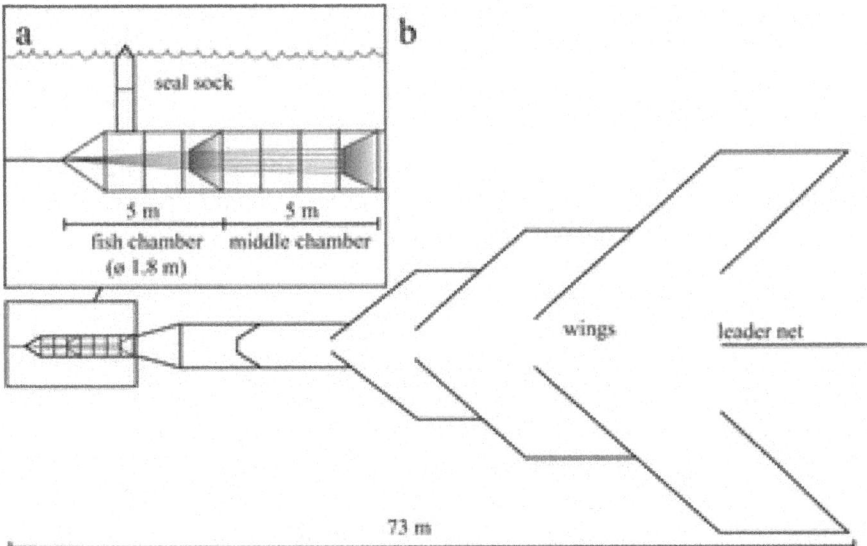

Figure 2: A type of by-catch reduction device, a seal sock, is attached to the fish chamber of a fyke net allowing the seal to have access to the surface.

(A) A side view of a seal sock and chambers. (B) A fyke net (seen from above) consists of a leader net, wings and chambers.

doi:10.1371/journal.pone.0127510.g002

The fisherman kept a logbook on the number, species and capture date of by-caught seals. During the testing of the sock in the years 2011–2013, the fisherman also recorded the sex and weight (kg) or age class (juveniles and adults) of the by-caught seals. Researchers took part to the field work (tagging live animals and taking samples from dead ones) during years 2011–2013 and they often had possibility to verify the reliability of the data. Although the fishermen's motivation to report by-catch has generally been low in Finland [18], the volume of by-catch reported in this study is high and comparable between the two periods (2008–2010 and 2011–2013). It is therefore likely that the volumes of by-catch detected in our study are not underrated. Weighed seals were further divided into age classes on the basis of an age-weight database (Natural Resources Institute Finland). Weighed ringed seals with a body weight > 50 kg were classified as adults (estimated age ≥ 4 years). Grey seal males with a body weight > 92 kg and females > 65 kg were categorized as adults (estimated age ≥ 5 years). We also used the seal socks for live-capturing ringed seals for satellite tagging (Oksanen et al., unpublished). We tested the effect of the seal sock on survival with Fisher's exact test. The effect of weight and capture month on survival of ringed seals when the sock was in use (2011–2013) was tested with binary logistic regression. Effects of weight and month on the grey seal survival were not tested due to the small sample size. We conducted statistical testing with IBM SPSS Statistics 20 software.

Ethics Statement

The use of a seal sock and animal handling was permitted by the game authorities (permit no. 2011/00082 and 2013/00197) and the Animal Experiment Board of Finland (no. ES AVI/1114/04.10.03/2011).

RESULTS

A total of 135 seals (30 live and 105 dead) were caught (4–6 fyke nets annually) during the study years in 2008–2013 (S1 Dataset). Of all the by-caught seals, 103 (76%) were ringed and 32 (24%) grey seals. During the years 2011–2013, 95% of the ringed (n = 40) and 67% of the grey seals (n = 18) were juveniles (Table 1). The mean weight of the by-caught and weighed ringed seals (n = 33) was 36 kg (SD 11). The weight of the ringed seals visiting fyke nets increased towards autumn (linear regression, R^2 = 0.469, F = 27.3, p<0.001; Table 1): ringed seals by-caught and weighed in May (n = 4) had a mean weight of 19 kg

(SD 9) (corresponding to young-of-the-year), whereas individuals by-caught in October–November (n = 7) weighed on average 44 kg (SD 10). The sex-ratio of the ringed seals was even (49% males and 51% females), but for the grey seals it was biased towards males (71% males and 29% females; Table 1). In fact, while the sex distribution of juvenile grey seals (n = 12) was even and they were caught mostly during summer (8 caught in May-July), the adults (n = 6) were mostly males (5 males, 1 unknown) and mostly captured during the autumn (5 caught in August-November: Table 1).

Table 1: Monthly variation in numbers, age and sex distribution and survival of the by-caught seals in fyke nets equipped with the seal sock

Month	Ringed seals					Grey seals				Total
	Total	Juveniles/ adults	Males/ females	Alive/ dead	Mean weight	Total	Juveniles/ adults	Males/ females	Alive/ dead	
May	6	6/0	3/3	2/4	19 ± 9	3	3/0	1/2	0/3	9
June	1	1/0	1/0	0/1	na	3	2/1	3/0	0/3	4
July	3	3/0	2/1	2/1	24 ± 8	3	3/0	1/2	0/3	6
Aug	4	4/0	2/2	4/0	34 ±8	2	1/1	2/0	0/2	6
Sept	16	15/1	7/9	13/3	38 ± 8	4	2/2	4/0	1/3	20
Oct	8	7/1	3/4[a]	5/3	45 ± 12	1	1/0	0/1	1/0	9
Nov	2	2/0	1/1	2/0	43 ± 1	2	0/2	1/0[a]	0/2	4
Total	40	38/2	19/20	28/12	36 ± 11	18	12/6	12/5	2/16	58

In years 2011–2013, annually 4–6 fyke nets equipped with the seal sock were set out for fishing in the Bothnian Bay. Monthly mean weight (kg) ± SD for ringed seals is reported for a total of 33 weighed individuals.
[a] Gender for one seal not recorded

doi:10.1371/journal.pone.0127510.t001

doi:10.1371/journal.pone.0127510.t001

Overall, the survival of seals in the fyke nets increased when the sock was used (Fisher's exact test, p<0.001). Altogether 77 dead seals were by-caught during the years when the seal sock was not used (2008–2010), whereas 30 live and 28 dead seals were caught when the seal sock was used (2011–2013) (Fig 3). The seal sock increased survival of ringed seals (p<0.001): of all caught ringed seals (n = 40), 83% found their way to the sock and 70% remained alive. However, the survival of grey seals was not increased (p = 0.308): only 17% of the grey (n = 18) seals found way to the sock, and 11% remained alive. The ringed seals that did not find their way into the sock were mostly small pups (5/7) having a mean weight of 16 kg (SD 9) and they were mostly captured in early summer. In fact, capture month was the only statistically significant predictor of ringed seal survival in logistic regression (model summary: $\chi^2(1) = 4.210$, p = 0.040, Nagelkerke $R^2 = 0.142$; parameter *month*: p = 0.048, coefficient = –0,405). By contrast, all the by-caught adult grey seals (n = 6) were caught in the middle chamber, while 8 out of 12 of juveniles were either in the fish chamber or in the sock. However, only 3 out of 8 grey seals reaching the fish chamber found their way into the sock.

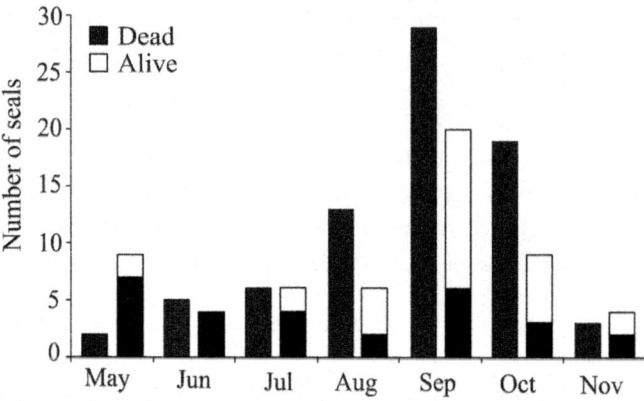

Figure 3: Monthly variation in the number of by-caught seals in fyke nets (4–6 fyke nets annually).

First bars: fyke nets without a seal sock (years 2008–2010). Second bars: fyke nets with a seal sock (2011–2013).

doi:10.1371/journal.pone.0127510.g003

DISCUSSION

The seal sock proved to be effective in reducing the by-catch mortality of ringed seals in fyke nets, but it did not perform as well with grey seals. According to our pilot test, the seal sock also seemed to work well in a so-called bottom fyke net (Flex R, Ab Scandi Net Oy), which we tested during the summer of 2013, and all of the captured seals (7 ringed seals and 1 grey seal) survived in the sock. Although the use of the seal sock increased survival of ringed seals in general, small pups caught during early summer survived poorly despite the sock. On the contrary, adult grey seals were by-caught only in the middle chamber, and it is likely that they could not reach the fish chamber through the narrow passage between the chambers (diameter 0.3 m). To overcome this problem, a sock could be inserted also into the middle chamber. However, only 3 out of 8 grey seals that reached the fish chamber found their way into the sock, which may indicate behavioral differences between the species. Ringed seals keep open breathing holes in the sea ice, which may be several meters thick in the Arctic [25], and swimming upwards in a narrow funnel might, therefore, be more typical behavior for them. In addition to reducing by-catch, the seal sock has potential applications in capturing seals alive and removing individuals repeatedly visiting the fishing gear. When the seal sock is used only for by-catch mitigation, it could have an opening at the surface enabling seals to climb out of the gear, as exclusion devices are usually designed to enable rapid exit of the animal from the fishing gear.

Our results illustrate that the majority of seals by-caught in fyke nets are young seals of both genders. This was especially evident in ringed seals, of which 95% were juveniles and equally of both genders, but also among grey seals most of the by-caught individuals were juveniles (67%). The dominance of juveniles in by-catch could to some extent be explained by their naivety, as juveniles of many seal species are more vulnerable to being by-caught [26,27]. Ringed seal pups are especially prone to get caught in fishing gear just after weaning in the early summer [28]. We also observed lower survival of ringed seals during early summer compared to autumn, and our results indicated that survival in the seal sock was more dependent on naivety of the young-of-the-year than weight of the seals. Although the overall sex ratio of grey seals was biased towards males, young grey seals of both sexes were by-caught especially in spring, and adult males in autumn. A similar temporal trend in by-caught grey seals has been reported in previous studies [29,30]. A strong male dominance has also been observed in grey seals visiting pontoon traps [10,21,22,31].

Our study indicates that by-catch particularly increases juvenile mortality, which may not reduce population growth and viability as much as a similar increase in adult—especially female—mortality would [32]. Nevertheless, by-catch mortality may constitute another significant source of anthropogenic mortality in the grey seal population, whose annual hunting pressure in the Finnish sea area was around 5% of the total population [33], and should therefore be taken into account in the population management. By-catch mortality also increases total mortality of the ringed seal population and mitigating it may become increasingly important for population management, as the population growth of this ice-dependent species is predicted to decrease due to climate change [12,34].

Knowledge of the by-catch mortality and its mitigation methods are important aspects of the sustainable management of seal populations. Reducing incidental mortality is also becoming increasingly important measure for the fisheries management to respond to growing demands for sustainable seafood. Several ecolabels take by-catch reduction into account [2]. For example ecolabel of the Marine Stewardship Council (MSC) is granted to fisheries that follow the sustainable fishery standards of the MSC. Our study introduces a novel and inexpensive tool, a seal sock, for mitigating by-catch of seals in coastal fyke net fisheries. In addition, the seal sock provides a practical and ethical method for the selective removal of seals repeatedly visiting fyke nets in areas where high seal abundance causes substantial losses to fisheries. It can also be used in scientific studies where seals need to be captured alive, for example telemetry studies.

ACKNOWLEDGMENTS

We would like to thank fishermen M. Viitanen, T. Matinlassi and M. Posti for their close co-operation during the study. We also wish to thank K. Kyyrönen for the fyke net graphics and L. Mehtätalo for the statistical consulting. Thanks are also due to M. Valtonen and anonymous reviewers for their valuable comments on the manuscript.

AUTHOR CONTRIBUTIONS

Conceived and designed the experiments: SMO MPA JO MK. Performed the experiments: SMO MPA JO MK. Analyzed the data: SMO. Contributed reagents/materials/analysis tools: SMO MPA JO MK. Wrote the paper: SMO MPA JO MK.

REFERENCES

1.   Lewison RL, Crowder LB, Read AJ, Freeman SA. Understanding impacts of fisheries bycatch on marine megafauna. Trends Ecol Evol. 2004;19: 598–604. doi: 10.1016/j.tree.2004.09.004

2.   Kirby DS, Ward P. Standards for the effective management of fisheries bycatch. Mar Policy. 2014;44: 419–426. doi: 10.1016/j.marpol.2013.10.008

3.   Soykan CU, Moore JE, Zydelis R, Crowder LB, Safina C, Lewison RL. Why study bycatch? An introduction to the Theme Section on fisheries bycatch. Endanger Species Res. 2008;5: 91–102. doi: 10.3354/esr005091

4.   Read AJ. The looming crisis: interactions between marine mammals and fisheries. J Mammal. 2008;89: 541–548. doi: 10.1644/07-mamm-s-315r1.1

5.   Bowen WD, Lidgard D. Marine mammal culling programs: review of effects on predator and prey populations. Mamm Rev. 2013;43: 207–220. doi: 10.1111/j.1365-2907.2012.00217.x

6.   Lunneryd S-G, Fjälling A, Westerberg H. A large-mesh salmon trap: a way of mitigating seal impact on a coastal fishery. ICES J Mar Sci. 2003;60: 1194–1199. doi: 10.1016/s1054-3139(03)00145-0

7.   Lehtonen E, Suuronen P. Mitigation of seal-induced damage in salmon and whitefish trapnet fisheries by modification of the fish bag. ICES J Mar Sci. 2004;61: 1195–1200. doi: 10.1016/j.icesjms.2004.06.012

8. Hemmingsson M, Fjälling A, Lunneryd S-G. The pontoon trap: Description and function of a seal-safe trap-net. Fish Res. 2008;93: 357–359. doi: 10.1016/j.fishres.2008.06.013

9. Hamilton S, Baker GB. Review of research and assessments on the efficacy of sea lion exclusion devices in reducing the incidental mortality of New Zealand sea lions*Phocarctos hookeri* in the Auckland Islands squid trawl fishery. Fish Res. 2015;161: 200–206. doi: 10.1016/j.fishres.2014.07.010

10. Lehtonen E, Suuronen P. Live-capture of grey seals in a modified salmon trap. Fish Res. 2010;102: 214–216. doi: 10.1016/j.fishres.2009.10.007

11. Harding K, Härkönen T. Development in the Baltic grey seal (*Halichoerus grypus*) and ringed seal (*Phoca hispida*) populations during the 20th century. Ambio. 1999;28: 619–627.

12. Sundqvist L, Härkonen T, Svensson CJ, Harding KC. Linking climate trends to population dynamics in the Baltic ringed seal: Impacts of historical and future winter temperatures. Ambio. 2012;41: 865–872. doi: 10.1007/s13280-012-0334-x. pmid:22851349

13. Finnish Game and Fisheries Research Institute. Itämeren hallikanta kasvaa edelleen [Internet]. 2014 [cited 2014 Oct 10]. Available:http://www.rktl.fi/tiedotteet/itameren_hallikanta_kasvaa.html

14. Varjopuro R. Co-existence of seals and fisheries? Adaptation of a coastal fishery for recovery of the Baltic grey seal. Mar Policy. 2011;35: 450–456. doi: 10.1016/j.marpol.2010.10.023

15. Kauppinen T, Siira A, Suuronen P. Temporal and regional patterns in seal-induced catch and gear damage in the coastal trap-net fishery in the northern Baltic Sea: Effect of netting material on damage. Fish Res. 2005;73: 99–109. doi: 10.1016/j.fishres.2005.01.003

16. Storm A, Routti H, Nyman M, Kunnasranta M. Hyljepuhetta— Alueelliset ja kansalliset näkökulmat ja odotukset merihyljekantojen hoidossa [Internet]. Finnish Game and Fisheries Research Institute. 2007 [cited 2014 Oct 10]. Available:http://www.rktl.fi/www/uploads/pdf/raportti396.pdf

17. Bruckmeier K, Höj Larsen C. Swedish coastal fisheries—From conflict mitigation to participatory management. Mar Policy. 2008;32: 201–211. pmid:18052860 doi: 10.1016/j.marpol.2007.09.005

18. Vanhatalo J, Vetemaa M, Herrero A, Aho T, Tiilikainen R. By-catch of grey seals (*Halichoerus grypus*) in Baltic fisheries—A Bayesian analysis of interview survey. PLoS One. 2014;9: e113836. doi: 10.1371/journal.pone.0113836. pmid:25423168

19. Lewison RL, Crowder LB, Wallace BP, Moore JE, Cox T, Zydelis R, et al. Global patterns of marine mammal, seabird, and sea turtle bycatch reveal taxa-specific and cumulative megafauna hotspots. Proc Natl Acad Sci U S A. 2014;111: 5271–5276. doi: 10.1073/pnas.1318960111. pmid:24639512

20. Ministry of Agriculture and Forestry. Management plan for the Finnish seal populations in the Baltic Sea [Internet]. 2007 [cited 2014 Oct 10]. Available:http://www.mmm.fi/attachments/mmm/julkaisut/julkaisusarja/2007/5sxiKHp2V/4b_Hylkeen_enkku_nettiin.pdf

21. Königson S, Fjälling A, Berglind M, Lunneryd S-G. Male gray seals specialize in raiding salmon traps. Fish Res. 2013;148: 117–123. doi: 10.1016/j.fishres.2013.07.014

22. Oksanen SM, Ahola MP, Lehtonen E, Kunnasranta M. Using movement data of Baltic grey seals to examine foraging-site fidelity: implications for seal-fishery conflict mitigation. Mar Ecol Prog Ser. 2014;507: 297–308. doi: 10.3354/meps10846

23. Suuronen P, Siira A, Kauppinen T, Riikonen R, Lehtonen E, Harjunpää H. Reduction of seal-induced catch and gear damage by modification of trap-net design: Design principles for a seal-safe trap-net. Fish Res. 2006;79: 129–138. doi: 10.1016/j.fishres.2006.02.014

24. Fjälling A. The estimation of hidden seal-inflicted losses in the Baltic Sea set-trap salmon fisheries. ICES J Mar Sci. 2005;62: 1630–1635. doi: 10.1016/j.icesjms.2005.02.015

25. Smith TG, Stirling I. The breeding habitat of the ringed seal (*Phoca hispida*). The birth lair and associated structures. Can J Zool. 1975;53: 1297–1305. doi: 10.1139/z75-155

26. Sipilä T. Conservation biology of Saimaa ringed seal (*Phoca hispida saimensis*) with reference to other European seal populations. Doctoral dissertation, The University of Helsinki, Finland. 2003. Available:https://helda.helsinki.fi/bitstream/handle/10138/22401/conserva.pdf?sequence=2

27. Björge A, Ölen N, Hartvedt S, Böthun G, Bekkby T. Dispersal and bycatch mortality in gray, *Halichoerus grypus*, and harbor, *Phoca vitulina*, seals tagged at the Norwegian coast. Mar Mammal Sci. 2002;18: 963–976. doi: 10.1111/j.1748-7692.2002.tb01085.x

28. Niemi M, Auttila M, Viljanen M, Kunnasranta M. Home range, survival, and dispersal of endangered Saimaa ringed seal pups: Implications for conservation. Mar Mammal Sci. 2013;29: 1–13. doi: 10.1111/j.1748-7692.2011.00521.x

29. Bäcklin BM, Moraeus C, Roos A, Eklöf E, Lind Y. Health and age and sex distributions of Baltic grey seals (*Halichoerus grypus*) collected from bycatch and hunt in the Gulf of Bothnia. ICES J Mar Sci. 2011;68: 183–188. doi: 10.1093/icesjms/fsq131

30. Kauhala K, Kurkilahti M, Ahola MP, Herrero A, Karlsson O, Kunnasranta M, et al. Age, sex and body condition of Baltic grey seals: Are problem seals a random sample of the population? Ann Zool Fennici. 2015;52: In press.

31. Mörner T, Malmsten J, Bernodt K, Lunneryd S-G. A study on the effect of different rifle calibres in euthanisation of grey seals (*Halichoerus grypus*) in seal traps in the Baltic Sea. Acta Vet Scand. 2013;55: 79. doi: 10.1186/1751-0147-55-79. pmid:24219864

32. Harding KC, Härkönen T, Helander B, Karlsson O. Status of Baltic grey seals: Population assessment and extinction risk. NAMMCO Sci Publ. 2007;6: 33–56. doi: 10.7557/3.2720

33. Kauhala K, Ahola M, Kunnasranta M. Demographic structure and mortality rate of a Baltic grey seal population at different stages of population change, judged on the basis of the hunting bag in Finland. Ann Zool Fennici 2012;49: 287–305. doi: 10.5735/086.049.0502

34. Meier HEM, Döscher R, Halkka A. Simulated distributions of Baltic Sea-ice in warming climate and consequences for the winter habitat of the Baltic ringed seal. Ambio. 2004;33: 249–256. pmid:15264604 doi: 10.1639/0044-7447(2004)033[0249:sdobsi]2.0.co;2

Chapter 9

LARVAL CONNECTIVITY AND THE INTERNATIONAL MANAGEMENT OF FISHERIES

Andrew S. Kough[1], Claire B. Paris[1], Mark J. Butler IV[2]

[1] Rosenstiel School of Marine and Atmospheric Sciences, University of Miami, Miami, Florida, United States of America

[2] Department of Biological Sciences, Old Dominion University, Norfolk, Virginia, United States of America

ABSTRACT

Predicting the oceanic dispersal of planktonic larvae that connect scattered marine animal populations is difficult, yet crucial for management of species whose movements transcend international boundaries. Using multi-scale biophysical modeling techniques coupled with empirical estimates of larval behavior and gamete production, we predict and empirically verify spatio-temporal patterns of larval supply and describe the Caribbean-wide pattern of larval connectivity for the Caribbean spiny lobster (Panulirus argus), an iconic coral reef species whose commercial value approaches $1 billion USD annually. Our results provide long sought information needed for international cooperation in the management of marine resources by identifying lobster larval connectivity and dispersal pathways throughout the Caribbean. Moreover, we outline how large-scale fishery management could explicitly recognize metapopulation structure by considering larval transport dynamics and pelagic larval sanctuaries.

INTRODUCTION

The lifecycle of most marine animals includes a dispersive planktonic larval stage lasting hours to months that connects scattered populations. Therefore, knowledge of larval connectivity is crucial for understanding population dynamics and sustainably managing marine taxa whose biogeographic distributions rarely coincide with political boundaries. Recent studies of larval connectivity employing natural or artificial tags [1]–[3], biophysical modeling[4]–[6], tracking of larval patches [7], and genetic analysis [8]–[10] have revealed surprising levels of population self-recruitment, eclipsing the long-held paradigm that marine populations are largely "open" and dependent upon an exogenous supply of larvae [11]. As compelling as these findings are, the ability to predict the actual dispersal of larvae from spawning grounds to nurseries remains a rare exception. Here, we describe how an empirically parameterized biophysical model provides estimates of larval supply and may be used to pinpoint larval origins, destinations, and pathways for one of the Caribbean›s most valuable marine species - the spiny lobster, *Panulirus argus*.

The Caribbean spiny lobster is a ubiquitous inhabitant of coral reefs and shallow tropical seas in the tropical West Atlantic. Commercial fishermen and recreational divers in over 30 Caribbean nations harvest lobsters, a resource valued at nearly $1 billion USD annually [12]. Like most marine animals, *P. argus* has a complex life cycle: adults inhabit coral reefs where they spawn, their planktonic larvae (phyllosoma) mature in the open sea and engage in diurnal and ontogenetic vertical migration during dispersal before returning to coastal nurseries in shallow, vegetated habitats [13]. Given the long pelagic larval duration (PLD) of this species (5–9 months) [14], larvae potentially disperse among lobster populations throughout the Caribbean [15]. Genetic analyses support the hypothesis of a single "pan-Caribbean" lobster metapopulation [16]–[18], indistinguishable within the Caribbean but distinct from a closely related species off the coast of Brazil [19].

Frequent and widespread dispersal of larvae can mask genetically distinct subpopulations, whereas demographic connectivity - the frequent (i.e., weeks to years) exchange of individuals within a metapopulation - is a fundamental ecological process relevant to the management of marine fisheries and protected areas [20]. Studies of demographic connectivity have largely focused on taxa with short PLDs (e.g., bivalves and reef fish) and though valuable scientific contributions, they likely bias our understanding of connectivity at the larger spatial scales most important for marine resource management [21]. Demographic connectivity among distant (>1000 km) populations is virtually undetectable given current tagging methods and genetic techniques [22], [23]. For this less tractable circumstance, biophysical modeling is a fast and affordable tool that is unhindered by the PLD of target species; moreover it permits the evaluation of hypothetical management strategies on larval connectivity within marine metapopulations [24].

To identify the origins, destinations, and dispersal corridors of spiny lobster larvae within and among Caribbean nations, we used an open source, multi-scale coupled biophysical larval transport model [25] built from an earlier configuration of a Lagrangian individual-based model[26]. The model has four components: 1) a GIS-based benthic module representing habitat for lobster spawning and recruitment, 2) a physical oceanographic module (Figure 1) containing daily 3-D current velocities from an array of hydrodynamic models, 3) a larval biology module depicting larval life history characteristics, and 4) a Lagrangian stochastic module that tracks the trajectory of individual larvae. We parameterized the model with data on spatio-temporal patterns of spiny lobster spawning and planktonic larval behavior, and then verified the model by comparing simulation results with empirical data on the spatio-temporal patterns of larval supply at four sites in the Caribbean (see Methods). Compared to other larval dispersal models created for spiny lobsters [27]–[33], our model uses the highest resolution, three-dimensional oceanographic circulation models and also larval behavior, both of which affect dispersal trajectories [34]. Our objectives were to employ this modeling system to investigate: (a) the demographic connectivity of spiny lobster larvae among Caribbean nations, (b) the international patterns of larval imports and exports, and (c) the relevancy of connectivity in designing Caribbean-wide networks of marine protected areas (MPAs). An unanticipated phenomenon also emerged from our modeling results: the predicted existence of pelagic larval nursery areas.

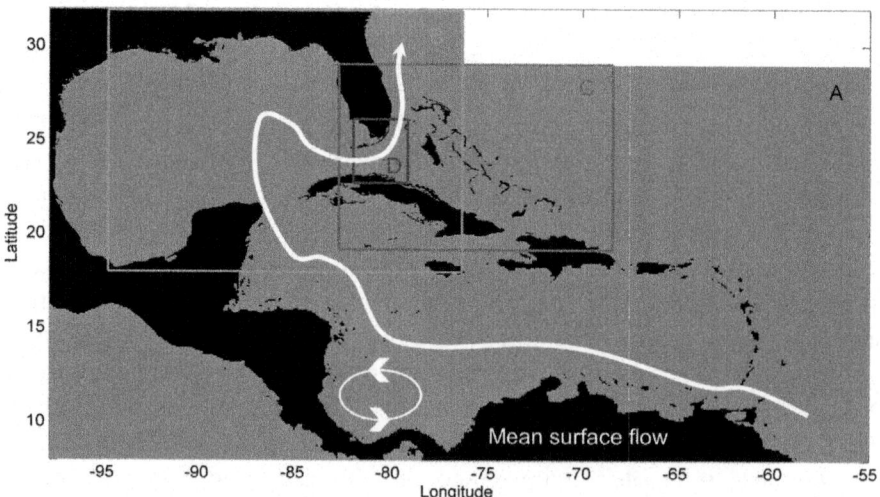

Figure 1: The hierarchy of nested circulation models used in the study and the conceptual mean Caribbean flow. The ocean circulation models used in reverse order of priority for use by the Lagrangian tracking module with their horizontal resolution and vertical depth bins in meters. A) HYCOM Global 1/12 degree: 0, 10, 20, 30, 50, 75, 100; B) GOM-HYCOM 1/25 degree: 0, 5, 10, 15, 20, 25, 30, 40, 50, 60, 70, 80, 90,

100; C) Bahamas ROMS 1/24 degree: 0, 2, 4, 8, 10, 20, 30, 40, 50, 55, 60, 80, 100; D) FLK-HYCOM 1/100 degree: 0, 5, 10, 30, 50, 75, 100. Mean surface flow after Fratantoni [76].

RESULTS

Model Verification

Two independent sets of empirical data on postlarval lobster settlement that were not used in the parameterization of our model [29], [35] were subsequently used to evaluate the final coupled system's performance. The model was compared against the monthly patterns of *P. argus* postlarval arrival at two sites in both Mexico and Florida, corresponding to four separate habitat polygons (sites) in the model (Figure 2). The simulated pattern of monthly arrival of postlarval lobsters was significantly correlated (p<0.05) with observed postlarval recruitment at two of the four sites and captured the peak in seasonal recruitment at all four sites (Figure 2). The model shows the fall peak in postlarval arrival in the Florida Keys, but does not show the spring peak (Long Key and Big Munson; Figure 2).

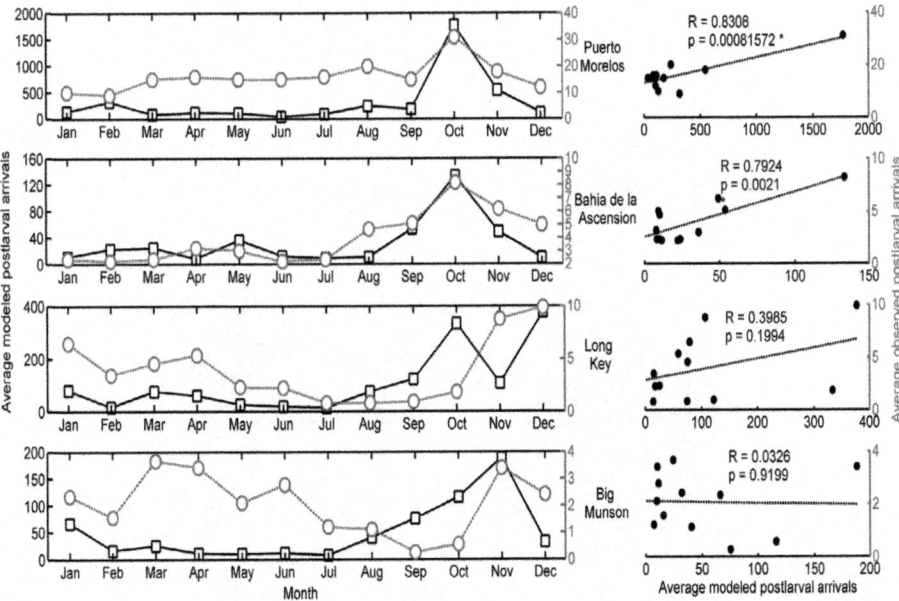

Figure 2: The seasonal pattern of observed postlarval arrival compared to model predictions. A comparison of the actual coastal arrival of *P. argus* postlarvae (red) as compared to model predictions (black) over four years at four different locations (Mexico: Bahia de Ascension, Puerto Morales; Florida: Long Key, Big Munson). The Florida observations[35] are of average postlarval arrivals per collector from 2004–2008. The

Mexican observations are averages from Briones-Fourzan [29]. The correlation between the modeled and the observed arrivals was significant (p<0.05) for Bahia de Ascension and Puerto Morales. The model also predicted the appropriate peak month of settlement in three locations, suggesting that the model can capture the temporal pattern of arriving larvae.

Connectivity Matrices

Our simulations reveal distinct flows of long-lived spiny lobster larvae among some regions of the Caribbean and pockets of larval retention within others (Figure 3). Probabilistic imports and exports of larvae from each of 261 sites show that the majority of larval exchanges transcend international boundaries when summarized by country (Figure 4). Nonetheless, domestic connectivity (i.e., self-recruitment of lobsters within a country) still dominates larval recruitment in some areas. For example, lobster populations in the Bahamas, Cuba, Nicaragua, and Venezuela are largely self-recruiting, whereas those in the Cayman Islands, Columbia, Honduras, Jamaica, Panama, and Puerto Rico depend largely on larval subsidies from outside their borders.

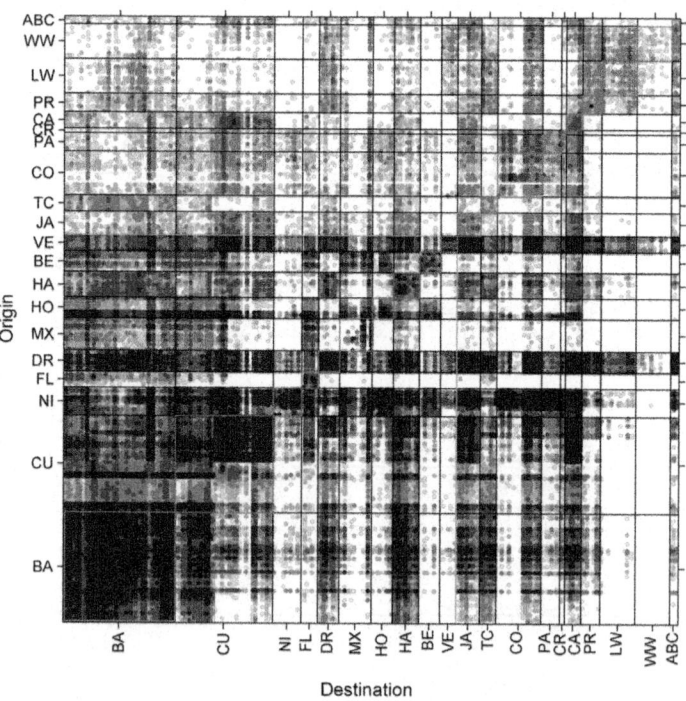

Figure 3: Connectivity matrix of spiny lobster (*P. argus*) larva. A simple matrix showing the number of larva migrating from place to place in a coupled biophysical model.

The origin of each larval connection is from the left (rows) and the destination of the larvae is at the bottom (column). Domestic connectivity (recruits that settled into their origin nation) follows the diagonal. The strength of connections among sites is a percentage of the total larval exchanged, and the grey shades represent five quantiles. The top 10 lobster fishery nations are separated by the green box. The results are from four years of Caribbean-wide lobster larval dispersal simulations among 261 habitat sites distributed into 39 countries whose abbreviations are: BA=Bahamas; CU=Cuba; NI=Nicaragua; FL=Florida; DR=Dominican Republic; MX=Mexico; HO=Honduras; HA=Haiti; BE=Belize; VE=Venezuela; JA=Jamaica; TC=Turks and Caicos; CO=Columbia; PA=Panama; CR=Costa Rica; CA=Cayman Islands; PR=Puerto Rico; LW=Leeward Islands (10 countries); WW=Windward Islands (9 countries); ABC=Aruba, Bonaire, and Curacao.

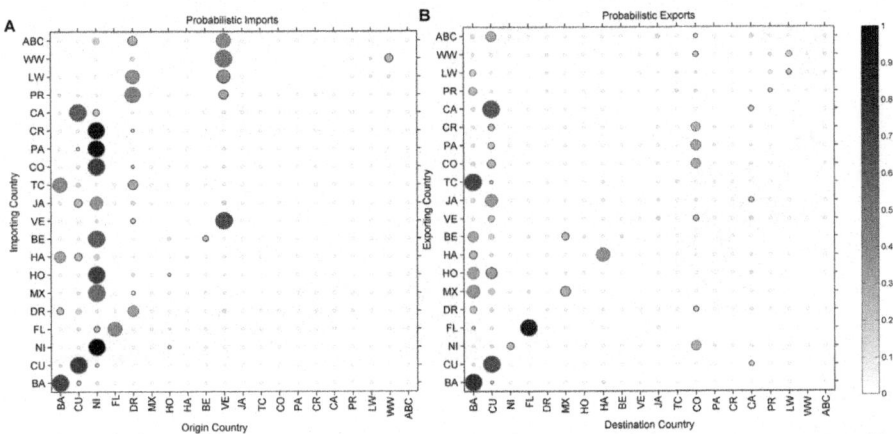

Figure 4: Probabilistic imports (A) and exports (B) of spiny lobster (*P. argus*) larva grouped by political boundaries. The probability for each instance is computed as: $Pij = Pj / \sum_{1}^{n} Pi$ where i=the country importing (A) or exporting (B), j=the origin (A) or the destination (B) country, and n=all countries. The size and shade of grey of the bubble represent the normalized probability, increasing with size and darkness. The three highest probabilities in each scenario are also colored in red, blue, and cyan, respectively.

Imbalanced International Exchange

Much like international trade, large disparities between larval imports and exports among countries abound in our simulations. We identified imbalances in the international exchange of lobster larvae by removing model predictions of domestic connectivity from the total larval supply and then compared the remaining difference in larval subsidies received and subsidies donated to the pan-Caribbean larval pool (Figure 5). This analysis reveals which countries harbor lobster populations that sustain populations elsewhere. The eastern Bahamas, southern Cuba, Dominican Republic, Nicaragua, and Venezuela export far more lobster larvae than those areas receive from the international community. In contrast, the western Bahamas, Cayman Islands, northern Cuba, Columbia, Florida Keys, Jamaica, and Panama are regions whose lobster populations receive more larvae from outside their boundaries than they donate to the Caribbean larval pool.

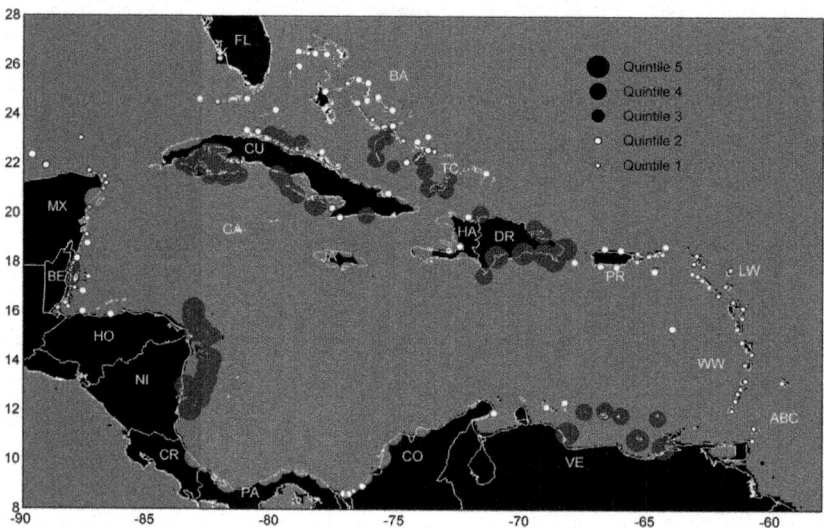

Figure 5: International larval exchange of spiny lobster (*P. argus*) larvae. The difference between larval exports and imports at a site (n=261), after removing self-recruitment. The size of the circle depicts the relative magnitude of the difference, grouped into 5 quantiles. The direction of the difference is shown as blue for positive (more larval exports) and red for negative (more larval imports). BA=Bahamas; CU=Cuba; NI=Nicaragua; FL=Florida; DR=Dominican Republic; MX=Mexico; HO=Honduras; HA=Haiti; BE=Belize; VE=Venezuela; JA=Jamaica; TC=Turks and Caicos; CO=Columbia; PA=Panama; CR=Costa Rica; CA=Cayman Islands; PR=Puerto Rico; LW=Leeward Islands (10 countries); WW=Windward Islands (9 countries); ABC=Aruba, Bonaire, and Curacao.

Connectivity and Marine Reserve Networks

Networks of MPAs have been proposed as a solution to ensure that demographic connectivity is maintained among marine animal metapopulations, with a recommendation that on average 20–30% of the coastal seas be set aside as MPAs [36]. We used our model to explore this recommendation specifically for spiny lobster in the Caribbean by designating various model sites as hypothetical MPAs and evaluated different networks of sites as if they were the sole sources of lobster larvae for the Caribbean (Table S1). Five MPA network scenarios were evaluated in simulations in which 40 habitat sites were designated as MPAs and selected in one of five ways: (1) *Random*: 40 sites individually and randomly selected from all those in the Caribbean, (2) *Stratified Random*: two randomly selected sites from each of the 20 countries, (3) *Self-Recruitment:* the top two self-recruiting sites per country, (4) *Long-distance Dispersal*: the top forty sites which successfully export larvae internationally in the Caribbean (5)*Maximum Export*: the top forty sites throughout the Caribbean with export imbalanced exchange (Figure 4). For these simulations the magnitude of larval production from each habitat site was fixed and uniform (unlike the more realistic and variable production used in our first set of simulations), which removed the effect of differences in local population size and focused on the effect of spatial arrangement of MPAs on biophysical connectivity networks. In each of the MPA scenarios, only the larval transport that originated from the 40 selected sites was considered, thus treating the system as a patchwork of MPAs.

The geographical location and connectivity characteristics of sites selected as MPAs altered patterns of spiny lobster larval dispersal and settlement (Table S1). Sites selected at random (scenarios 1 and 2; bootstrapped 1000 times to create averages) produced less successful larval connectivity than sites selected based on their merit as international (scenarios 4 and 5) or domestic (scenario 3) larval exporters. Simulations focusing on preserving domestic connectivity caused a near universal increase in larval recruitment across the Caribbean, although smaller than the ideal internationally managed scenario. Thus, by taking into consideration the complex patterns of connectivity for a species, we can add specificity to the general recommendation that a certain proportion of the sea requires protection to sustain marine fishery resources.

Pelagic Larval Nurseries

An unexpected pattern in larval distribution within the open ocean also appeared in our simulations. When we examined the oceanic pathways travelled (i.e., sum of PLD spent in each oceanic locale) by successfully settling larvae in contrast to the paths taken by larvae that are eventually lost from the system, zones emerged that could be described as "pelagic larval nurseries". That is, regions in the open Caribbean Sea where lobster larvae from around the Caribbean spend much of their planktonic existence before later settling into coastal benthic nurseries. These larval nurseries include relatively large regions offshore of Nicaragua, southern Cuba, and the central Bahamas as well as smaller areas north of Cuba and southeast of Hispaniola (Figure 6). We evaluated the role of larval behavior in creating these pelagic nurseries by conducting an additional simulation without ontogenetic vertical migration (OVM) by larvae, thus simulating passive larval dispersal. The segregation between the regions of concentration was accentuated when larvae drifted passively (Figure 6), indicating that the larval nursery zones were governed primarily by physical oceanographic features, not OVM behavior specific to spiny lobsters. Thus, these pelagic larval nurseries are potentially relevant to the pelagic retention of other Caribbean species, not just spiny lobster.

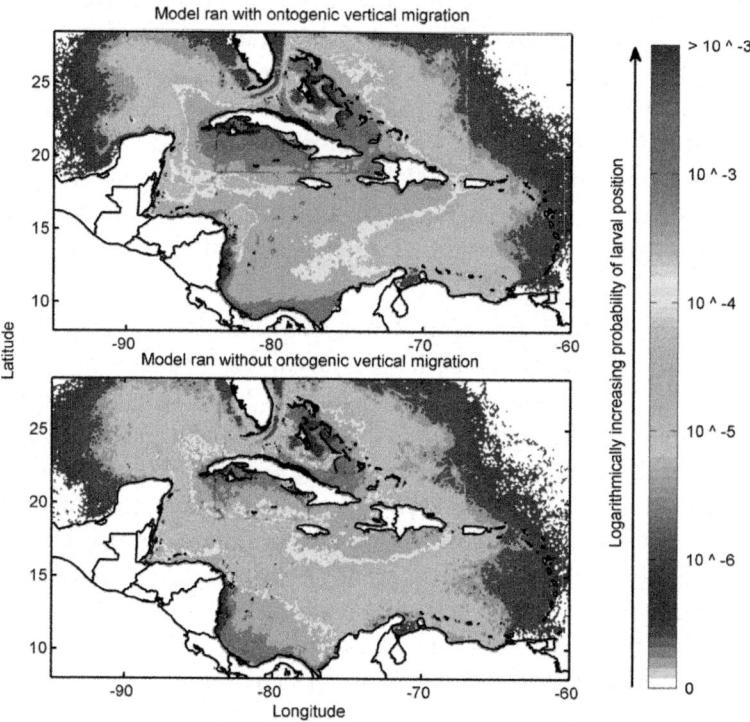

Figure 6: Probabilistic modeled spiny lobster (*P. argus*) larval concentrations. The probability density distributions represent pelagic nursery habitat within the Caribbean Sea for successfully recruiting spiny lobster larvae. The output location was recorded on a ten day frequency and added into a 0.1°×0.1°gridcell. Blue areas were relatively devoid of successfully dispersing larvae; warmer colored regions had more larval trajectories pass through them, increasing logarithmically from blue to red. The most important pelagic nursery zones for larvae are represented in red-orange. The areas of highest mean flow through the Caribbean represent a distinct, inter-linked larval 'graveyard'. Simulations were conducted with (A; n=54,186,756 larval locations) and without (B; n=68,675,786 larval locations) ontogenetic vertical migration.

DISCUSSION

Managing marine fisheries organisms as if they were constrained within geopolitical boundaries is not working as fisheries worldwide are in decline [37], [38]. For example, in regions where the spiny lobster *P. argus* are most abundant and thus heavily fished, adult stocks have declined by 30% or more over the past two decades despite spirited management [39]–[42]. For many species, an approach to fisheries management that acknowledges dispersal dynamics with estimates of larval connectivity is needed and now possible.

When we used MPAs in our model to "protect" specific locales that tend to export larvae internationally, those simulations yielded the highest successful settlement of lobster larvae throughout the Caribbean. Certain regions contribute disproportionately to the wider Caribbean larval pool, so maintaining the health of spawning stocks in those countries should be an international priority. One strategy for doing so, similar to the trade of "carbon credits" outlined in article 6 of the Kyoto protocol [43], would be to assign each nation "larval credits" based on regional larval export production. Nations that absorb disproportionally more larvae from the international larval pool bear an ethical responsibility and financial incentive to assist in the preservation of spawning stocks in other areas best suited for exporting larvae. Such non-traditional management recommendations are likely to be met with skepticism and their implementation difficult considering the political and economic realities of international agreements and the needs of local communities [44]. Yet scientific evidence suggests that populations of many marine animals persist in an intricate web of metapopulations that are often linked across geopolitical boundaries by larval connectivity and should be managed accordingly.

Just as preserving pathways between habitat fragments is essential for sustaining many terrestrial species [45], intact connectivity corridors for marine organisms may be needed. Our results suggest that larval corridors may exist in the open ocean that regularly concentrate and nurture pelagic larvae during their ontogenetic journey to coastal habitats (Figure 6). In contrast, the prevailing Caribbean current that snakes through the Caribbean Basin appears to be a "graveyard" for larval lobsters. Its high mean flows (Figure 1) presumably entrain and then wash larvae into the North Atlantic where few will survive (Figure 6). This stands in contrast to the view that larvae harness strong currents to successfully disperse long distances [11]. Our simulations with and without larval behavior indicated that the pelagic nursery zones we identified were stable and likely maintained by oceanographic features. Thus, our findings for *P. argus* are likely to be robust despite differences in larval origins, destinations, and avenues of dispersal that invariably differ among taxa with dissimilar dispersive traits [46]. If so, then the existence of pelagic nurseries for larvae has implications beyond lobsters and may constitute consideration as oceanic "essential fish habitat" [47]. Protection of these open ocean larval habitats from potentially deleterious processes (e.g., pollution from oil spills, coastal runoff, and vessel discharges) may be considerations for the long-term sustainability of marine species with dispersive larvae.

Although an adequate flow of larvae among sub-populations is crucial for the sustainability of marine resources, the arrival of larvae at a site does not necessarily equate to successful recruitment. Whereas larval supply and later recruitment are correlated for some species of spiny lobster and in some areas [39]; [48]–[50], unsuitable nursery habitats decouple the relationship between larval supply and juvenile recruitment in others [51], [52]. The transition from pelagic larva to benthic juvenile and on to adulthood is dependent on a variety of post-settlement processes [53], many of which are site-specific and not accounted for in models like ours that assume homogeneous and static habitat quality. Other studies indicate that phenotype-environment mismatches between settlers from one region into another can also contribute to post-settlement mortality and be a barrier to population connectivity [54]. Thus, the integration of biophysical larval dispersal models with spatially-explicit and dynamic depictions of benthic habitat conditions that drive benthic population dynamics [55], [56] are a logical next step in the development of predictive large-scale metapopulation models.

Advances in computing, genetics, and oceanographic remote sensing are yielding tools useful in addressing questions about the connectivity of marine metapopulations that were unfathomable only a decade ago. The dispersal of long-lived larvae is a complex function of temporally unstable hydrodynamics and ontogenetically variable larval behavior. Therefore, models that do not capture these essential system traits or whose results are not verified with empirical data will be misleading. Management of marine resources should benefit from new tools such as biophysical modeling that quantify larval connectivity and thus can be used to help guide policy. For example, the establishment of MPA networks in ecologically relevant areas that maximize larval production and connectivity among disparate populations will maximize population viability in both self-recruiting regions as well as regions dependent upon larvae from elsewhere. Our findings with respect to spiny lobster connectivity in the Caribbean suggest that international management agreements that recognize the existence of marine metapopulations, focus on rebuilding and sustaining adequate spawning stocks [57], and protect sensitive coastal and pelagic nurseries [42] represent a scientifically sound policy for sustainable management of many marine resources.

METHODS

Focusing on the Caribbean's most valuable fishery resource as a model system, we investigated larval dispersal through the use of an open-source coupled biophysical larval transport model, specifically parameterized using empirical data collected for *P. argus* (Table S2). Our model adheres to the recommended practices for Lagrangian biophysical modeling laid forth in North *et al* [58], while also incorporating empirical data for biological parameterization. Empirical estimates of spawning population (this study), laboratory and field observations of larval vertical migration in the water column [59], and postlarval sensory behavior [60] were used to parameterize the early life history traits of *P. argus*. Each of four submodules was specifically parameterized for spiny lobster larvae.

The Lagrangian Stochastic Module

The Lagrangian stochastic module drives the coupled biophysical Connectivity Modeling System (CMS). It uses a 4th order Runga-Kutta integration scheme [25] in both time and space to improve the accuracy of simulated larval trajectories as is best practice [61]. For each particle, the next position along the trajectory was calculated during each integration time-step of 2700 seconds, comparable to a previous experiment using spiny lobster that used a time-step of 4500 seconds [59]. The trajectories resulting from the modeled time-step and turbulence are smooth and relatively free of artifacts. Submesoscale turbulent movement was accounted for with stochastic turbulent diffusion during each time-step [25], calculated by multiplying a random number between 0 and 1 by the square root of twice the diffusivity coefficient (0.1 m^2/s) divided by the time-step. We ran simulations starting daily from January 1, 2004 until December 31, 2007, tracking larval flow for over 4 years. Details on the coupled biophysical algorithms and modeling approach can be found in Paris *et al.* [25].

The Physical Oceanographic Module

The physical oceanographic module contains the various oceanographic models that provide the currents with which to move larvae. These currents vary as depth changes from the surface down to 100 m, which is the likely maximum depth utilized by lobster phyllosoma [59]. A hierarchy of ocean circulation models are nested offline in the physical oceanographic module, allowing a Caribbean-wide simulation scale (−100:−55 degrees longitude West and 8:32 degrees latitude North) while not compromising resolution in areas with advanced local circulation models (Figure 1). Four different ocean circulation models were nested together for this study: 1/12 degree HYCOM+NCODA Global Hindcast Analysis [62] provided the base, followed by the higher resolution HYCOM+NCODA Gulf of Mexico 1/25° Analysis (GOM10.04)[63], a 1/24th degree ROMS model of the Bahamas [64], and the fine scale 900 meter resolution FLKeys-HYCOM [65].

The GIS-based Benthic Module

The GIS-based benthic module determines where larvae can settle and the location, quantity, and timing of larval release. It is directly coupled to the particle tracking module and is accessed during each integration time step. It consists of 261 habitat sites (polygons - vector GIS data) that are a combination of settlement habitat and a sensory envelope reflecting the threshold at which lobster postlarvae can detect and move to settlement habitat (Figures S1–S20 in File S1). Further information on polygon theory is in Paris *et al* [6]. The 18km sensory envelope for this study was constructed based on the sensory abilities of spiny lobster postlarvae [60]. Postlarvae are the highly mobile, non-feeding, settlement stage of spiny lobsters and are capable of detecting nursery habitat cues over similarly long distances [66]. Lobster benthic habitats were delineated based on data from the Millennium Reef Project [67]. Larvae were released from the nearest non-land location to the center of each habitat site.

The daily timing and magnitude of lobster spawning and thus larval release from each habitat site was estimated as a function of lobster density, sex ratio, size, and fecundity. First, the relative abundance of adult lobsters within each Caribbean country was estimated from FAO fishery landing statistics and an independent mail survey. Data was gathered from the top 10 lobster fishing nations that make up ≅95% of the fishery in the Caribbean. We assumed that the FAO [40] fishery landing statistics are an indicator of relative adult lobster abundance due to the overexploited nature of spiny lobster fisheries. However, these are fishery dependent data with unknown bias (e.g., under reporting of total catch) that may well vary among countries in an unpredictable manner. If so, then the magnitude of larval release in our model and our conclusions would be similarly biased. Unfortunately, there are no other data sources available and these data for *P. argus*, the most prized fishery in the Caribbean, are among the very best for any Caribbean species. The data we obtained on the relative abundance of lobsters among nations was then supplemented by a mail survey distributed to lobster scientists and fishery managers around the Caribbean with intimate knowledge of their local jurisdiction [68]. These data sources provided fine-scale resolution of the timing of spawning, the sex ratio, and the size-structure of adult male and female lobsters, which we used to determine fecundity [69] per habitat site.

Using these data, we scaled the larval production per habitat site per day proportional to the total annual egg production in the Caribbean (Figure 7A). These estimates of total *P. argus* egg production per year in the Caribbean were then divided into monthly patterns of spawning for each region based on the FAO and survey data (Figure 7B). The total spawned per month and site was further divided into each day because *P. argus* does not spawn synchronously. Finally, we scaled these empirical estimates so as to restrict the annual release of particles in the model to approximately 40,000,000; of which 38,000,000 were distributed to the 10 countries representing 95% the fishery and the remaining 2,000,000 particles distributed equally throughout the rest of the habitat sites with less accurately known lobster population structure. The annual value of 40,000,000 particles was found *a priori* to saturate movement paths in the model, after accounting for mortality (Figure 8). The end result is a daily release of larvae that varied in magnitude proportional to the total fishery, constructed with the local size, population, and spawning patterns when known for each of the 261 habitat sites.

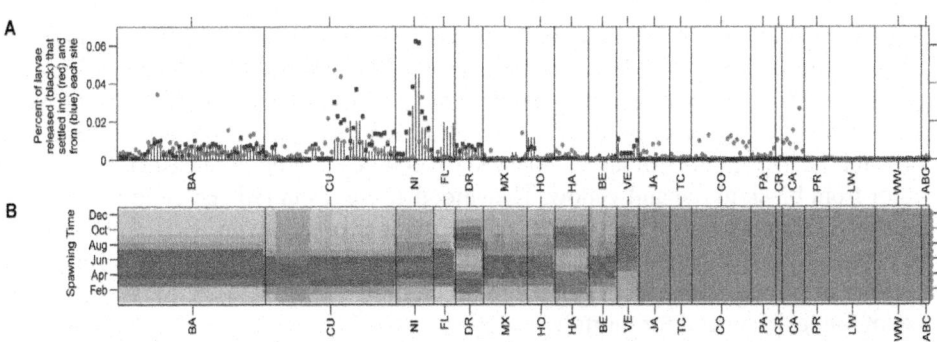

Figure 7: Simulation larval release, settlement, and seasonality. The details of the timing and magnitude of the simulated releases and the larvae received at each habitat site (n=261). The annual release (black lines), the larvae successfully received (red circles), and larvae donated (blue squares) at each habitat site as a percentage of the total (A). The annual timing of spawning at each site (B). The monthly effort increases from cyan to a peak of spawning occurring in red for locations with dynamic reproductive seasons. A uniform spawning pattern was used in locations that did not have empirical data on spawning time.

A modified pattern was used to test for idealized MPA placement, which assumed that each habitat site could hold the same climax population size and have the same reproductive potential. This had timing structured as in the original release, but allocated an equal number of particles to each site, rather than scaling population size based on survey and fishery data.

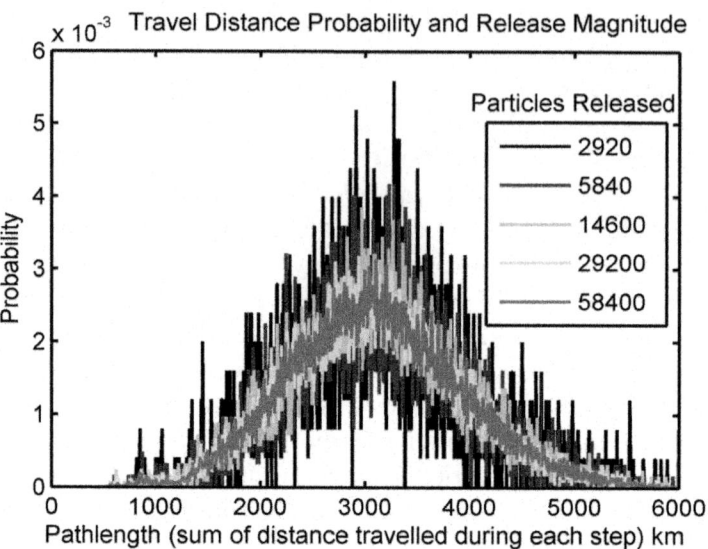

Figure 8: Larval release magnitude and movement pathlength. The probability of dispersal distances for larval releases from a central Caribbean release location (-68°W,14°N). The X-axis is the pathlength (sum of distances moved during each time-step) traveled by each larva binned into 5km increments, and the Y-axis is the probability. The number of larvae released (over 4 years of daily releases) increases in color from black to red and yellow. The smoother curves in red and yellow reflect the stochastic saturation, and suggest the proper number of larvae needed to release daily to probabilistically describe potential lobster larvae dispersal. These values reflect the number of larvae from a single site with no mortality, and had to be multiplied to account for each site and for mortality.

The Larval Biology Module

The larval biology module accounts for the early life history traits of spiny lobster including PLD, larval competency period, and ontogenetic vertical migration. Lobster larvae display distinct patterns of vertical distribution throughout ontogeny, which greatly alters which currents they are exposed to and therefore their dispersal. To reproduce this behavior, CMS assigns larvae probabilistically to different depth bins [25]. In the present simulations, individual larvae may reside during each time-step within one of five depth ranges (0–20 m, 20–40 m, 40–60 m, 60–80 m, and >80 m) with an age-dependent probability. During each time step, the depth bin is assigned randomly from the age-specific distribution [59]. However, larvae are not allowed to travel more than one depth bin per time step. Older larvae (>3 months old) have a higher chance of being deeper than younger larvae. These probabilities were determined through a combination of plankton trawls and laboratory experiments described in another study [59]. The mean PLD of lobster larvae was observed to be (±1 SD) 174±22 d, based on data from laboratory rearing of *P. argus* from egg to postlarval stage [14]. Larvae in the model metamorphose to postlarvae within a competency period (152 to 196 d) and postlarvae are recorded as 'settled' if they enter a benthic nursery habitat site (habitat module) within this competency period; if suitable habitat was not encountered within the competency period they 'die' and are removed from the simulation.

Mortality is a key parameter in biophysical modeling [70]. There is no evidence that vertebrate plankton mortality rates are similar to that of invertebrate plankton, however there is growing evidence that mortality changes throughout ontogeny for both coral [71] and fish [72]. To impose mortality, we used a half-life function to reflect varying survivorship as a function of larval duration. There are no known mortality rates for *P. argus* phyllosoma, thus we used an estimate for another spiny lobster (*P. cygnus*) used in Feng *et al* [28], based on trawl surveys that had diminishing returns of later stage larva [73], suggesting abundance based mortality of 85–90%. The cumulative mortality imposed on the larva in our model is ≈ *ca.* 90%, including advective mortality resulting from not reaching settlement habitat.

The Verification of Our Model

The verification of our model lends credence to its results. Whereas the backbone of coupled bio-physical models are ocean circulation models whose physical dynamics have been validated and peer reviewed, the biological predictions of larval dispersal models should also be verified [74] but few are. Our verification of the model predictions is based on correlations between model predictions and empirical observations of recruitment into relatively small ca. ≈50km² habitat patches following the dispersal of larvae over thousands of kilometers during their 5–9 month PLD (Figure 2). There is precedent for using postlarval collector seasonal settlement trends to verify a Lagrangian model [28], and predictable seasonal patterns are vital for fishery management. Correlating the spatial concentration of observed pelagic larval or juvenile patches with modeled outputs has been done in smaller scale studies [5], [7], [33],[75], but is prohibitively costly and difficult to do at a Caribbean-scale which our model is based on.

Sensitivity analyses of some parameters for which empirical data are lacking or based on laboratory studies (e.g., mortality, PLD, age of competency) could potentially improve the accuracy of our model [6]. Incorporating specific biological traits, for example vertical migrations, into a model alters outputs.

For example, Briones-Fourzan *et al* [29] used stochastic perturbations of a particle backtracking simulation to investigate potential origins of postlarvae arriving on the Mexican Quintana Roo coast, without having data on ontogenic vertical migrations. In comparison with their findings, our results suggest diminished larval supply to Mexico from the Lesser Antilles Caribbean Islands and the Venezuelan corridor, while increasing the supply of larva from Central America and Hispaniola (Figure S21 in File S1). This was expected since the vertical migratory behavior of the actively moving larvae increases retention [59]. A simulation that we conducted without larval OVM nor adult population structure did not capture the seasonal recruitment pattern evident in the empirical data (Figure S22 in File S1), and is more similar to the connectivity described in Briones-Fourzan *et al* [29], suggesting that additional biological parameterization could further improve model performance.

ACKNOWLEDGMENTS

We are grateful to Drs. Laurent Cherubin and Villy Kourafalou for access to their oceanographic models. The Connectivity Modeling System code development was aided by Judith Helgers.

AUTHOR CONTRIBUTIONS

Conceived and designed the experiments: ASK MJB CBP. Performed the experiments: ASK CBP MJB. Analyzed the data: ASK CBP MJB. Contributed reagents/materials/analysis tools: CBP MJB. Wrote the paper: ASK CBP MJB.

REFERENCES

1. Becker B, Levin L, Fodrie F, McMillan P (2007) Complex larval connectivity patterns among marine invertebrate populations. Proc Natl Acad Sci USA 104: 3267–3272. doi: 10.1073/pnas.0611651104

2. Almany GR, Berumen ML, Thorrold SR, Planes S, Jones GP (2007) Local replenishment of coral reef fish populations in a marine reserve. Science 316: 742–744. doi: 10.1126/science.1140597

3. Hamilton SL, Casellea JE, Malone DP, Carr MH (2010) Incorporating biogeography into evaluations of the Channel Islands marine reserve network. Proc Natl Acad Sci USA 107: 18272–18277. doi: 10.1073/pnas.0908091107

4. Cowen RK, Paris CB, Srinivasan A (2006) Scaling of connectivity in marine populations. Science 311: 522–527. doi: 10.1126/science.1122039

5. Hidalgo M, Gusdal Y, Dingsør DE, Hjermann D, Ottersen G, et al.. (2011)

A combination of hydrodynamical and statistical modeling reveals non-stationary climate effects on fish larvae distributions. Proc R Soc B. DOI: 10.1098/rspb.2011.0750.

6. Paris CB, Cowen RK, Claro R, Lindeman KC (2005) Larval transport pathways from Cuban spawning aggregations (Snappers; *Lutjanidae*) based on biophysical modeling. Mar Ecol Prog Ser 296: 93–106. doi: 10.3354/meps296093

7. Paris CB, Cowen RK (2004) Direct evidence of a biophysical retention mechanism for coral reef larvae. Limnol Oceanogr 49: 1964–1979. doi: 10.4319/lo.2004.49.6.1964

8. Planes S, Jones GP, Thorrold SR (2009) Larval dispersal connects fish populations in a network of marine protected areas. Proc Natl Acad Sci USA 106: 5693–5697. doi: 10.1073/pnas.0808007106

9. Saenz-Agudelo P, Jones GP, Thorrold SR, Planes S (2011) Connectivity dominates larval replenishment in a coastal reef fish metapopulation. Proc R Soc B. DOI: 10.1098/rspb.2010.2780.

10. Puebla O, Bermingham E, McMillan WC (2012) On the spatial scale of dispersal in coral reef fishes. Mol Ecol. DOI: 10.1111/j.1365-294X.2012.05734.x.

11. Roberts CM (1997) Connectivity and management of coral reefs. Science 278: 1454–1457. doi: 10.1126/science.278.5342.1454

12. Ehrhardt NM, Puga R, Butler MJ IV (2010) In:Fanning L, Mahon R, McConney P, editors. Towards Marine Ecosystem-Based Management in the Wider Caribbean. Amsterdam, NL: Amsterdam Univ. Press, 157–175.

13. Lipcius RN, Cobb JS (1994). In:Phillips BF, Cobb JS, Kittaka JK, editors. Spiny lobster management. Oxford, UK : Fishing news books, 1–30.

14. Goldstein JS, Matsuda H, Takenouchi T, Butler MJ IV (2008) The complete development of larval Caribbean spiny lobster, *Panulirus argus*, in culture. J Crustacean Biology 28: 306–327. doi: 10.1163/20021975-99990376

15. Ehrhardt NM (2005) Population dynamic characteristics and sustainability mechanisms in key Western Central Atlantic spiny lobster, *Panulirus argus*, fisheries. Bull Mar Sci 76: 501–525.

16. Silberman JD, Sarver SK, Walsh PJ (1994) Mitochondrial DNA variation and population structure in the spiny lobster *Panulirus argus*. Mar Biol 120: 601–608. doi: 10.1007/bf00350081

17. Naro-Maciel E, Reid B, Holmes KE, Brumbaugh DR, Martin M, et al. (2011) Mitochondrial DNA sequence variation in spiny lobsters:

population expansion, panmixia, and divergence. Mar Biol 158: 2027–2041. doi: 10.1007/s00227-011-1710-y

18. Diniz FM, Maclean N, Ogawa M, Cintra IHA, Bentzen P (2005) The hypervariable domain of the mitochondrial control region in Atlantic spiny lobsters and its potential as a marker for investigating phylogeographic structuring. J Mar Biotechnol 7: 462–473. doi: 10.1007/s10126-004-4062-5

19. Tourinho JL, Sole-Cava AM, Lazoski C (2012) Cryptic species within the commercially most important lobster in the tropical Atlantic, the spiny lobster *Panulirus argus*. Mar Biol 159: 1897–1906. doi: 10.1007/s00227-012-1977-7

20. Kritzer JP, Sale PF (2004) Metapopulation ecology in the sea: from Levins' model to marine ecology and fisheries science. Fish and Fisheries 5: 131–140 DOI: 10.1111/j.1467-2979.2004.00131.x.

21. Pelc RA, Warner RR, Gaines SD, Paris CB (2010) Detecting larval export from marine reserves. Proc Natl Acad Sci USA 107: 18266–18271. doi: 10.1073/pnas.0907368107

22. Lowe WH, Allendorf FW (2010) What can genetics tell us about population connectivity? Mol Ecol. DOI: 10.1111/j.1365-294X.2010.04688.x.

23. Waples RS, Punt AE, Cope JM (2008) Integrating genetic data into management of marine resources: how can we do it better? Fish and Fisheries 9: 423–449. doi: 10.1111/j.1467-2979.2008.00303.x

24. Botsford LW, White JW, Coffroth MA, Paris CB, Planes S, et al.. (2009) Connectivity and resilience of coral reef metapopulations in marine protected areas: matching empirical efforts to predictive needs. Coral Reefs. DOI: 10.1007/s00338-009-0466-z.

25. Paris CB, Helgers J, Van Sebille E, Srinivasan A (2013) The Connectivity Modeling System: A probabilistic modeling tool for the multi-scale tracking of biotic and abiotic variability in the ocean. Environ Modell Softw. DOI: 10.1016/j.envsoft.2012.12.006.

26. Paris CB, Cowen RK, Lwiza KMM, Wang DP, Olson DB (2002) Objective analysis of three-dimensional circulation in the vicinity of Barbados, West Indies: Implication for larval transport. Deep Sea Res 49: 1363–1386. doi: 10.1016/s0967-0637(02)00033-x

27. Stockhausen WT, Lipcius RN (2001) Single large or several small marine reserves for the Caribbean spiny lobster? Mar Freshw Res 52: 1605–1614.

28. Feng M, Caputi N, Penn J, Slawinski D, de Lestang S, et al. (2011)

Ocean circulation, Stokes drift, and connectivity of western rock lobster (*Panulirus cygnus*) population. Can J Fish Aquat Sci 68: 1182–1196. doi: 10.1139/f2011-065

29. Briones-Fourzan P, Candela J, Lozano-Alvarez E (2008) Post-larval settlement of the spiny lobster *Panulirus argus* along the Caribbean coast of Mexico: patterns, influence of physical factors, and possible sources of origin. Limnol Oceanogr 53: 970–985. doi: 10.4319/lo.2008.53.3.0970

30. Griffin DA, Wilkin JL, Chubb CF, Pearce AF, Caputi N (2001) Ocean currents and the larval phase of Australian western rock lobster, *Panulirus cygnus*. Mar Freshw Res 52: 1187–1199. doi: 10.1071/mf01181_co

31. Rudorff CA, Lorenzzeti JA, Gherardi DF, Lins-Oliveira JE (2009) Modeling spiny lobster larval dispersion in the Tropical Atlantic. Fish Res 96: 206–215. doi: 10.1016/j.fishres.2008.11.005

32. Chiswell SM, Booth JD (2008) Sources and sinks of larval settlement in *Jasus edwardsii* around New Zealand: where do larvae come from and where do they go? Mar Ecol Prog Ser 354: 201–217. doi: 10.3354/meps07217

33. Incze L, Xue H, Wolff N, Xu D, Wilson C, et al. (2010) Connectivity of lobsters (*Homarus americanus*) populations in the coastal Gulf of Maine: part II. Coupled biophysical dynamics. Fisheries Oceanography 19: 1–20. doi: 10.1111/j.1365-2419.2009.00522.x

34. Sponaugle S, Paris CB, Walter KD, Kourafalou V, d'Alessandro E (2012) Observed and modeled larval settlement of a reef fish in the Florida Keys. Mar Ecol Prog Ser 453: 201–212. doi: 10.3354/meps09641

35. Muller R, Matthews T, FWC collector data 2005–22009, (FWC, Marathon, FL 33001 USA).

36. Gaines SD, White C, Carr MH, Palumbi SR (2010) Designing marine reserve networks for both conservation and fisheries management. Proc Natl Acad Sci USA 107: 18286–18293. doi: 10.1073/pnas.0906473107

37. Beddington JR, Agnew DJ, Clark CW (2007) Current problems in the management of marine fisheries. Science 316: 1713–1716. doi: 10.1126/science.1137362

38. Pauly D (2009) Beyond duplicity and ignorance in global fisheries. Scientia Marina 73: 215–224. doi: 10.3989/scimar.2009.73n2215

39. Ehrhardt NM, Fitchett MD (2010) Dependence of recruitment on parent stock of the spiny lobster, *Panulirus argus*, in Florida. Fisheries Oceanography 19: 434–447. doi: 10.1111/j.1365-2419.2010.00555.x

40. Food and Agriculture Organization (2006) Fifth regional workshop

on the assessment and management of the Caribbean spiny lobster. Available:ftp://ftp.fao.org/docrep/fao/010/a1518b/a1518b00. pdf Accessed 2013 Jan 1.

41. Chavez EA (2009) Potential production of the Caribbean spiny lobster (*Decapoda, Palinura*) fisheries. Crustaceana 82: 1393–1412. doi: 10.116 3/001121609x12481627024373

42. Cruz R, Bertelsen RD (2008) The Spiny Lobster (*Panulirus argus*) in the Wider Caribbean: A Review of Life Cycle Dynamics and Implications for Responsible Fisheries Management. Proc Gulf Carib Fish Inst 61: 433–446.

43. United Nations (1998) Kyoto protocol to the United Nations framework convention on climate change. Available: http://unfccc.int/resource/docs/convkp/kpeng.pdf. Accessed 2013 Jan 1.

44. Smith MD, Lynham J, Sanchirico JN, Wilson JA (2009) Political economy of marine reserves: understanding the role of opportunity costs. Proc Natl Acad Sci USA 107: 18300–18305. doi: 10.1073/pnas.0907365107

45. Wikramanayake E, Dinerstein E, Seidensticker J, Lumpkin S, Pandav B, et al. (2011) A landscape-based conservation strategy to double the wild tiger population. Conservation Letters 00: 1–9. doi: 10.1111/j.1755-263x.2010.00162.x

46. Largier JL (2003) Considerations in estimating larval dispersal distances from oceanographic data. Ecol Appl 13: S71–S89. doi: 10.1890/1051-0761(2003)013[0071:cieldd]2.0.co;2

47. National Oceanographic Atmospheric Administration (2002) Fishery Conservation and Habitat. Available: http://www.habitat.noaa.gov/pdf/efhregulatoryguidelines.pdf. Accessed: 2012 Jun 18.

48. Phillips BF (1986) Prediction of commercial catches of the western rock lobster *Panulirus cygnus* george. Canadian Journal of Fisheries and Aquatic Sciences 43: 2126–2130. doi: 10.1139/f86-261

49. Caputi N, Brown RS, Chubb CF (1995) Regional prediction of the western rock lobster, *Panulirus cygnus*, commercial catch in Western Australia. Crustaceana 68: 245–256. doi: 10.1163/156854095x00142

50. Chavez EA, Chavez-Hildalgo A (2013) The ecological importance of larval dispersal pathways of connectivity amongst Western Caribbean spiny lobster stocks. Proc 12th Intl. Coral Reef Symp: In Press.

51. Lipcius RN, Stockhausen WT, Eggleston DB, Marshall LS, Hickey B (1997) Hydrodynamic decoupling of recruitment, habitat quality and adult abundance in the Caribbean spiny lobster: source–sink dynamics?

Mar Freshw Res 48: 807–815. doi: 10.1071/mf97194

52. Butler MJ, IV, Herrnkind WF (1997) A test of recruitment limitation and the potential or artificial enhancement of spiny lobster populations in Florida. Can J Fish Aquat Sci 54: 452–463. doi: 10.1139/cjfas-54-2-452

53. Pineda J, Reyns NB, Starczak VR (2009) Complexity and simplification in understanding recruitment in benthic populations. Popul Ecol 51: 17–32. doi: 10.1007/s10144-008-0118-0

54. Marshall DJ, K Monro, M Bode, MJ Keough, S Swearer (2010) Phenotype-environment mismatches connectivity in the sea. Ecol Lett. doi: 10.1111/j.1461-0248.2009.01408.x.

55. Butler MJ IV, Dolan T, Hunt JH, Herrnkind WF, Rose K (2005) Recruitment in degraded marine habitats: a spatially-explicit, individual-based model for spiny lobster. Ecol App 15: 902–918. doi: 10.1890/04-1081

56. Butler MJ IV (2003) Incorporating ecological process and environmental change into spiny lobster population models using a spatially-explicit, individual-based approach. Fisheries Res 65: 63–79. doi: 10.1016/j.fishres.2003.09.007

57. Steneck RS, Paris CB, Arnold SN, Ablan-Lagman MC, Alcala AC, et al. (2009) Managing outside the box: coalescing connectivity networks to build resilience in coral reef ecosystems. Coral Reefs 28: 367–378. doi: 10.1007/s00338-009-0470-3

58. North EW, Gallego A, Petitgas P, Adlandsvik B, Bartsch J, et al.. (2009) Manual of recommended practices for modeling physical – biological interactions during fish early life history. ICES Cooperative Research Report 295.

59. Butler MJ IV, Paris CB, Goldstein JS, Matsuda H, Cowen RK (2011) Behavior constrains the dispersal of long-lived spiny lobster larvae. Mar Ecol Prog Ser 422: 223–237. doi: 10.3354/meps08878

60. Goldstein JS, Butler MJ IV (2009) Behavioral enhancement of onshore transport by post-larval Caribbean spiny lobster (*Panulirus argus*). Limnol Oceanogr 54: 1669–1678. doi: 10.4319/lo.2009.54.5.1669

61. Brickman D, Ådlandsvik B, Thygesen UH, Parada C, Rose K, et al. (2009) Particle Tracking *in* Modelling physical–biological interactions during fish early life (North EW, Gallego A, Petitgas P, eds.) ICES Cooperative Research Report. 295: 9–13.

62. Dataset from HYCOM Consortium, HYCOM+NCODA Global 1/12°. Available:http://tds.hycom.org/thredds/global_combined/glb_analysis_

catalog.html. Accessed: 2013 Apr 16.

63.   Dataset from HYCOM Consortium, HYCOM+NCODA Gulf of Mexico 1/25° Analysis (GOMl0.04). Available: http://tds.hycom.org/thredds/ GOMl0.04/expt_20.1.html. Accessed: 2013 Apr 16.

64.   Cherubin LM (2013) High-resolution simulation of the circulation in the Bahamas and Turks and Caicos Archipelagos. Progress in Oceanography: In press.

65.   Kourafalou VH, Kang H (2012) Florida Current meandering and evolution of cyclonic eddies along the Florida Keys Reef Tract: Are they interconnected? J Geophys Res: doi:10.1029/2011JC007383.

66.   Jeffs AG, Montgomery JC, Tindle CT (2005) How do spiny lobster post-larvae find the coast? N Z J Mar Freshwater Res 39: 605–617. doi: 10.1080/00288330.2005.9517339

67.   Andréfouët S, Muller-Karger FE, Robinson JA, Kranenburg CJ, Torres-Pulliza D, et al.. (2004) In:10th ICRS. Global assessment of modern coral reef extent and diversity for regional science and management applications: a view from space. Okinawa, Japan: Japanese Coral Reef Society 1732–1745.

68.   International collaborators who assisted us by completing surveys (2008) or providing information on adult lobster population structure within their countries include: Karl Aiken (Jamaica), James Azueta (Belize), Julio Baisre and Raul Cruz (Cuba), Richard Beaver (Florida), Nelson Ehrhardt (Nicaragua), Alejandro Herrera (Dominican Republic), Lester Gittens (Bahamas), Nilda Jimenez (Puerto Rico), Kathy Lockhart (Turks and Caicos), Alicia Medina (Honduras), Renaldy Navarro (Nicaragua), Hazel Oxenford (Barbados), Paul Phillip (Grenada), Juan Posada (Venezuala), Martha Prada (Columbia), Lionel Reynal (Martinique and Guadeloupe), Maria Romero, Christine Shing (British Virgin Islands), and Eloy Sosa (Mexico).

69.   Bertelsen RD, Matthews TR (2001) Fecundity dynamics of female spiny lobster (*Panulirus argus*) in a south Florida fishery and Dry Tortugas National Park lobster sanctuary. Mar Freshwater Res 52: 1559–1565.

70.   Houde E, Bartsch J (2008) Mortality *in* Modelling physical–biological interactions during fish early life (North EW, Gallego A, Petitgas P, eds.) ICES Cooperative Research Report. 295: 27–42.

71.   Graham EM, Baird AH, Connolly SR (2008) Survival dynamics of scleractinian coral larvae and implications for dispersal. Coral reefs 27: 529–539. doi: 10.1007/s00338-008-0361-z

72.   Paris CB (2009) Fate of reef fish larvae trough ontogeny: advection or

true mortality? Theme Session T: Death in the sea, Proceedings of the 2009 Annual Science Conference, September 21–25 2009, Berlin, ICES CM 2009/T: 13, 22.

73. Rimmer DW, Phillips BF (1979) Diurnal migration and vertical distribution of phyllosoma larvae of the western rock lobster *Panulirus cygnus*. doi: 10.1007/BF00386590.

74. Metaxas A, Saunders M (2009) Quantifying the "bio-" components in biophysical models of larval transport in marine benthic invertebrates: advances and pitfalls. Biol Bull 216: 257–272.

75. Vikebø FB, Ådlandsvik B, Albretsen J, Sundby S, Stenevik EK, et al.. (2011) Real-Time Ichthyoplankton Drift in Northeast Arctic Cod and Norwegian Spring-Spawning Herring. PLoS One. doi:10.1371/journal. pone.0027367.

76. Fratantoni DM (2001) North Atlantic surface circulation during the 1990's observed with satellite tracked drifters. J Geophys Res 106: 22067–22093. doi: 10.1029/2000jc000730

Chapter 10

ESTUARINE FISHERIES COMMUNITY-LEVEL RESPONSE TO FRESHWATER INFLOWS

James M. Tolan[1]

[1]Texas Parks and Wildlife Department, Coastal Fisheries Division, Natural Resource Center 2501, Unit 5846, Corpus Christi, TX, USA

INTRODUCTION

Every estuary needs freshwater inflow (FWI) to maintain proper salinity regimes, nutrient loading, and sediment inputs to support its geographically unique levels of biological productivity [1-4]. Watershed elevations and soil types determine surface and groundwater flows into estuaries, and these flows have source, timing, and velocity components that can be significantly affected by anthropogenic alterations at the landscape level. It is estimated that approximately 60% of the global storage of freshwater is now contained behind reservoirs and dams [5] and 77% of the total water discharge from 139 of the largest river systems in the northern hemisphere are either strongly or moderately affected by dams, interbasin transfers, and surface water withdrawals [6]. Hydrologic modifications of estuarine watersheds influence wetland and open-water salinity patterns, nutrients, sediment fertility, bottom topography, dissolved oxygen, and concentrations of xenobiotics [7]. Because demand for freshwater is only expected to increase as population continues to grow [8], it is incumbent upon resource managers to examine the environmental effects and biological consequences of hydrologic alterations within coastal ecosystems [9-11].

Resource-based approaches seek to link freshwater inflows to a number of fishery species generally considered valuable by society [12]. The optimization model utilized by Powell and Matsumoto [13] uses a series of relationships between monthly inflows and the catch of a number of commercially and recreationally important finfish (red drum *Sciaenops ocellatus*, black drum *Pogonias cromis*, spotted seatrout *Cynoscion nebulosus*, southern flounder *Paralichthys lethostigma*), crustaceans (blue crab*Callinectes sapidus*, white shrimp *Litopenaeus setiferus*, brown shrimp *Farfantepenaeus aztecus*, pink shrimp *F. duorarum*) and mollusks (eastern oyster *Crassostrea virginica*) to arrive at a set of targeted monthly freshwater inflows to maintain healthy ecological conditions in estuaries. The goal of this method, which jointly considers the salinity tolerance range of each of the target organisms and limits the inflow volume solution by imposing numerous process constraints (such as fishery biomass and harvest ratios; monthly, bi-monthly, and yearly freshwater volumes; upper and lower bounds for salinity; nutrient and sediment loading; see [13]), is to estimate the minimum amounts of FWI needed to maintain historical fisheries production. Although the inflow-harvest equations were originally based on fishery-dependent commercial catch records, recent modeling efforts have incorporated a greater proportion of fishery-independent data sources [14].

A problem with the resource-based approach is that it focuses on adults, which are harvestable. Although these are estuarine dependent species, the adults are widely distributed along salinity gradients [15]. A number of transient taxa (sensu, facultative estuarine-dependent, see [16]) which recruit from offshore spawning areas are known to have size-specific use patterns within shallow habitats of the oligohaline-to-freshwater portions of estuaries [17-20]. Inflows, especially those large pulses associated with flooding events, can displace seaward the boundary between the brackish and freshwater interface, on a kilometers-to-estuary wide scale. Taxa with specific nursery habitat requirements could therefore be restricted from ingress into portions of the estuary, potentially altering an important habitat for juvenile nekton.

The goal of this study was to expand the focus of interest of the resource-based approach beyond the limited number of fisheries target species and to include juvenile stages of fisheries species to examine the functional role of FWI in shaping the total nekton assemblage structure in estuaries. The approach was to perform an analysis of a long-term, state agency, bag seine monitoring program. Bag seine samples are fishery independent and contain juvenile stages of fishery species. There are three FWI gradients examined; within estuaries from river to sea, among estuaries along a climatic gradient, and over time as changes in freshwater inflows, in the form of flood pulses and drought events, dramatically alter the salinity structure of the estuary.

MATERIALS AND METHODS

Study Area

Texas' coastline extends along 600 km of open Gulf of Mexico shoreline and contains 3,420 kilometers of bay-estuary-lagoon shoreline. This is a biologically rich and ecologically diverse region of the state, supporting more than 247,576 hectares of fresh, brackish, and salt marshes. Within the state, over 305,600 km of rivers and streams coalesce into 15 major river systems, and these rivers empty into seven major estuaries (Figure 1). All seven estuaries have similar geomorphic structure and physiography, yet each is quite diverse hydrologically. This is primarily due to a climatic gradient influencing freshwater inflows. This gradient of decreasing rainfall from northeast to southwest is one of the most distinctive features of the coastline (Table 1). Along this gradient, rainfall decreases by a factor of two, yet inflow decreases by almost two orders of magnitude. The Laguna Madre, a hypersaline lagoon, has a negative inflow balance because this estuary lacks any major riverine inflow and evaporation normally exceeds precipitation. The net effect is a gradient of estuaries with similar physical characteristics but greatly differing salinity regimes.

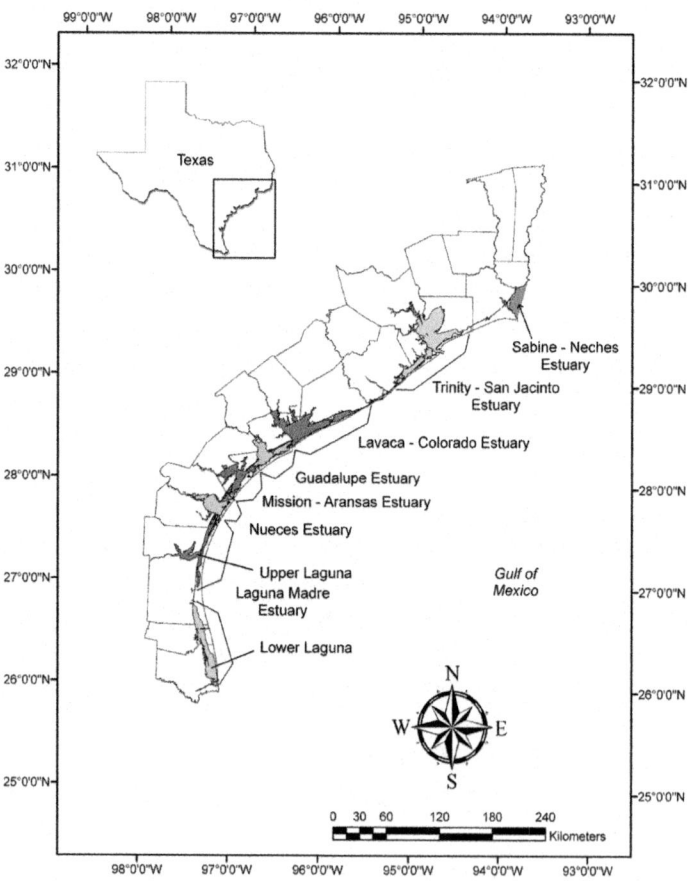

Figure 1: Map of Texas showing the location along the coast of each of the major estuarine systems.

Table 1: Climatic gradient in Texas estuaries, listed from north to south. Drainage basin size (Drainage), bay surface area (Area), and Habitat (SAV = submerged aquatic vegetation) characteristics from USEPA (1999). Average annual Inflow Balance, Rainfall, and Salinity characteristics (1941-1999) from the Texas Water Development Board; http://www.twdb.state.tx.us/data/bays_estuaries/bays_estuary_toc.asp

Estuary	Drainage (km²)	Area (km²)	Rainfall (cm y⁻¹)	Inflow (10⁶m³ y⁻¹)	Salinity (PSU)	Wetlands (km²)	SAV (km²)
						Habitat	
Sabine-Neches	45 705	243	142	16 897	8	967	-
Trinity-San Jacinto	57 900	1399	112	14 000	16	1594	73
Lavaca-Colorado	111 890	1158	102	3801	18	348	28
Guadalupe	26 330	551	91	2664	16	271	65
Mission-Aransas	7860	453	81	265	19	393	85
Nueces	43 350	433	76	298	29	121	53
Laguna Madre	29 695	3658	69	-893	36	1825	773

From the Louisiana border to the Trinity-San Jacinto estuary the coastline is characterized by marshy plains with low, narrow beach ridges, and from there to the border of Mexico the coastline is characterized by long barrier islands and large, shallow lagoons. Barrier islands are parallel to the mainland along the coast, and between the barrier islands and the mainland are lagoons. These lagoons are interrupted with drowned river valleys that form the bays and estuaries. Inlets through the barrier island connect the Gulf of Mexico to these lagoons, with each lagoon opening into a large primary bay. There is typically a constriction between the primary and secondary (and in some cases tertiary) bays. While ungauged coastal watershed runoff can locally influence estuarine salinity, most inflow into each bay is supplied primarily by just one or two gauged rivers draining hydrologically isolated watersheds.

BAG Seine Community Structure Data

Starting in January 1992 and continuing through the present, 20 replicate bag seine samples have been collected each month within each major estuary system along the Texas coast. Sampling locations are randomly selected from a grid system of one minute latitude and one minute longitude, with no selected grid sampled more than once per calendar month. For each sample, a bag seine (18.3 m X 1.8 m, 1.9 cm stretch nylon multifilament; central bag, 1.8 m wide, 1.3 cm stretch mesh) is pulled parallel to the shore for 15.2 m [21].

The surface area sampled is estimated using the distance pulled and length of extension of the bag seine. All fish and invertebrates collected in each sample are identified, enumerated, and measured. Total catch of each taxon is standardized and expressed as catch per hectare. Prior to each bag seine collection, surface salinity in Practical Salinity Scale were measured with either handheld Hydrolab or YSI multiprobes calibrated to the manufacturers specifications.

Data Analysis

The experimental unit defined for this study was each bay system as a whole, as each of the major estuaries along the Texas coast can be defined by the underlying hydrologic gradient. To assemble the bag seine collections into a time series, catch data for each taxa from the monthly replicate samples were summed across estuaries individually, and reported as total catch per month. Salinity records were similarly transformed into a time series, although the replicate values were first averaged across each estuary and reported as mean salinity per month. For each estuary, a categorical 'Inflow Condition' variable was defined by evaluating the average salinity time series. Salinities above the 85[th]percentile were deemed indicative of 'Drought' conditions and values below the 15th percentile representative of high flow or 'Flood' conditions. Values between these two extremes were identified as 'Normal' flows. The Laguna Madre Estuary was further sub-divided into an Upper and Lower components, providing for eight estuaries under investigation (see Figure 1). This sub-division is based on a natural sand sheet or land bridge (the Land Cut) connecting Padre Island with the mainland [21]. The Gulf Intracoastal Waterway bisects this extensive sand flat, thus connecting the two lagoons via the Land Cut.

Community analyses were performed using Primer-E (Version 6.0) software [22]. A matrix of Bray-Curtis Distance similarities between each total catch per month sample was created. Catch data was initially transformed [Log_{10} + 1] to down-weight the most abundant taxa. Significant differences in rank similarities between groups of samples were then tested by Analysis of Similarity (ANOSIM). In the ANOSIM procedure, the probability of *a priori* groupings of samples is estimated by repeated permutations of the original data matrix. Values of the R statistic can range from -1 to 1, although R will usually fall between 0 and 1 with R values > 0.4 indicating higher degrees of discrimination among groups. The *a priori* factors tested with the ANOSIM procedure were the external factors associated with each sample (e.g., season of collection and inflow condition) within a common estuary. The entire collection was then merged across estuaries, and then tested for differences in community structure among estuaries. The SIMPER (SIMilarity PERcentages) routine was used to examine the contribution of individual species to the community structure seen among the *a priori* factors. Similarities among the samples are graphically represented with non-metric multidimensional scaling (MDS) ordinations [23]. Although outcomes of the ANOSIM are not dependant on MDS ordinations, the ordinations are presented here as they are a helpful way of visualizing patterns in the data. Stress values indicate how well the two-dimensional plot represents relationships among samples in the multidimensional space. Stress values < 0.15 indicate a good fit. MDS ordinations may be arbitrarily rotated so axes are not labeled.

RESULTS

From 1992 through 2006, bag seine sampling resulted in 28,786 individual collections from the eight major estuarine systems. This time series of 180 months revealed dramatically fluctuating mean salinities throughout the study period (Table 2). Despite dramatic differences in total inflows across the coastal hydrologic gradient, temporal inflow patterns were generally similar across the coast. The timing of extended flooding conditions (i.e., on the order of 10 months during 1992 and again in 1997) or droughts (the majority of the calendar years of 2000 and 2001) were similar within each estuary (Figure 2).

Table 2: Salinity summary statistics by estuary for the study period January 1992 through December 2006

	n	Mean	Std. Dev.	Min.	Max.
Estuary					
Sabine-Neches	3599	6.73	6.23	0.0	32.0
Trinity-San Jacinto	3597	16.52	9.21	0.0	41.0
Lavaca-Colorado	3599	18.47	9.57	0.0	40.0
Guadalupe	3594	16.62	11.36	0.0	45.0
Mission-Aransas	3600	18.13	9.70	0.0	41.0
Nueces	3600	28.66	7.42	0.0	59.0
Upper Laguna Madre	3599	35.44	10.64	0.0	78.0
Lower Laguna Madre	3598	32.04	7.98	0.0	64.0

The bag seines recorded 3,583,061 individuals from 387 unique taxa. Analysis of Similarity of the entire collection showed that in each estuary, community structure was significantly different across seasons (Table 3), and these seasonal differences were repeated annually across all inflow categories. The greatest disparity in community composition involved comparisons across opposite seasons (e.g., winter vs. summer, spring vs. fall), with significant pairwise comparison R values ranging from 0.609 – 0.971 (see Table 3). While seasonal differences in communities were quite evident, there appears to be little correspondence between community structure and synoptic-scale inflow events (Figure 3). This general disconnect between shallow water nekton assemblages and inflows was evident in every estuary along the Texas coast (Figure 4). The only estuaries to display significant community-level differences across the different inflow conditions were from opposite ends of the salinity spectrum. The Sabine-Neches estuary (mean salinity approximately 7) had significantly different community compositions during a drought relative to flood conditions (R = 0.520, $p < 0.001$). Greater abundances of white shrimp (7 fold increase), brown shrimp (17 fold increase), pinfish *Lagodon rhomboides* (6 fold increase), white mullet *Mugil curema* (12 fold increase), spotted seatrout (17 fold increase), and sheepshead minnow *Cyprinodon variegatus* (9 fold increase) were recorded during the periods of elevated salinities. The Lower Laguna Madre (mean salinity 32) also had significantly different community compositions during drought conditions (Drought vs. Normal comparison, R = 0.253, $p < 0.001$; Drought vs. Flood comparison, R = 0.235, $p < 0.001$), although the elevated salinities in this estuary during drought conditions (mean salinity > 40) led to lower abundances of some of these same taxa. Substantial decreases in brown shrimp (5 fold), white shrimp (15 fold), and Atlantic croaker*Micropogonias undulatus* (7 fold), as well as lower abundances of rainwater killifish *Lucania parva* (4 fold decrease) and red drum (2 fold decrease) were noted during extended low inflow conditions.

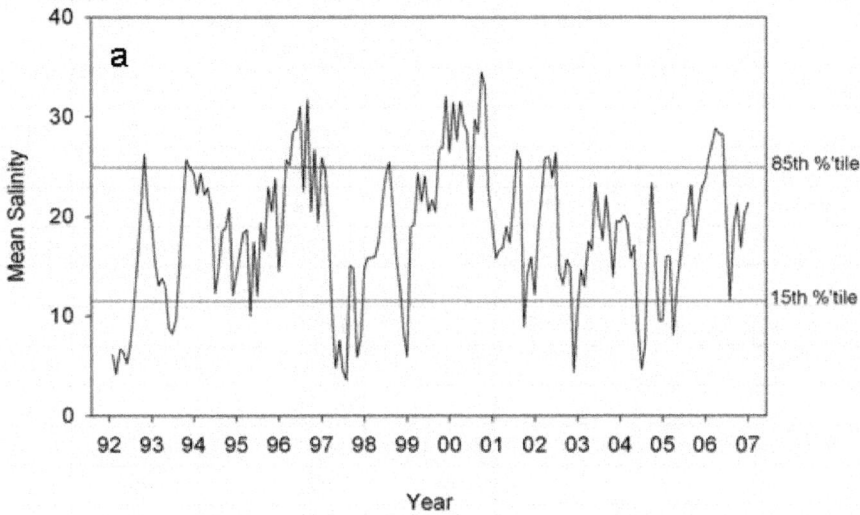

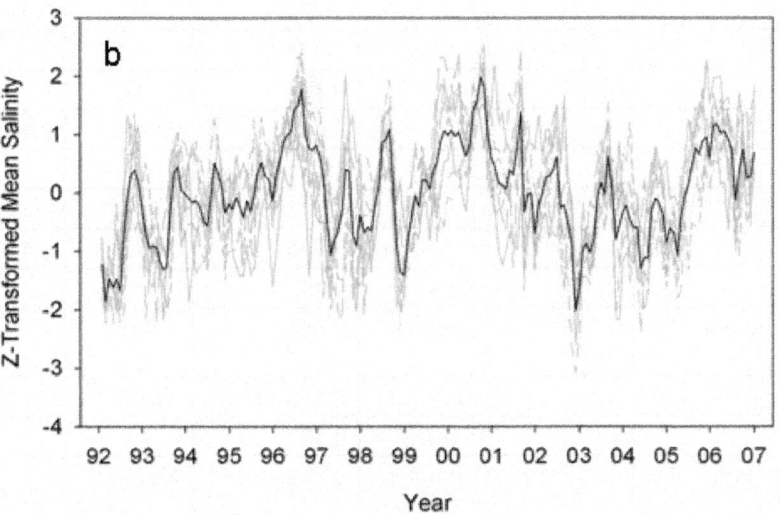

Figure 2: Estuarine-wide mean salinity time series during the study period of 1992 through 2006: (a) from a representative estuary (Lavaca-Colorado), and (b) from all eight major estuaries. Salinities in (b) are Z-transformed (not labeled individually for clarity), and a mean line added to aid in interpretation.

Table 3: Analysis of Similarity results of community structure within each estuary across seasons. Global R by Estuary, *** = $p < 0.001$, pairwise comparison R values by season (significant pairwise R values > than the Global R in bold)

	Global R		Seasonal Pairwise Comparison R		
Estuary					
			Winter	Spring	Summer
		Spring	0.464	-	
Sabine-Neches	0.669***	Summer	0.908	0.566	-
		Fall	0.836	0.834	0.501
		Spring	0.628	-	
Trinity-San Jacinto	0.649***	Summer	0.900	0.553	-
		Fall	0.815	0.786	0.362
		Spring	0.576	-	
Lavaca-Colorado	0.695***	Summer	0.971	0.697	-
		Fall	0.820	0.796	0.391
		Spring	0.573	-	
Guadalupe	0.661***	Summer	0.897	0.660	-
		Fall	0.756	0.786	0.411
		Spring	0.582	-	
Mission-Aransas	0.677***	Summer	0.848	0.652	-
		Fall	0.763	0.845	0.502
		Spring	0.565	-	
Nueces	0.647***	Summer	0.802	0.627	-
		Fall	0.700	0.873	0.450
		Spring	0.493	-	
Upper Laguna Madre	0.516***	Summer	0.651	0.420	-
		Fall	0.515	0.659	0.381
		Spring	0.382	-	
Lower Laguna Madre	0.532***	Summer	0.609	0.572	-
		Fall	0.629	0.764	0.302

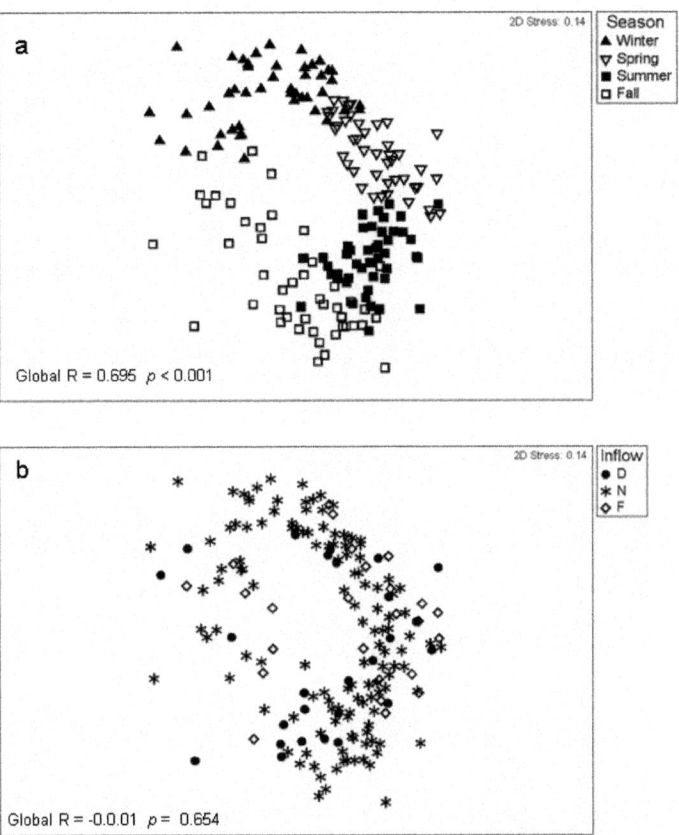

Figure 3: Multidimensional scaling (2D) configuration of bag seine community structure from a representative estuary (Lavaca-Colorado) overlaid with (a) Season, and (b) Inflow Condition. Season of collection defined as: Winter (Dec, Jan, Feb); Spring (Mar, Apr, May); Summer (Jun, Jul, Aug); and Fall (Sep, Oct, Nov). Inflow Condition designations; D = Drought, N = Normal, F = Flood. Global R values for each Analysis of Similarity test included.

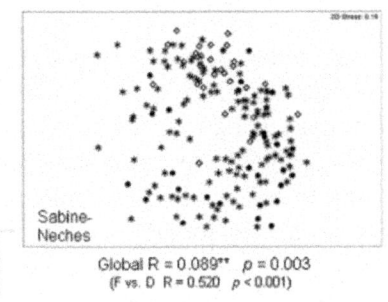

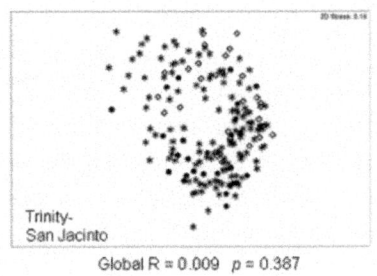

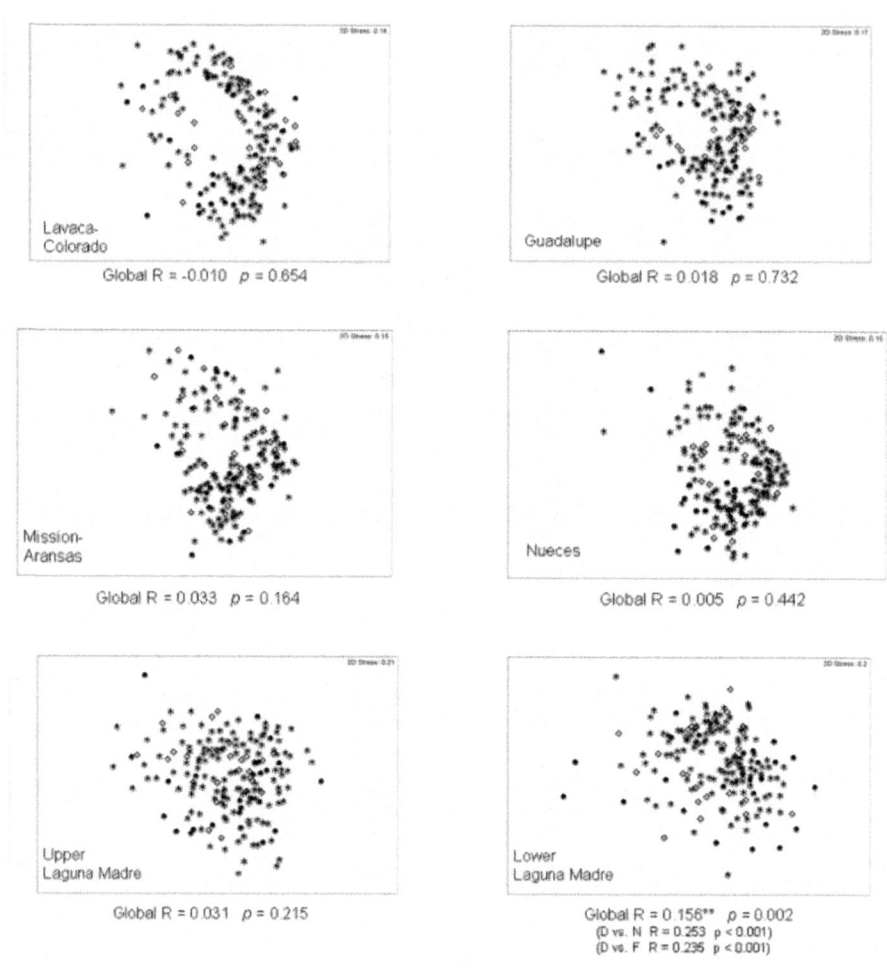

Figure 4: Multidimensional scaling (2D) configurations of bag seine derived community structure for each Texas estuary identifying the categorical Inflow Condition, with the Global R values for the seasonal Analysis of Similarity test. Inflow Condition symbols follow Figure 3.

While estuary-specific assemblages do not appear to be responding to synoptic inflow events, the inclusive role of salinity on overall community structure can be seen in Table 4. Across the estuaries, patterns of community structure roughly matched the NE-SW salinity gradient present on the Texas coast, with the freshest estuaries on the upper coast having significantly different communities than the more saline estuaries found on the lower coast. The middle coast estuaries (Lavaca-Colorado, Guadalupe, and Mission-

Aransas) showed the greatest degree of overlap in their community structure. Of the hundreds of taxa that constitute the nekton recorded with the bag seines, differences among the estuaries could be explained by examining only a fraction of this total. Abundance levels of 36 taxa accounted for the vast majority of the individuals found in each estuary, ranging from a low of 99.5 % in the Lavaca-Colorado system to a high of 100% in the Nueces Estuary (Table 5). Of the commercially and recreationally important species currently used for TxEMP modeling, only blue crab, white shrimp, and brown shrimp contributed substantially to nekton community structure patterns. Other taxa used for TxEMP either contributed little to the overall community (e.g., red drum ranked no higher than 12th from any estuary; spotted seatrout no higher than 14th) or were identified as a characteristic component from only a single estuary (southern flounder contributed to at least 1% of the community structure only in the Sabine-Neches estuary). Black drum were not identified as a significant component from any estuary. Community structure across the salinity gradient of estuaries appears to be driven by the relative proportion of only a few oligohaline (e.g., Atlantic croaker, bay anchovy *Anchoa mitchilli*, and Gulf menhaden *Brevoortia patronus*) and polyhaline to metahaline taxa (pinfish, Gulf killifish *Fundulus grandis*, sheepshead minnow, longnose killifish *F. similis,* and grass shrimp *Palaemonetes* spp.). Ubiquitous euryhaline taxa that were identified at equivalent ranks across the coastal salinity gradient included blue crab, striped mullet *Mugil cephalus*, spot *Leiostomus xanthurus*, brown shrimp, white shrimp, and silversides *Menidia* spp.

Table 4: Matrix of pairwise comparison R values for the Analysis of Similarity test of community structure among estuaries. Global R = 0.455, $p < 0.001$; significant pairwise R values > than the Global R in bold

Estuary	Sabine-Neches	Trinity-San Jacinto	Lavaca-Colorado	Guadalupe	Mission-Aransas	Nueces	Upper Laguna Madre
Trinity-San Jacinto	0.299	-					
Lavaca-Colorado	0.324	0.081	-				
Guadalupe	**0.686**	0.423	0.347	-			
Mission-Aransas	**0.555**	0.283	0.213	0.295	-		
Nueces	**0.710**	0.371	0.335	0.418	0.116	-	
Upper Laguna Madre	**0.875**	**0.806**	**0.727**	0.403	**0.579**	**0.679**	-
Lower Laguna Madre	**0.784**	**0.578**	**0.527**	0.324	0.381	0.336	**0.485**

Table 5: Rank order of the nekton taxa contributing to the top 90% of average similarity within each estuarine assemblage. Species identified by an asterisk (*) represent commercially or recreationally important target species currently used in TxEMP modeling. Blank entries represent taxa recorded from each estuary, but their overall contribution to the community in that estuary is less than 1%

Species	Sabine-Neches	Trinity-San Jacinto	Lavaca-Colorado	Guadalupe	Mission-Aransas	Nueces	Upper Laguna Madre	Lower Laguna Madre
Micropogonias undulatus	1	2	4	12	13	17		14
Callinectes sapidus*	2	1	2	5	3	2	7	5
Anchoa mitchilli	3	5	10	11	11	13	8	17
Brevoortia patronus	4	3	11	16				
Menidia beryllina/ peninsulae	5	4	3	2	2	5	2	8
Mugil cephalus	6	6		8	10	10	10	9
Litopenaeus setiferus*	7	10	6	15	12	14		13
Palaemonetes spp.	8	7	1	1	1	6	3	12
Leiostomus xanthurus	9	8	9	9	8	7	11	6
Farfantepenaeus aztecus*	10	9	5	6	7	8	9	3
Lagodon rhomboides	11	12	8	3	4	3	6	1
Sciaenops ocellatus*	12	16	15	14	14	16	16	18
Paralichthys lethostigma*	13							
Cynoscion arenarius	14	17	18					
Fundulus similis		13	7	7	6	4	4	4
Fundulus grandis		14	13	10	9	9	5	15

Cynoscion nebulosus*				17	17	18	14	
Gobiosoma bosc				18		19		
Eucinosto-mus argen-teus				19	18			16
Cyprinodon variegatus		11	12	4	5	1	1	2
Mugil curema		15	14	13	15	15	15	11
Citharichthys spilopterus		18	19					
Callinectes similis		19				12		10
Menticirrhus americanus		20	16					
Arius felis		21	17					
Lucania parva							12	
Farfante-penaeus duorarum					16	11		7
Syngnathus scovelli							13	
Percent Total Abundance	99.7	99.7	99.6	99.5	99.8	100	99.8	99.7

DISCUSSION

Long-term data sets are fundamental to an understanding of factors that regulate system level processes, because the inherent complexity and variability of open natural systems make it difficult to establish causal relationships between and among the important components. These data are needed to ensure that the environmental conditions which potentially can lead to dramatic fluctuations in observed nekton abundance levels are recorded at least once and preferably several times [24]. Decadal-scale continuous records of biological data utilizing uniform sampling strategies are the exception rather than the rule for most estuarine and coastal realms [25]. Many of the estuarine studies that do take into account the spatial and temporal aspects of the physical environment often utilize commercial catch and effort records [26-30], and these catch per unit effort (CPUE) indices of abundance are not without problems. Technological advances and external economic factors which can directly affect actual effort are either poorly documented, or often entirely dismissed. Circumventing some of the inherent problems associated with fishery-dependent CPUE indices, TPWD utilizes fisheries-independent sampling methodologies to assemble long-term data sets of estuarine biotic and abiotic structure [21]. Besides providing resource managers with uniform information that is reliably documented and collected under standardized sampling designs and techniques, these long-term data sets offer the antithesis to short-term management solutions dictated by monetary constraints that emphasize research and monitoring projects of limited temporal and spatial duration [31].

The current TxEMP methodology which uses salinity as a proxy for FWI to establish inflow–species spatial relationships has demonstrated varying levels of correspondence between abundance and salinity gradients [14, 32-33]. In estuaries receiving substantial inflows that facilitate defined salinity zonations (e.g., Trinity-San Jacinto and Guadalupe), peak densities of many target species were spatially correlated with specific salinity zones. Conversely, in estuaries receiving lower amounts of inflows (e.g., Nueces and Laguna Madre), well defined salinity zonations were either dramatically compressed into the upper-most reaches of the estuary, or absent altogether, and consequently these same target species were far less associated with their recognized salinity preferences. Expanding the spatial scale beyond individual estuaries and using the bag seine information to encompass the entire nekton community, the present analysis shows that a much lower degree of correspondence exists between the synoptic-scale FWI signal and community assemblage. While this general lack of correspondence between the motile nekton and FWI may seem contradictory to the reported positive flow effects on fisheries abundance [29-30, 34-35], similar neutral responses by fisheries to FWI have been reported in other studies conducted at equivalent spatial scales as was used for this study. For example, in East Bay, Florida, Livingston et al. [2] found that river flow and primary production were associated mainly with changes in the communities at the lower trophic levels (herbivores and omnivore), whereas the carnivores (e.g., spotted seatrout, southern flounder, and red drum) were associated primarily with other animal trophic interactions. Their study showed that salinity changes were only indirectly involved in biological interactions at the highest trophic levels. Similarly, Griffiths [36] showed that yellowfin bream *Acanthopagrus australis* (the functional equivalent of pinfish used for this study) and striped mullet were generally resilient to salinity perturbations in Shellharbour Lagoon, Australia. Neutral responses to fluctuating salinities is not exclusive to finfish, as both Kimmerer [37] reporting on California bay shrimp *Crangon franciscorum* (functional equivalent of white shrimp) in the northern San Francisco Estuary, California, and Rozas et al. [11] working with brown shrimp in Breton Sound, Louisiana, each showed a general de-coupling of abundance levels and hydrologic conditions. Similarly, increases in temporal scales have also revealed a general de-coupling between abundance levels and FWI, as both spot and Atlantic croaker did not correlate with year-to-year variation of river discharge in Apalachicola Bay, Florida [38]. Weinstein et al. [39] also showed that shallow-water fish assemblages in the Cape Fear River estuary were not affected by annual differences in river discharge.

The repeatability of species assemblage composition and abundances from year to year across the salinity spectrum in the estuarine systems along the Texas coast is one of the most prominent features of this study. From Figure 4, it is clear that an orderly seasonal succession in abundance and species composition of the dominant components confirms the many published accounts of annually repeating community structure from a variety of locations [40-43]. A common theme found within these studies is that the identification of quite specific arrival times, or dates of first occurrence within each season of recruitment, can be shown for a number of taxa, regardless of the hydrologic conditions within an estuary at the time of recruitment. Interannual variations in these dates of first occurrence are typically small, suggesting that temporal stability of assemblage structure may be more closely related to temperature [40] or seasonal photoperiods than to salinity. The current analysis shows that the greatest disparity in community composition, regardless of any underlying salinity level difference, involved comparisons across opposite seasons. These seasonal differences were steadfastly replicated year after year, in spite of the dramatically different levels of freshwater inflows producing temporally unpredictable flood and drought conditions. During these environmental extremes, no wholesale changes in community composition were noted; only changes in the relative abundances within a set of common taxa.

Absent from the current analysis is a recognition of the role of physical habitat in structuring nekton community compositions (reviewed in [44]). From Table 1, it is quite clear that major differences in the areal extent of fringing wetlands and submerged aquatic vegetation exists among the eight estuaries under investigation, and despite these obvious differences, the major nekton components of each community assemblage are, for the most part, the same limited suite of taxa (Table 5). Many estuarine organisms have increased (sometimes dramatically) abundances in areas closest to the freshwater source, and these same oligo- and mesohaline areas are noted for supporting much of the wetland habitats cited in Table 1. Even though the direct effect that FWI has on wetlands and the species that use them has not been definitely demonstrated [12], there is very good evidence that these relationships exist, at least for some size-groups or life history stages [7, 18, 36, 45]. Transient groups of young-of-the-year clupeiforms (Gulf menhaden, bay anchovy), perciforms (Atlantic croaker, spot, red drum, pinfish, spotted seatrout, and both species of mullets) and pleuronectiforms (southern flounder, bay whiff) that are spawned in deeper estuarine, nearshore, or offshore waters have all been shown to enter the shallow portions of estuaries and occur in very high densities in their recruitment and residence periods [17-19, 34,47]. Those nekton communities from the fresher, upper coast estuaries with large amounts of surrounding wetlands supported greater proportions of Atlantic croaker, bay anchovy, Gulf menhaden, and striped mullet than did the more southern estuaries with less fringing marsh. In the more saline estuaries, where seagrasses generally replace fringing marsh systems as one of the dominant structured habitats, the nekton communities identified by the bag seines were characterized by increases in cyprinidontiforms (sheepshead minnows and longnose killifish), grass shrimp, pinfish. Except for pinfish, all these taxa are estuarine-residents that do not recruit from nearshore or offshore spawning grounds, therefore the intermediate linkage between freshwater inflows and physical habitats may not be as important for these populations to be successful.

Many investigations have suggested that variability in estuarine production can be attributed either directly or indirectly to the fertilizing effects of freshwater input [1, 10, 48-49]. This estuarine 'agricultural model' is based on a mechanistic link between nutrient loading and increased phytoplankton production, ultimately leading to increased fisheries yields. Although Sutcliffe's [48] arguments have been disputed on interpretation and statistical grounds [46, 50], the concept persists and is fundamental to the implementation of the TxEMP methodology. Relating the flow effects to animal populations requires trophic transfer up the food web, and numerous studies have focused on the relative importance of 'top-down' vs. 'bottom-up' control of aquatic food webs [51-54]. A distinct dichotomy in the response of FWI controlling factors within estuarine systems appears to be that the herbivores and omnivores are more directly linked to physical and chemical controls associated implicitly with primary production ('bottom-up' regulation), whereas the carnivores (primary, secondary, and tertiary) are more closely associated with 'top-down' biological factors [2, 38]. Examination of the target taxa used for the TxEMP inflow-species relationships (Table 5) shows that all the vertebrates fall into the tertiary carnivore class, and the epibenthic macroinvertebrates are classified by as primary and secondary carnivores [2]. Omnivorous taxa that are more likely to have a more direct trophic linkage to the effects of FWI included pinfish, spot, striped mullet, white mullet, hardhead catfish *Arius felis*, and Gulf pipefish *Syngnathus scovelli*. Spot, a bottom-feeding perciform characteristic of the assemblage structure in every estuary, are potentially an important linkage between inflows and production because of their ability to regulate benthic invertebrates [55]. The connection between the benthos and FWI associated production appears show a much stronger mechanistic link [38, 56-57], although the benthic environment is represented only by eastern oyster within the current modeling paradigm. Thus, evaluating the biological effects of FWI within Texas estuaries is currently dependent upon taxa that empirically have been shown to display the weakest mechanistic couplings.

The time steps involved in TxEMP FWI modeling are on the order of a calendar month, whereas the nekton appear to be operating on the order of months (the seasonal signal was clearly evident in every estuary across the coast) to a calendar year (the repeating pattern of seasonal compositions common to each estuary resulted in the circular configuration of the samples seen in Figure 3a). Conversely, the drought and flood FWI signal common to the entire Texas coast is due by climate-level drivers operating at fundamental frequencies of approximately 11, 5, and 3.5 years [58]. While all of the commercially and recreationally important finfish used to quantify optimal FWI are long lived species and can contribute a number of different year-classes to the nekton community, the macroinvertebrates used by TxEMP all have life spans less than even the shortest frequency inflow signal driver. The shrimp species are all considered annual species [15], with maximum life spans from 18 months to 2 years, whereas blue crabs are reported to have a life span approaching 3 years. For the shrimp species abundant throughout the Texas coast, physical timing of their recruitment periods appear to be more in synchrony with species-specific temperature ranges instead of estuarine salinity requirements. Brown shrimp recruit to the estuary from February through May, while white shrimp typically show up from June through October. Once recruited from offshore spawning areas, white shrimp juveniles can migrate farther into the less saline waters of the upper estuary because they are more tolerant of lower salinities than the other shrimp species. This pattern is evidenced by the higher rank abundance values for white shrimp seen in the less saline upper coast estuaries, whereas rank values for brown shrimp were higher in the more saline lower coast estuaries. While the interaction of available habitat and salinity tolerance levels can therefore aid in successful recruitment, the annual frequency in shrimp spawning does not appear to be closely tied to the multiyear to decadal frequencies of the inflows. These observations conform to the conclusions of Allen and Barker [19], in that 'responses of populations to major changes in the estuarine environment are more strongly expressed as alterations in the magnitude than in the timing of habitat utilization'.

Because estuarine fishes have evolved to exploit one of the more physiologically challenging environments, it should not too surprising that they do not appear to be dramatically responding to the synoptic-scale inflow events that are currently used to quantify 'ecological health'. Even when utilizing species ranks to adjust for any gross differences in relative abundance among estuaries (e.g., biomass), the present analysis reveals that in each estuary, the contributions of a very limited number of taxa were strikingly similar. To evaluate the functional role of freshwater inflow into estuaries and determine estuarine FWI needs for the future, incorporating more sensitive 'measuring stick' organisms are recommended. One way this could be accomplished is to incorporate a greater range of trophic structure, utilizing some of the lower trophic level taxa that constituted a majority portion of the community assemblage (e.g., Gulf menhaden, bay anchovy, Gulf killifish, striped mullet, sheepshead minnow), or base the FWI modeling on taxa that appear to show a definite salinity response (e.g., Atlantic croaker, longnose killifish, white mullet, pinfish). Still another more challenging option would be to move down the trophic food web and index measures of 'estuarine health' to benthic taxa that show more direct mechanistic linkages to FWI.

ACKNOWLEDGEMENTS

I am sincerely indebted to the more than twenty years of field staff and technicians at the Coastal Fisheries Division of Texas Parks and Wildlife Department for their diligent collection of the biotic and abiotic parameters used for this study. This project was never explicitly funded by research grants, but I gratefully acknowledge the continued support of Sportfish Restoration Funds, without which the time for data synthesis and interpretation would not have been possible.

REFERENCES

1. Flint, R.W. Long-term estuarine variability and associated biological response. Estuaries 1985;8 158-169.

2. Livingston, R.J., N. Xufeng, F.G. Lewis, III, and G.C. Woodsum. Freshwater input to a Gulf estuary: long-term control of trophic organization. Ecological Applications 1997;7 277-299.

3. Logeragan, N.R., and S.E. Bunn. River flows and estuarine ecosystems: implications for coastal fisheries from a review and case study of the Logan River, southeast Australia. Australian Journal of Ecology 1999;24 431-440.

4. Alber, M. A conceptual model of estuarine freshwater inflow management. Estuaries 2002;25 1246-1261.

5. Vörösmarty, C.J., and D. Sahagian. Anthropogenic disturbance of the terrestrial water cycle. BioScience 2000;50 753-765.

6. Dynesius, M., and C. Nilsson. Fragmentation and flow regulation of river systems in the northern third of the world. Science 1994;266 753-762.

7. Sklar, F.H., and J.A. Browder. Coastal environmental impacts brought about by alterations to freshwater flow in the Gulf of Mexico. Environmental Management 1998;22 547-562.

8. Rijsberman, F.R. Water scarcity: Fact or fiction? Agricultural Water Management 2006;80 5-22.

9. Skreslet, S. Freshwater outflow in relation to space and time dimensions of complex ecological interactions in coastal waters. In: S. Skreslet (ed.) The role of freshwater outflow in coastal marine ecosystems. Berlin, Germany: Springer-Verlag; 1986. p3-12.

10. Mallin, M.A., H.W Paerl, J. Rudek, and P.W. Bates. Regulation of estuarine primary productivity by watershed rainfall and river flow. Marine Ecology Progress Series 1993;93 199-203.

11. Rozas, L.P., T.J. Minello, I. Munuera-Fernandez, B. Fry, and B. Wissel. 2005. Macrofaunal distributions and habitat change following winter-spring releases of freshwater into the Breton Sound estuary, Louisiana (USA). Estuarine, Coastal and Shelf Science 65:319-336.

12. Longley, W.L., editor. Freshwater inflows to Texas bays and estuaries: Ecological relationships and methods for determination of needs. Texas Water Development Board and Texas Parks and Wildlife Department, Austin, Texas. 1994.

13. Powell, G.L., and J. Matsumoto. Texas estuarine mathematical programming model: a tool for freshwater inflow management. In: K.R. Dyer and R.J. Orth (eds.) Changes and fluxes in estuaries. Fredensborg, Denmark: Olsen and Olsen; 1994. p401-406.

14. Pulich, W. Jr, J. Tolan, W.Y. Lee, and W. Alvis. Freshwater inflow recommendation for the Nueces Estuary. http://www.tpwd.state.tx.us/ landwater/water/conservation/freshwater_inflow/nueces/ (accessed 20 Feb 2010).

15. Patillo, M.E., T.E. Czapla, D.M. Nelson, and M.E. Monaco. Distribution and abundance of fishes and invertebrates in Gulf of Mexico estuaries, Volume II: Species life history summaries. ELMR Report Number 11. NOAA/NOS Strategic Assessment Division, Silver Spring, MD. 1997.

16. Able, K.W. A re-examination of fish estuarine dependence: Evidence for connectivity between estuarine and ocean habitats. Estuarine, Coastal and Shelf Science 2005;64 5-17.

17. Weinstein, M.P., and H.A. Brooks. Comparative ecology of nekton residing in a tidal creek and adjacent seagrass meadow: community

composition and structure. Marine Ecology Progress Series 1983;12 15-27.

18. Rogers, S.G., T.E Targett, and S.B. Van Sant. Fish-nursery use in Georgia salt-marsh estuaries: the influence of springtime freshwater conditions. Transactions of the American Fisheries Society 1984;113 595-606.

19. Allen, D.M., and D.L. Barker. Interannual variation in larval fish recruitment to estuarine epibenthic habitats. Marine Ecology Progress Series 1990;63 113-125.

20. Hare, J.A., and K.W. Able. Mechanistic links between climate and fisheries along the east coast of the United States: explaining population outbursts of Atlantic croaker (Micropogonias undulatus). Fisheries Oceanography 2007;16 31-45.

21. Martinez-Andrade, F., P. Campbell, and B. Fuls. Trends in relative abundance and size of selected finfishes and shellfishes along the Texas Coast: November 1975-December 2003. Management Data Series No. 232, Texas Parks and Wildlife Department, Coastal Fisheries Division. Austin, Texas. 2005.

22. McKee, D.A. Fishes of the Laguna Madre: A guide for anglers and naturalists. College Station, Texas: Texas A&M University Press. 2008.

23. Clarke, K.R., and R.M. Warwick. Change in marine communities: an approach to statistical analysis and interpretation, 2nd edition. Plymouth : PRIMER-E. 2001.

24. Kruskal, J.B. Multidimensional scaling by optimizing goodness of fit to a non-metric hypothesis. Psychometrika 1964;29 1-27.

25. Rose, K.A., and J.K. Summers. Relationships among long-term fisheries abundances, hydrographic variables, and gross pollution indicators in northeastern U.S. estuaries. Fisheries Oceanography 1992;1 281-293.

26. Wolfe, D.A., M.A. Champ, D.A Flemer, and A.J. Mearns. Long-term biological data sets: their role in research, monitoring, and management of estuarine and coastal marine systems. Estuaries 1987;10 181-193.

27. Summers, J.K., T.T. Polgar, J.A. Tarr, K.A. Rose, D.G. Heimbuch, J. McCurley, R.A. Cummins, G.F. Johnson, K.T. Yetman, and G.T. DiNardo. Reconstruction of long-term time series for commercial fisheries abundance and estuarine pollution loadings. Estuaries 1985;8 114-124.

28. Pearson, T.H., and P.R.O. Barnett. Long-term changes in benthic populations in some west European coastal areas. Estuaries 1987;10 220-226.

29. Houde, E.D., and E.S. Rutherford. Recent trends in estuaries fisheries-predictions of fish production and yield. Estuaries 1993;16 161-176.

30. Jassby, A.D., W.J. Kimmerer, S.G. Monismith, C. Armor, J.E. Cloern, T.M. Powell, J.R. Schubel, and T.J. Vendlinski. Isohaline position as a habitat indicator for estuarine populations. Ecological Applications 1995;5 272-289.

31. Diop, H., W.R. Keithly, Jr., R.F. Kazmierczak, Jr., and R.F. Shaw. Predicting the abundance of white shrimp (Litopenaeus setiferus) from environmental parameters and previous life stage. Fisheries Research 2007;86 31-41.

32. Champ, M.A. Monitoring: Painting a moving train. Sea Technology 1986;27 73.

33. Pulich, W. Jr, J. Tolan, W.Y. Lee, and W. Alvis. Freshwater inflow recommendation for the Nueces Estuary. http://www.tpwd.state.tx.us/landwater/water/conservation/freshwater_inflow/nueces/ (accessed 20 Feb 2010).

34. Tolan, J.M., W.Y. Lee, G. Chen, and D. Buzan. Freshwater inflow recommendation for the Laguna Madre Estuary system. Texas Parks and Wildlife Department. Austin, Texas. 2004.

35. Copeland, B.J. Effects of decreased river flow on estuarine ecology. Journal of the Water Pollution Control Federation 1966;38 1831-1839.

36. Weinstein, M.P., and M.P. Walters. Growth, survival and production in young-of-year populations of Leiostomous xanthurus Lacépéde residing in tidal creeks. Estuaries 1981;4 185-197.

37. Griffiths, S.P. Factors influencing fish composition in an Australian intermittently open estuary. Is stability salinity-dependent? Estuarine, Coastal and Shelf Science 2002;52 739-751.

38. Kimmerer, W.J. Effects of freshwater flow on abundance of estuarine organisms: physical effects or trophic linkages? Marine Ecology Progress Series 2002;243 39-55.

39. Kobylinski, G.J., and P.F. Sheridan. Distribution, abundance, feeding and long-term fluctuations of spot, Leiostomus xanthurus, and croaker, Micropogonias undulates, in Apalachicola Bay, Florida, 1972-1977. Contributions in Marine Science 1971;22 149-161.

40. Weinstein, M.P., S.L. Weiss, and W.F. Walters. Multiple determinants of community structure in shallow marsh habitats, Cape Fear River estuary, North Carolina, USA. Marine Biology 1980;58 227-243

41. McGovern, J.C., and C.A. Wenner. Seasonal recruitment of larval and juvenile fishes into impounded and non-impounded marshes. Wetlands 1990;10 203-221.

42. Witting, D.A., K.W. Able, and M.P. Fahay. Larval fishes of a Middle Atlantic Bight estuary: assemblage structure and temporal stability. Canadian Journal of Fisheries and Aquatic Sciences 1999;56 222-230.

43. Hagan, S.M., and K.W. Able. Seasonal changes of the pelagic fish assemblage in a temperate estuary. Estuarine, Coastal and Shelf Science 2003;56 15-29.

44. Tolan, J.M. Larval fish assemblage response to freshwater inflow: a synthesis of five years of ichthyoplankton monitoring within Nueces Bay, Texas. Bulletin of Marine Science 2008;82 275-296.

45. Minello, T.J. Nekton densities in shallow estuarine habitats of Texas and Louisiana and the identification of essential fish habitat. In L. Benaka (ed.) Fish habitat: Essential fish habitat and habitat restoration. American Fisheries Society, Symposium 22. Bethesda, MD. 1999. p43-75.

46. Sinclair, M., G.L. Bugden, C.L. Tang, J.C. Therriault, and P.A. Yeats. Assessment of effects of freshwater runoff variability on fisheries production in coastal waters. In S. Skreslet (ed.) The role of freshwater outflow in coastal marine ecosystems. Berlin, Germany: Springer-Verlag. 1986. p139-160.

47. Martino, E.J., and K.W. Able. Fish assemblage across the marine to low salinity transition zone of a temperate estuary. Estuarine, Coastal and Shelf Science 2003;56 969-987.

48. Sutcliffe, W.H., Jr. Some relation of land drainage, particulate matter, and fish catch in to eastern Canadian bays. Journal of the Fisheries Research Board of Canada 1972;29 357-362.

49. Cloern, J.E., A.E. Alpine, B.E. Cole, R.L.J Wong, J.F. Arthur, and M.D. Ball. 1983. River discharge controls phytoplankton dynamics in the northern San Francisco Bay estuary. Estuarine, Coastal and Shelf Science 1983;16 415-429.

50. Drinkwater, K.F., and R.A. Myers. Testing predictions of marine fish and shellfish landings from environmental variables. Canadian Journal of Fisheries and Aquatic Sciences 1987;44 1568-1573.

51. McQueen, D.J., R.S. Johannes, J.R. Post, T.J. Stewart, and D.R.S. Lean. Bottom-up and top-down impacts on freshwater pelagic community structure. Ecological Monographs 1989;59 289-309.

52. Menge, B.A. Community regulation: under what conditions are bottom-up factors important on rocky shores. Ecology 1992;73 755-765.

53. Flinkman, J., E. Aro, I. Vuorinen, and M. Viitasalo. Changes in northern Baltic zooplankton and herring nutrition from 1980s to 1990s: top-down and bottom-up processes at work. Marine Ecology Progress Series 1998;165 127-136.

54. Micheli, F. Eutrophication, fisheries, and consumer-resource dynamics in marine pelagic ecosystems. Science 1999;285 1396-1398.

55. Killam, K.A., R.J. Hochberg, and E.C. Rzemiem. Synthesis of basic life histories of Tampa Bay species. Tampa Bay National Estuary Program, Technical Publication Number 10-92. 1992.

56. Nilsson, P., B. Jönsson, I.L. Swanberg, and K. Sundbäck. Response of a marine shallow-water sediment system to an increased load of inorganic nutrients. Marine Ecology Progress Series 1991;71 275-290.

57. Montagna, P.A., and R.D. Kalke. The effect of freshwater inflow on meiofaunal and macrofaunal populations in the Guadalupe and Nueces Estuaries. Estuaries 1992;15 307-326.

58. Tolan, J.M. El Nino-Southern Oscillation impacts translated to the watershed scale: salinity patterns along the Texas Gulf Coast, 1982 to 2004. Estuarine, Coastal and Shelf Science 2007;72 247-260.

Chapter 11

FISHERIES AND BIODIVERSITY IN THE UPPER GULF OF CALIFORNIA, MEXICO

Gerardo Rodríguez-Quiroz[1], Eugenio Alberto Aragón-Noriega[2], Miguel A. Cisneros-Mata[3] and Alfredo Ortega Rubio[4]

[1]Centro Interdisciplinario de Investigaciones para el Desarrollo Integral Regional, Unidad Sinaloa

[2]Centro de Investigaciones Biológicas del Noroeste, Unidad Sonora

[3] Instituto Nacional de Pesca, SAGARPA, Centro Regional de Investigaciones Pesqueras de Guaymas

[4]Centro de Investigaciones Biológicas del Noroeste, Unidad La Paz, México

INTRODUCTION

The Upper Gulf of California (UGC) has been recognized by its high primary productivity and abundant fishing (Aragon-Noriega & Calderon-Aguilera, 2000). Sediments and nutrients from the Colorado River, and complex hydrodynamics render this as an important site for spawning, mating and nursing for numerous species of commercial and ecological importance (Cudney & Turk, 1998; Ramirez-Rojo & Aragón-Noriega 2006). Temperature, salinity and abundance of nutrients in this region vary depending on fresh water runoff from the Colorado River (Alvarez-Borrego et al., 1975; Hernández-Ayón et al., 1993; Lavín & Sánchez, 1999).

Commercial fishing of high market value resources such as shrimp takes place in the UGC by artisanal or small scale, and industrial fishing. Artisanal fishing is done on relatively small (30 feet) fiber glass boats or artisanal boats with outboard motors, usually operated by two fishers; their primary fishing gear is drift gillnets, which they use to catch croakers, Spanish mackerel and even shrimp. This type of fishing is carried out by cooperatives and individual fishers from the three ports of the UGC: Puerto Peñasco and El Golfo de Santa Clara, in the State of Sonora, and San Felipe, in Baja California. Because marine resources in the region are migratory, fisheries are seasonal generating bursts of accumulated fishing effort over a few months depending on availability of species (see Cudney & Turk 1998). Increasing demand of economically important species has motivated a steady rise in fishing effort and use of gear and fishing practices jeopardizing critical species such as totoaba, Totoaba macdonaldi, an endemic croaker declared under risk of extinction (Cisneros-Mata et al., 1995), and the rare vaquita, Phocoena sinus. Vaquita are accidentally caught in all kinds of gillnets used in the Upper Gulf (D'Agrosa et al., 1995; Blanco 2002).

Vaquita is the world's smallest cetacean; it is endemic to the Upper Gulf of California and has the most restricted distribution range of all marine mammals (Jaramillo-Legorreta et al., 2007). In a situation of increased mortality in fishing activities, the reasons why vaquita is under risk of extinction are its historical small population size (Jaramillo-Legorreta et al., 2007) and possibly reduced habitat (Lavin et al., 1999) with decreased flow of Colorado river inflow (Hanski, 1998; Fagan et al., 2005). They occur only in the northern quarter of the Gulf of California, Mexico, mainly north of 30°45' N and west of 114°20' W with a highly productive core area of about 2,235 km2, between San Felipe and Rocas Consag archipelago, a small upwelling spot where they have being seen feeding (Rojas-Bracho et al., 2006).

The high productivity of the upper-most portion of Gulf of California maintains a diverse number of marine species which interact with the vaquita. This porpoise is known to feed on grunt, Orthopristis reddingi, and ronco croaker, Bairdiella icistia, as well as different species of market squid (Vaquita Marina, 2007). Vaquita competes with dolphins and rays for several food items in the area; historically it was predated by sharks and killer whales (Barlow, 1986); at present shark predation and competition with ray species has lowered because of reduced abundance of these species possibly due to over fishing in the region (Rojas-Bracho et al., 2006).

The Upper Gulf of California and Colorado River Delta was declared a Biosphere Reserve (henceforth, Reserve) on June 10, 1993; it has an extension of 934,756 hectares including marine and terrestrial environments (Diario Oficial de la Federación [DOF], 1993; Fig. 1). The Reserve was implemented to protect species inhabiting that region, some of which are commercially important, endemic or under risk of extinction (Instituto Nacional de Ecología [INE], 1995; van Jaarsveld et al., 1998). A management program was designed to promote sustainable use of the biodiversity and landscape (SEMARNAT, 1995; RojasBracho et al., 2006; Aragón-Noriega et al., 2010).

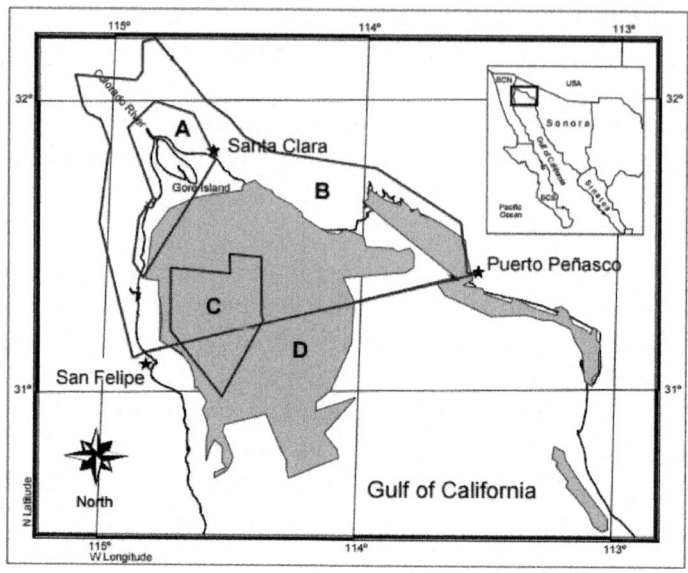

Figure 1: The Upper Gulf of California. The thin line depicts the Biosphere Reserve declared in June 1993; the shaded area represents the vaquita refuge declared in December 2005.A) Core Zone, B) Buffer Zone in the Biosphere Reserve of the Upper Gulf of California, C) Vaquita Refuge Area, D) Shadowed are fishing grounds.

The Reserve hosts an important number of species with high commercial value. Such is the case of the gulf croaker, Cynoscion othonopterus, an endemic fish species that arrives massively into the UGC where it spawns during winter; and the blue shrimp, Litopenaeus stylirostris, which is highly priced in local and international markets (Rodriguez-Quiroz et al., 2010).

Implementation of Biosphere Reserve in June 1993 limits the use of gillnets (> 15cm mesh size) and fishing effort (up to 2,100 artisanal boats) to protect, most of all, totoaba and vaquita, considered under risk of extinction (INE, 1995; Greenberg, 2005). The most recent additional measure to protect vaquita and its habitat was a declaration in December 2005 of a Refuge to further limit fishing activities (Fig. 1). The Refuge, located in the western side of the UGC, comprises an area of 1,263.85 km² and is divided into two polygons: Polygon A – northern portion-, within the Reserve and with a surface of 897.09 km²; and Polygon B – southern portion-, outside the Reserve and with a surface of 266.76 km². The Refuge was declared in the most likely distribution range of vaquita and includes a 65 km² zone where gillnets and trawl nets are prohibited (DOF, 2005).

Management of the Reserve and the Refuge imply a series of actions to achieve both protection of critical species and use of commercially important species. Consequently, fishing in the Upper Gulf becomes an economic activity with environmental implications. Conservation measures in the Reserve and the Refuge pose a challenge because they were designed to minimize negative impacts of fishing on vaquita. Several studies have been conducted in the area to determinate an average of accidental captures of vaquita in each fishing season since 1985; the most conclusive data was produced by D'Agrosa et al. (2000); who reported 39 vaquitas/year as by-catch before 1995. For those years less than 600 artisanal boats were in use, 1/3 of the boats registered in 2007 in the three fishing communities. Because of its critical condition it has been estimated that the maximum catch rate to avoid mid-term extinction is one vaquita per year (DOF, 2004). Mexican legislation recognizes that it is through participation of human communities affected by these measures that agreements can be achieved (see Palumbi et al., 2003). Therefore, solving this challenge will require a clear definition of common goals in fisheries management and conservation, expressed in a single policy (Davis, 2005).

Several management measures have been implemented both to protect vulnerable marine species and fishing resources. Amongst such measures we have: no-take zones (Mangel, 1998), subsidies (Munro & Sumaila 2002), buy-out of fishing gear and boats (Clark et al., 2005), fishing rights (Gonzalez-Laxe, 2006), and individual transferable quotas (ITQ) (Townsend et al., 2006). In the Gulf of California, some of these measures have been implemented to reduce fishing effort and protect soft bottom biological communities: buyout of shrimp trawlers[1]; most recently, an ITQ program started aimed at rebuilding fishing stocks in the Gulf of California[2].

In this work we identify and analyze the most important artisanal fisheries of the Upper Gulf of California, which are in continuous interaction with the vaquita. We propose a scheme to reduce vaquita by-catch as a fishery management and biological conservation policy.

METHODS

Basic information used for the present analysis was generated in a series of studies conducted by World Wildlife Fund and Centro de Investigaciones Biológicas del Noroeste in the Upper Gulf of California during 2005. Additional information on fishing sites by species came from a previous report (Cudney & Turk, 1998). Artisanal fisheries data spanning since 1999 to 2007 were collected from official records in the ports of San Felipe, El Golfo de Santa Clara and Puerto Peñasco. Further information was gathered form a survey based on direct interviews to 146 artisanal fishers in those three fishing ports. Questionnaires were designed to compute direct cost structure during fishing operations, as well as fishing sites by species. A section of the survey was specifically designed to ascertain what types of activities alternative to artisanal fishing might be implemented in the Upper Gulf. Following Cochran (1989) we estimated sample size (n) for fishermen interviews:

$$n = \frac{\dfrac{Z^2 q}{E^2 p}}{1 + \dfrac{1}{N}\left[\dfrac{Z^2 q - 1}{E^2 p}\right]}$$

(1)

where: Z= CI=95%; p and q = Equation distribution; E= 6% Precision level; N= Fishermen community size. Following Greenberg (1993), local fishermen at each port were randomly selected.

From artisanal fishing landing records declared by fishers in local government fishery offices we obtained the following: Capture site, species, weight of landings and first-hand or "beach" economic value of landings by species.

Artisanal catch by species was processed and spatially represented in a geographic information system (GIS), identifying fishing sites within the Refuge (Fig. 1), overlapping the vaquita refuge polygon through the use of a ArcView 3.2 software using a 2002 Conica Lambert projection in maps by fishery and community in the vaquita polygon. Relative size (percentage) of the fishing activities zones in the vaquita refuge were obtained from the overall projected fishing sites.

To assess impact of fishery management measures oriented to protect the vaquita population, we arbitrarily assumed reductions in the number of artisanal boats for the three fishing communities. We establish two scenarios according to Gerrodette et al. (2011) who considered the number of vaquitas in 254 individuals. A deterministic model to describe vaquita population is defined as (Haddon, 2001):

$$P_t + 1 = P_t + rP_t(1 - \frac{P_t}{K}) - q_t f_t P_t$$

(2)

where for year t, P is vaquita population size, r is the per capita population growth rate, K is carrying capacity, q is catchability, and f is artisanal fishing effort in the Upper Gulf of California. Here we define catchability as the number of vaquitas incidentally caught per boat in artisanal fisheries in a given year and C = qfP is the total number of vaquitas caught in that year. Because of recent efforts to reduce vaquita incidental mortality during the first decade of 2000, we assume that these efforts have resulted in a proportional reduction in q starting in 2011. This model can thus be utilized to assess incidental mortality of vaquitas under different scenarios with respect to the number of active artisanal boats in a period of 15 years (2011-2025). For simplicity reasons, demographic and environmental stochasticity and their impacts in the vaquita population dynamics are neglected in our model, although we are keenly aware of their potential effects. Due to chance events alone, demographic stochasticity at very low population numbers might drive populations of marine mammals such as vaquita to extinction (Burgman et al., 1993).

RESULTS

Fisheries Analysis

A total of 2,554 catch reports by artisanal fishers were compiled and analyzed for the three fishing communities of the Upper Gulf. Additionally, a total of 146 fishers were interviewed. Based on catch volume and beach economic value, six artisanal fisheries are the most important in the Upper Gulf: Shrimp Litopenaus stylirostris, curvina Cynoscion othonopterus, bigeye croaker Micropogonias megalops, Spanish mackerel Scomberomorus spp., rays (several species) and sharks (several species) (Table 1). Due to its high value, shrimp represents the largest gross income to artisanal fishers. The curvina is the second most economically important species for fishers of El Golfo de Santa Clara, bigeye croaker for fishers of San Felipe, and rays for those of Puerto Peñasco.

Table 1: Catch (Kg) and value ($US) of the main species in artisanal fisheries landed in the three ports of the Upper Gulf of California during 2007

Species	Golfo de Santa Clara		Puerto Peñasco		San Felipe	
	Catch	Value	Catch	Value	Catch	Value
Shrimp	279	3'847,477	53	588,833	399	5'176,521
Curvina	1,552	806,283	43	23,843	677	376,272
Bigeye croaker	508	212,350	87	24,044	726	201,748
Spanish mackerel	888	870,345	58	54,029	95	76,769
Rays	106	91,677	121	89,713	244	180,988
Sharks	3	2,539	26	19,146	24	17,799

There are 2,100 small boats working in the Upper Gulf of California and, as discussed later in this work, fishers use different kinds of gillnets to fish for a variety of species. Table 2 shows the number of artisanal boats officially registered in each community. The number of artisanal boats is greater for El Golfo de Santa Clara, where artisanal fishing is virtually the only economic activity. Two types of fisheries concentrate the largest authorized fishing effort, shrimp with 606 artisanal boats and fishes (curvina, bigeye croaker, Spanish mackerel, sharks and rays), with 882 artisanal boats. The greatest number of authorized artisanal boats for both shrimp and fishes are registered in San Felipe and El Golfo de Santa Clara.

Table 2: Authorized artisanal fishing vessels by group of species in the three ports of the Upper Gulf of California. Source: Federal government offices in the communities of the Upper Gulf of California. *Curvina, bigeye croaker, Spanish mackerel, rays

Species	San Felipe	El Golfo de Santa Clara	Puerto Peñasco
Clams	15	12	39
Jumbo squid			4
Shrimp	318	232	56
Snails	1		42
Fishes *	295	412	175
Swimming crab	11	39	229
Mullet	10	76	8
Octopus	2		40
Sharks	10	26	69
Total	662	797	662

Our survey data and GIS analysis showed that fishing is conducted within the Vaquita Refuge Area and in the Biosphere Reserve. Approximately 62% of the total catch in the Upper Gulf of California was caught in the marine protected areas. Approximately 77% of the marine area of the Biosphere Reserve and the entire Vaquita Refuge Area are used for fishing (Fig. 2). In the Vaquita Refuge, 97% of the total area is fished for shrimp, 94% is fished for corvine, 85% is fished for shark, 79% is fished for bigeye croaker and 69% is fished for Spanish mackerel. In the Biosphere Reserve, 56% of the total area is fished for shrimp, 55% is fished for corvine, 44% is fished for bigeye croaker, 39% is fished for shark and 30% is fished for Spanish mackerel.

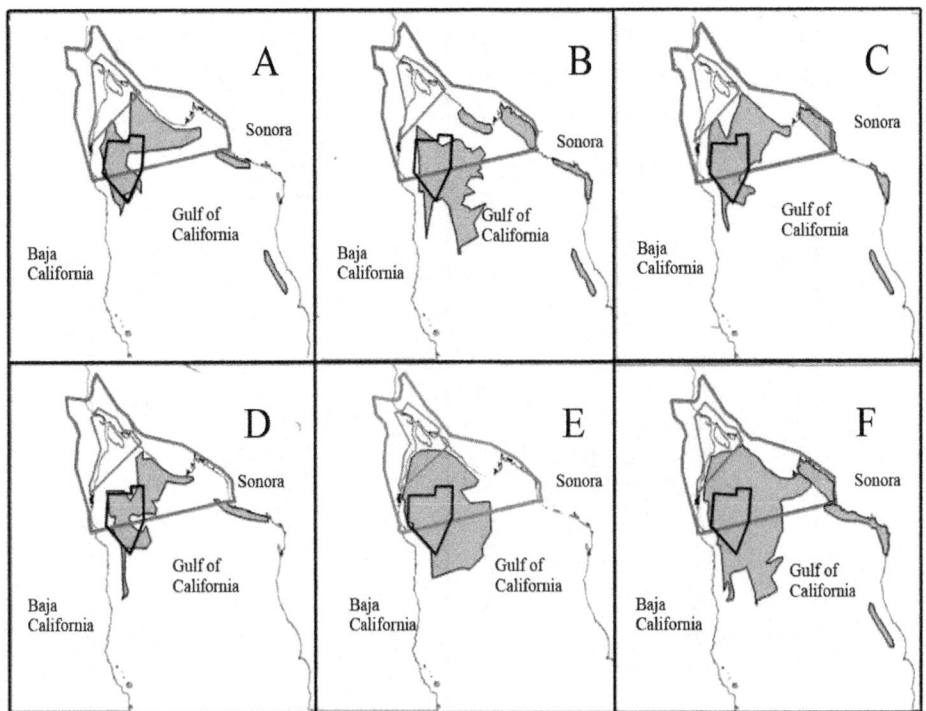

Figure 2: Spatial distribution of artisanal fisheries as compared with the vaquita refuge declared in the Upper Gulf of California. A) bigeye croaker, B)sharks and rays , C) shrimp, D) Spanish mackerel, E) curvina, F) all fisheries.

Fishermen from Puerto Peñasco fish close to the Sonoran shoreline. 75% of the capture occurs inside the Biosphere Reserve and the fishermen fish in 20% of the northern area of the Vaquita Refuge. Fishermen from Golfo de Santa Clara carry out their fishing inside the marine protected areas and they fish in about half of the Vaquita Refuge Area. San Felipe fishermen fish near the Baja California shoreline in the UGC from the core zone to Puertecitos, which covers the entire Vaquita Refuge Area and 70% of the Biosphere Reserve (Fig. 3).

Within the Refuge, curvina represents the greatest annual catch with ~3,000 mt, followed by bigeye croaker with ~1,240 mt; other species amount to 2,192 mt (Table 3). In terms of economic value, shrimp represents 80% of ca. $US 1.7 million total gross incomes in the marine protected areas; sharks and rays represent the lowest gross income with only ~$US 385,000.

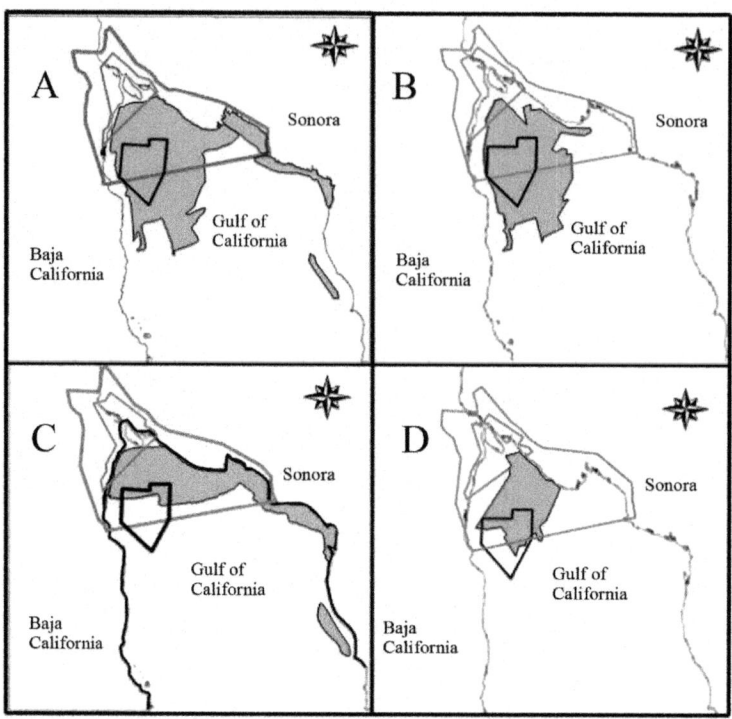

Figure 3: Spatial distribution of the artisanal fisheries in the UGC by community. A) all communities; B) San Felipe; B) Puerto Peñasco; C) El Golfo de Santa Clara.

Table 3: Catch, first-hand value and operation costs by group of species in the artisanal fisheries of the Upper Gulf of California in 2007. Source: Local fishery offices in the communities. *US thousands of dollars

Indicator	Shrimp	Bigeye croaker	Curvina	Rays	Spanish mackerel	Sharks	Total
Catch (metric tons)	459	938	2,957	429	1,239	65	6,087
Value of catch *	4,791	528	2,765	357	1,812	66	10,319
Costs of catch *	3,073	165	489	29	131	9	3,897
Gross profit *	1,718	363	2,276	328	1,681	57	6,423
Return rate (%)	36	69	82	92	93	86	62

Total annual catch in the marine protected areas from 1999 to 2007 were ~5,506 mt with a first-hand economic value of $US 8'563,000. The operation costs spent to obtain that total catch were $US 2'666,000 for a total gross income of $US 5'897,000, or a mean return rate of 68%. During that period we registered an increased fishing effort; gross profits provided high incomes.

Therefore, estimated opportunity costs for artisanal fishers giving up their activities in the vaquita Refuge amount to ca. $US 1.7 million per year (Table 4).

Table 4: Catch (metric tons), first-hand value and operation cost by group of species in the artisanal fisheries of the Upper Gulf of California inside the vaquita refuge and the Biosphere Reserve from 1995 to 2007. Source: Local fishery offices in the communities. *US thousands of dollars

	Catch	Value of catch*	Costs of catch*	Gross profit*	Return rate (%)
1995	2,510	3,444	569	2,874	83
1996	2,354	2,323	1,185	1,138	49
1997	5,466	4,327	1,640	2,688	62
1998	6,450	9,451	1,945	7,505	79
1999	7,536	9,625	2,390	7,234	75
2000	6,786	9,907	2,726	7,181	72
2001	6,050	9,181	2,994	6,186	67
2002	7,492	12,882	3,131	9,750	76
2003	5,029	8,755	3,321	5,435	62
2004	5,407	8,082	3,493	4,589	57
2005	5,888	12,660	3,609	9,051	71
2006	4,525	10,356	3,755	6,601	64
2007	6,087	10,319	3,897	6,423	62
Average	5,506	8,563	2,666	5,897	68

Social Analysis

In our study, opinions of fishers can be interpreted as guidelines of a comprehensive strategy to achieve the purposes of the Reserve (Table 5). When asked what their activity would be if the most important fishery to them were closed, 56% of fishers responded that they would continue to fish anyway: 22% claim that they would fish on the same, and 34% other species. These responses were mostly responded by El Golfo de Santa Clara fishermen, who do not have enough employment alternatives as compared to Puerto Peñasco and San Felipe fishermen. A total of 23.8% expected an economic aid and 19.6% would ask for something else such as a credit for a new business or local employment (plumber, carpenter, construction, etc).

Table 5: Response of fishers of the Upper Gulf of California to the question: If the most important fishery to you was closed, what would you ask in return?

Option	Percentage	Frequency
Permit for another fishery	34.3	49
Economic compensation	16.1	23
Payment of permit cost	7.7	11
Continue fishing the same	22.4	32
Other	14.7	21
Nothing	4.9	7

When we posed the question: If you were asked to stop fishing, what would you do? A large number of them responded that they would switch to the tourism and trade sector (49%), 6% would like to work in aquaculture and maquilas, 25.2% in another fishery (clams, oysters, etc.) or the same, and the remaining 20.1% would seek employment in domestic duties (Table 6).

Table 6: Response of fishers of the Upper Gulf of California to the question: If you were asked to stop fishing, what would you do?

Option	Percentage	Frequency
Tourism	24.5	34
Trade	24.5	34
Work in a private sector	5.8	8
Other activity in fisheries	7.2	10
Would not stop fishing	18.0	25
Other	20.1	28

Vaquita Recovery Analysis

Our model was fitted to reported mean annual vaquita abundance, effort (number of artisanal boats), and incidental mortality (Fig. 4). For the four years where there is information on vaquita population size (P) and incidental catch (C), we estimated q =C/(fN) and computed a weighted q (= 0.00011374) which was used for other years in the calculations. Using a fixed value for r (= 0.09531; Barlow, 1986) we fitted our model using least squares as criterion and found K = 4,640 vaquitas.

Our scenarios showed that it could take a large period of time for the vaquita population to recover to its 2010 size In scenario 1 (Fig. 4a) and according to response from fishermen (cf. table 6), starting in year 2011 we reduced the number of boats to 506, which is the quantity that would continue fishing and maintained constantly through year 2025. We observed that the number of vaquitas could recover in a relatively short time and could continue slow growth to over 349 individuals in 2025. In scenario 2 (Fig. 4b) we considered a 15% per year reduction of the artisanal fleet through year 2018 when the fleet reaches 506 boats and maintained it constant until year 2025. The model predicted a fast decrease in the vaquita population numbers until year 2015, stabilized thereafter and even showed a recovery of 6.3% per year through 2025 when the vaquita population reaches 242 individuals.

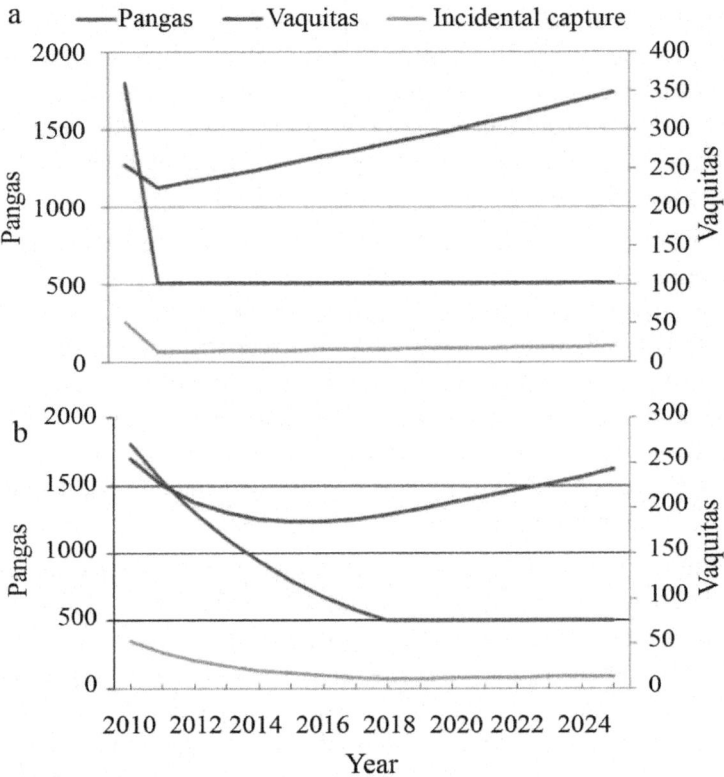

Figure 4: Scenarios for vaquita recovery if artisanal boats are reduced in the Upper Gulf of California: a) 75% in year 2011; b) 15% yearly until 2018 and then maintained constant through year 2025.

Again, we caution that our model neglected consideration of potentially important ecological and biological aspects such as demographic and environmental stochasticity at low population numbers, as well as environmental forcing (Burgman et al., 1993). It has been well established that in theory an age-structured population will experience a sharp increase in extinction risk, even if growing geometrically, when subjected to stochastic shocks in its vital rates (Burgman et al., 1993; Caswell, 1989).

DISCUSSION

Our study showed that both the Upper Gulf of California Biosphere Reserve and the recently declared vaquita Refuge are important grounds for artisanal fishing. Catch of shrimp, curvina and bigeye croaker have the largest distribution range in the Reserve and Polygon B of the Refuge. Shrimp generates the most important income for artisanal fishers. Our survey data indicates that 98% of artisanal fishers of El Golfo de Santa Clara and 100% of San Felipe fish on shrimp due to its high commercial value, gross revenues, and availability during the fishing season (September to January). This result is an important challenge to the fulfillment of goals of the Reserve and the vaquita Refuge, moreover because the number of registered artisanal boats is higher than recommended when the refuge was declared (DOF, 2005).

Operation costs determine to a great extent where fishing is conducted in the Upper Gulf of California and they depend mostly upon the distance of fishing sites to the ports and seasonal distribution of natural resources. San Felipe is the fishing port nearest to the recently declared Refuge; fishers residing in this port work in that vicinity throughout the year. Although El Golfo de Santa Clara holds the greatest number of registered permits and artisanal boats, fishers from this port do not fish near the Refuge because of high operation costs related to travel distance. Fishers of Puerto Peñasco fish near the sonoran coast to reduce operation coasts.

The high number of artisanal boats working in the Upper Gulf represents a clear threat to the vaquita (Rojas and Jaramillo 2001, Blanco 2002). The bulk of artisanal fisheries are done using gillnets to catch curvina (100%), shrimp (93%), Spanish mackerel (68%), bigeye croaker and rays (44%), and sharks (10%) (Vidal et al., 1994; D'Agrosa et al., 1995; Vidal, 1995). Gillnet mesh size varies from 5.7cm to 17.8cm; the highest vaquita mortality has been registered in gillnets with 11.43cm mesh size (Ortiz, 2002). These nets cover a great proportion of the water column where they are set and left for various hours (D'Agrosa et al., 1995). Length of these nets varies from 99 to 1,485 meters, the most common measuring between 594 and 990 meters with a mean height of 5.4 to 18 meters (Walsh et al., 2004). Because of the mesh size used, curvina and bigeye croaker fishing represent the biggest fishery-induced potential impact to the vaquita population (D'Agrosa et al., 2000; RojasBracho et al., 2006).

To succeed, a strategy to protect vaquita from artisanal fishing should consider social aspects such as attachment of fishers to their activity. It is clear that 25% of the fishermen would not stop fishing because that is the only activity they feel comfortable with and have done for years. In that context, some fisheries must be assessed considering the species value and impact to the environment, and it would regulate fishing enforcement allowing a specific number of fishermen and fishing tools by species to fish within the Reserve and the vaquita Refuge. The capture in the Biosphere Reserve and the Vaquita Refuge maintains a steady level of production with important economic incentives, which make it attractive to fishermen despite recent restrictions on their activity. However, the continued recruitment of new fishermen to the area will not enhance the welfare of the existing fishermen and there is no guarantee that the fishery will be sustained over the coming years (Ponce et al. 2006).

Our model results indicate that the vaquita population has never been too large (K < 5,000 individuals) and thus the importance of management actions to account for reductions of incidental kill in fishing activities, the only proven source of anthropogenic mortality. According to our scenarios, a fisheries management and vaquita conservation strategy can consider the reduction of the artisanal fleet by 15% every year until it reaches 506 boats in year 2018. Without compromising the vaquita population and the fishers in the region, this analysis could serve as a reference point to the buy-out program implemented since 2007, Parallel to this yearly reduction of the fleet, an integral strategy should incorporate a program to compensate and promote fishermen phase-out, including investments in equipment (sport fishing vessels with sustainable gears and refrigerated vehicles), infrastructure (storages, docks, freezers, added value of fishing products) and training in new activities. To support this program and according to table 4, ~$US 6'423,000 should be invested in the initial years to finance fishing opportunity cost to prevent a massive return of fishermen to the activity in cases where the buy-out program fails in the short term. Jobs must be accompanied by education and long-term provision for new qualifications, because many of the fishermen do not have technical business skills or experience in tourism administration or other activities. Also needed is a periodic evaluation of criteria set out in all implemented actions to increase chances of the vaquita recovery (Aragón-Noriega et al., 2010).

Through implementation since 2007 of an integral vaquita recovery program 242 artisanal boats and 340 fishing permits have been bought-out; in addition, 190 permits have been converted to alternative fishing gears. Also, a shrimp farm was rebuilt and in agreement with artisanal fishers whom will exchange their gear and permit, this could represent an additional 150 to 180 fishing boats phased-out of the region (Rojas-Bracho et al. 2010)

CONCLUSION

Our contribution is significant because we now have more information about the habitat utilized by commercially important species. We have also collected and analyzed information that can be used to elucidate which kinds of fisheries represent more risk to endangered species. Not all fisheries are necessarily a threat to biodiversity. A better understanding of the fisheries can help determine which fisheries are the most important to consider when developing a conservation strategy (Rodriguez-Quiroz et al., 2010).

Conservation success of vaquita must be based on agreements which dignify inhabitants of the Upper Gulf. Governments of all levels and conservation organizations should promote development of the region. We must strive to improve the quality of life of fishers while recovering the endangered vaquita considering socio-economic, ecological and institutional factors. Our calculations can serve as basis of a gradual compensation scheme to reduce artisanal fishing in the vaquita refuge through a buy-out scheme in the long run.

Success of most fisheries management policies to conserve species is contingent upon vulnerability of the species, size of the protected area and viable, equitable economic alternatives to fishers. Contrary to this view and given the critical situation of the vaquita, clearly enforcement of the Refuge as a no-take zone by itself will not suffice to save vaquita from extinction.

Continuation of the recently implemented integral program that includes measures for sustainable fishing, economic alternatives, and buy-out of artisanal fishing units must be guaranteed. A constant monitoring of the program must be put in place in an adaptive manner so as to ensure efficiency of interventions.

ACKNOWLEDGMENTS

We thank Javier de la Cruz for aid in field work and analyzing profit information. Helen Regan and Mauricio Ramirez provided helpful comments to an earlier version of this work. Mary López processed the catch information and produced the GIS figures. WWF-México provided partial funding for field work. GRQ thanks CONACYT (112401) and COTEPABEIPN (347) for a scholarship during his doctoral studies. EAAN thanks CONACYT Grant-48445.

REFERENCES

1. Álvarez Borrego, S.; Flores Báez, B.P. & Galindo Bect. L. (1975). Hidrología del Alto Golfo de California II. Condiciones durante invierno, primavera y verano. Ciencias Marinas, Vol. 2, pp. 21-36

2. Aragón-Noriega, E.A. & Calderon-Aguilera, L.E. (2000). Does damming the Colorado River affect the nursery area of blue shrimp Litopenaeus stylirostris (Decapoda:Penaeidae) in the Upper Gulf of California?.

International Journal of Tropical Biology and Conservation, Vol. 48, pp. 867-871

3. Aragón-Noriega, E.A.; Rodríguez-Quiroz, G.; Cisneros-Mata, M.A. & Ortega-Rubio, A. (2010). Managing a protected marine area for the conservation of critically endangered Vaquita Phocoena sinus (Norris, 1958) in the Upper Gulf of California. International Journal of Sustainable Development & World Ecology, Vol. 17, No. 5, pp. 410-416

4. Barlow, J. (1986). Factors affecting the recovery of Phocoena sinus, the Vaquita or Gulf of California harbor porpoise. U.S. National Marine Fisheries Service. Admistrative Report No. 86- 37, December 1986, pp. 19

5. Blanco, M.L. (2002). Pobreza y explotación de los recursos pesqueros en el Alto Golfo de California, In: Manejo de Recursos Pesqueros, Reunión Temática Nacional, R.E. Morán Angulo, M.T. Bravo, S. Santos Guzmán & J.R. Ramírez Zavala, (Ed.), 318-338, Editorial Universidad Autónoma de Sinaloa. Culiacán, México.

6. Burgman, M.A.; Ferson S. & Akçakaya, H.R. (1993). Risk assessment in conservation biology. Chapman & Hall, N.Y.

7. Caswell, H. (1989.) Matrix Population Models. Sinauer Associates, Inc. Sunderland, Massachusetts. 328 pp.

8. Clark, C.W.; Munro, G.R. & Sumaila, U.R. (2005). Subsidies, buybacks, and sustainable fisheries. Journal of Environmental Economics and Management, Vol. 50, 47-58

9. Cisneros-Mata, M.A.; Montemayor-López, G. & Román-Rodríguez, M.J. (1995). Life history and conservation of Totoaba macdonaldi. Conservation Biology, Vol. 9, pp. 806-814

10. Cochran, GW. (1989). Sampling Techniques, New York, Willey and Sons

11. Cudney, R. & Turk, P.J. (1998). Pescando entre mareas del Alto Golfo de California. Centro intercultural de Estudios del Desiertos y Océanos. Puerto Peñasco, Sonora, Mexico. 166 pp.

12. D'agrosa, C.; Vidal O. & Gram, W.C. (1995). Mortality of the vaquita Phocoena sinus in gillnet fisheries during 1993-1994. Reports of the International Whaling Commission (special issue), Vol. 16, pp. 283-291

13. D'agrosa, C.; Lennert-Cody, C.E. & Vidal, O. (2000). Vaquita bycatch in Mexico's artisanal gillnet fisheries: driving a small population to extinction. Conservation Biology, Vol. 14, pp. 1110-1119

14. Davis, G.E. (2005). Science and society: marine reserve design for the California Channel Islands. Conservation Biology, Vol. 19, pp. 1745-

1751

15. Diario Oficial de la Federación. (1993). Decreto por el que se declara área natural protegida con el carácter de Reserva de la Biosfera, la región conocida como Alto Golfo de California y Delta del Río Colorado. Diario Oficial de la Federación, junio de 1993.

16. Álvarez Borrego, S.; Flores Báez, B.P. & Galindo Bect, L. (2004). Acuerdo mediante el cual se aprueba la actualización de la Carta Nacional Pesquera y su anexo. Diario Oficial de la Federación, Marzo del 2004.

17. Aragón-Noriega, E.A. & Calderon-Aguilera, L.E. (2005). Programa de protección de la vaquita dentro de área de Refugio ubicada en la porción occidental del Alto Golfo de California. Diario Oficial de la Federación, Septiembre del 2005.

18. Fagan, W.F.; Kennedy, C.M. & Unmank, P.J. (2005). Quantifying rarity, losses and risks for native fishes of the lower Colorado River Basin: Implications for conservation listing. Conservation Biology, Vol. 19, pp. 1872-1882

19. Gerrodette, T.; Taylor, B.L.; Swift, R.; Rankin, S.; Jaramillo-Legorreta, A.M. & Rojas-Bracho, L. (2011). A combined visual and acoustic estimate of 2008 abundance, and change in abundance since 1997, for the vaquita, Phocoena sinus. Marine Mammal Sceince, Vol. 27, pp. 79-100

20. Gonzalez-Laxe, F. (2006). Transferability of fishing rights: The Spanish case. Marine Policy, Vol. 30, pp. 379-388

21. Greenberg, J.B. (1993). Local preferences for develop, In: Marine community and Biosphere Reserve: crises and response in the Upper Gulf of California, TR McGuire & JB Greenberg, (Ed.), p. 168. Occasional paper number 2. BARA: University of Arizona

22. Greenberg, J.B. (2005). Neoliberal reforms and the political ecology of fishing in the Upper Gulf of California, In: Las Dimensiones humanas en el estudio y conservación del Golfo de California, G.D. Danemann, (Ed.), 9-18, Pronatura Noroeste, Ensenada, Baja California, México

23. Haddon, M. (2001). Modelling and quantitative methods in fisheries. Chapman and Hall/CRC

24. Hanski, I. (1998). Metapopulation dynamics. Nature, Vol. 396, pp. 41-49

25. Hernández-Ayón, J.B.; Galindo-Bect, M.S.; Flores-Báez, B.P. & Álvarez-Borrego, S. (1993). Nutrient concentrations are high in the turbid waters of the Colorado River Delta. Estuarine, Coastal and shelf Science, Vol. 37, pp. 593-602

26. Instituto Nacional de Ecología. (1995). Programa de Manejo: Áreas Naturales Protegidas, Reserva de la Biosfera Alto Golfo de California y Delta del Río Colorado. pp. 94. México, D.F.

27. Jaramillo-Legorreta, A.M.; Rojas-Bracho, L.; Brownell, R.L.; Read, A.J.; Reeves, R.R.; Ralls, K. & Taylor, B.L. (2007). Saving the vaquita: Immediate action, not more data. Conservation Biology, Vol. 21, pp. 1653-1655

28. Lavin, M.F. & Sanchez, S. (1999). On how the Colorado River affected the hydrography of the Upper Gulf of California. Continental Shelf Research, Vol. 19, pp. 1545-1560

29. Mangel, M. (1998). No-take areas for sustainability of harvested species and a conservation invariant for marine reserves. Ecology Letters, Vol. 1, pp. 87-90

30. Munro, G. & Sumaila, R. 2002. The impact of subsidies upon fisheries management and sustainability: the case of the North Atlantic. Fish and Fisheries, Vol. 3, pp. 233-250

31. Ortiz, I. (2002). Impacts of fishing and habitat alteration on the population dynamics of the vaquita Phocoena sinus. Master Thesis. School of Aquatic and fishery Sciences. University of Washington, USA.

32. Palumbi, S.R.; Gaines, S.D.; Leslie, H. & Warner, R.R. (2003). New wave: high-tech tools to help marine reserve research. Frontiers in ecology and the environment, Vol. 1, pp. 73- 79

33. Ponce, D.G.; Arreguín, F & Beltrán, L.F. (2006). Indicadores de sustentabilidad y pesca: casos en Baja California Sur, México, In: Desarrollo sustentable: ¿Mito o realidad?, M.L.F. Beltrán, J Urciaga & A Ortega, (Eds.), 183-221. Centro de Investigaciones Biológicas del Noroeste, S.C, México

34. Ramirez-Rojo, R.A. & Aragon-Noriega, E.A. (2006). Postlarval ecology of the blue shrimp Litopenaeus stylirostris and brown shrimp Farfantepenaeus californiensis in the Colorado River Estuary. Ciencias Marinas, Vol. 32, pp. 45-52

35. Rodríguez Quiroz, G.; Aragón Noriega, E.A.; Valenzuela Quiñónez, W. & Esparza Leal, H.M. (2010). Artisanal fisheries in the conservation zones of the Upper Gulf of California. Revista de Biología Marina y Oceanografía. Vol. 45, No. 1, pp. 89-98

36. Rojas-Bracho, L. & Jaramillo-Legorreta, A. (2001). Vaquita Marina, In: Sustentabilidad y pesca responsable en México. Instituto Nacional de la Pesca, (Ed.), 963-981 Secretaria de Agricultura, Ganadería, Desarrollo Rural, Pesca y Alimentación, México

37. Rojas-Bracho, L.;   Reeves R.R. & Jaramillo-Legorreta, A. 2006. Conservation of the vaquita Phocoena sinus. Mammal Rev. Vol. 36, pp. 179-216

38. Rojas-Bracho, L.; Jaramillo-Legorreta, A.M.; Taylor, B.; Barlow, J.; Gerrodette, T.; Tregenza, N.; Swift, R. & Akamatsu, T. (2010). Assessing Trends in Abundance for Vaquita using Acoustic Monitoring: Within Refuge Plan and Outside Refuge Research Needs. Report of Vaquita Expedition 2008 and Current Conservation Actions. Paper SC/62/SM5 presented to the IWC Scientific Committee, 11 pp.

39. SEMARNAT. (1995). Programa de manejo. Áreas Naturales Protegidas 1. Reserva de la Biosfera del Alto Golfo de California y Delta del Río Colorado. Diciembre. SEMARNAT/CONANP.

40. Tognelli, M. F.; Silva-Garcia, C.; Labra, F.A. & Marquet, P. A. (2005). Priority areas for the conservation of coastal marine vertebrates in Chile. Biological Conservation, Vol. 126, pp. 420-428

41. Townsend, R.E.; Mccoll, J. & Young, M.D. (2006). Design principles for individual transferable quotas. Marine Policy, Vol. 30, pp, 131-141

42. Van Jaarsveld, A.S.; Freitag, S.; Chown, S. L.; Muller, C.; Koch, S.; Hull, H.; Bellamy, C.; Kruger, M.; Endrody-Younga, S.; Mansell, M.W. & Scholtz, C. H. (1998). Biodiversity assessment and conservation strategies. Science, Vol. 279, pp. 2106- 2108

43. Vaquita Marina. (2007). Todo sobre la vaquita marina. 17 Jun 2007, Available from: http://www.vaquitamarina.org/todo_sobre.php.

44. Vidal, O.; Van Waerebeek, K. & Findley, L.T. (1994). Cetaceans and gillnets fisheries in Mexico, Central America and the wider Caribbean: a preliminary Review. Rep. Int. Whal. Comm (special issue), Vol. 15, pp. 221-233

45. Vidal, O. (1995). Population biology and exploitation of the vaquita Phocoena sinus. Rep. Int. Whal. Comm (special issue), Vol. 16, pp. 247-272

46. Walsh, P.; Grant, S.M.; Winger, P.D. & Blackwood, G. (2004). An investigation of alternative harvesting methods to reduce the by-catch of vaquita porpoise in the Upper Gulf of California shrimp gillnet fishery. Centre for Sustainable Aquatic Resources, Fisheries and marine institute of Memorial University of Newfoundland, Canada. 32p.

Chapter 12

OPTIMUM FISHERIES MANAGEMENT UNDER CLIMATE VARIABILITY: EVIDENCE FROM ARTISANAL MARINE FISHING IN GHANA

Wisdom Akpalu[1], Isaac Dasmani[2, 4], and Ametefee K. Normanyo[3, 4]

[1]United Nations University—World Institute for Development Economics Research (UNU-WIDER), University of Ghana, Legon-Accra, Ghana

[2]Economics Department , University of Cape Coast, University Post Office, Cape Coast, Ghana

[3]Ho Polytechnic, Ho, Ghana

[4]Center for Environmental Economics Research & Consultancy (CEERAC), Accra, Ghana

ABSTRACT

In most developing coastal countries, the artisanal fisheries sector is managed as a common pool resource. As a result, such fisheries are overcapitalized and overfished. In Ghana, in addition to anthropogenic factors, there is evidence of rising coastal temperature and its variance, which could impact the environmental carrying capacity of the fish stock. This study investigates the effect of climate variation on biophysical parameters and yields. Our results indicate that the rising temperature is decreasing the carrying capacity. As a result, an optimum tax on harvest must reflect climate variability, as well as the congestion externality.

INTRODUCTION

In spite of the plethora of policies aimed at sustaining capture fish stocks around the world, evidence abound that most stocks are heavily overexploited [1,2]. In developing coastal countries where fishery sectors directly employ significant numbers of people and regulations are generally inadequate, food security and sustainable livelihoods are directly threatened [3,4]. In Sub-Saharan Africa, for example, the fisheries sector directly employs close to three million people and additional 7.5 million people are engaged in fish processing and trading.

In addition, it is estimated that in Africa the current annual revenue from capture fishery (US$2 billion) generates a multiplier effect of 2.5 times (US$5 billion) through trickle-up linkages [5]. The high number of fishers in coastal developing countries is due to a growing poverty trap.

In Ghana, artisanal and semi-industrial fishing are the most important direct and indirect employment generating activities within the entire coastal zone. The artisanal sector supported about 1.5 million people (about 9% of the total population) and landed about 70%–80% of total marine catches in 1996 [6]. The artisanal and semi-industrial fisheries are managed as unregulated common pool resources (CPR), hence are overcapitalized resulting in biological overfishing (*i.e.*, declining catch per unit effort (CPUE)). The existing regulations include a ban on the use of light aggregation equipment, which involves shinning light in the ocean, when the moon is out, to attract fish and increase harvest; a ban on the use of mesh sizes smaller than an inch in stretch diagonal; and a ban on the use of explosives in fishing. These regulations aim at limiting fishing efforts which is on the rise. For example, after a sharp increase in artisanal catch per unit effort between 1989 and 1992, it declined from 1992 through 2008 although fishing techniques improved and the number of crew per boat also increased. Within the same period, available data shows the annual coastal temperature and its variance has been on the rise. Since pelagic stocks targeted by artisanal fishers feed on planktons that depend on seasonal upwelling, it is likely that the rising coastal temperature is impacting the catch per unit effort. Although favorable upwelling can increase with global warming, the rising temperature could impact other environmental conditions for spawning, recruitment, or larval development, among others [7,8].

To reduce fish catches to sustainable levels, an optimum market-based policy instrument such as a tax on cost per unit effort or harvest is necessary. However, the efficacy of such a policy instrument hinges on the knowledge of the biophysical dynamics of the stocks. Two recent studies have shown that a fish stock could be potentially depleted if the biodynamic is misperceived, even if catch policies exist [9,10]. Using time series data on artisanal marine fishing in Ghana (1972–2007), this study (1) extends the existing surplus production function to account for the impact of changes in atmospheric temperature and its variance on the environmental carrying capacity of artisanal fish stock; (2) estimates the biophysical parameters employing the generalized maximum entropy (GME) estimators, which addresses the classical linear regression problems of endogeneity, multi-collinearity, and limited observations; and (3) estimates the optimum tax necessary to internalize congestion externality and the climate impact on fish yield; and forecasts the local atmospheric temperature as well as discusses its implication for the optimum tax. The results showed that the rising temperature yields negative biological response by decreasing the carrying capacity. In addition, a univariate analysis of the annual coastal temperature indicated that it will continue to rise at least in the near future. As a result, the tax rate must be set high enough to account for the increasing temperature in order to protect the artisanal fish stock.

The remainder of the paper is organized as follows. Section 2 presents the optimal control model for the optimal tax, and this is followed by incorporating the atmospheric forcing in the surplus production function in Section 3. Section 4 contains the empirical model and discussion on the estimation method. Section 5 provides the preliminary results and the final section, Section 6, concludes the paper.

THE MODEL FOR OPTIMUM TAX

To briefly outline the model for obtaining the optimum tax, following Akpalu [11], suppose a fishery is managed as a CPR. Let the biomass (x) of the fish stock grow according to a logistic function g(x,k), where k is a constant environmental carrying capacity $g_x(\cdot) > 0$ and $g_{xx}(\cdot) \leq 0$. For analytical convenience let the logistic growth function be $g(x,k) = rx\left(1 - \frac{x}{k}\right)$, where r is intrinsic growth rate. Furthermore, let c(x) and p be cost per unit harvest and price per kg of fish, respectively. In addition, assume future benefits and costs are discounted at a positive rate, δ. The value function of the entire fishery is given by Equation (1) and the stock dynamic Equation (2).

$$V(x,H) = \max_{H} \int_0^\infty \left(pH - c(x)H\right)e^{-\delta t}dt$$

(1)

$$\dot{x} = rx\left(1 - \frac{x}{k}\right) - H, \text{ with } H = \sum_{i=1}^{n} h_i$$

(2)

where $\dot{x} = \frac{dx}{dt}$, H is aggregate harvest, and h_i is the harvest of one economic agent (i). The corresponding current value Hamiltonian of the programme is

$$\Gamma(x,H,\mu) = \left(pH - c(x)H\right) + \mu\left(rx\left(1 - \frac{x}{k}\right) - H\right)$$

(3)

where μ is the scarcity value of the fish stock.

From the maximum principle, assuming an interior solution exists, the first order condition with respect to harvest (H) is

$$\frac{\partial \Gamma(.)}{\partial H} = p - c(x) - \mu = 0 \tag{4}$$

Equation (4) simply stipulates that in an inter-temporal equilibrium harvest must be at a level that equates net marginal benefit (*i.e.*, p–c(x)) to the scarcity value of the stock (*i.e.*, μ). If p–c(x)>μ, harvest has to be at its maximum. On the other hand it must be set to zero if p–c(x)<μ. The corresponding costate equation is

$$\dot{\mu} - \delta\mu = -\frac{\partial H(.)}{\partial x} = c_x H - \mu r\left(1 - \frac{2x}{k}\right) \tag{5}$$

Equation (5) implies that, in dynamic equilibrium, the interest earnable on the net marginal benefit from harvesting one kilogramme of fish today (*i.e.*, δμ) must equate the sum of the capital gain from conserving that kilogramme of fish (*i.e.*, $\dot{\mu}$) and some stock effect (*i.e.*, $-c_x H + \mu r(1 - 2xk^{-1})$). In steady state $\dot{x} = \dot{\mu} = 0$ so that Equations (4) and (5) become

$$p - c(x) = -c_x \left(rx\left(1 - \frac{x}{k}\right)\right)\left(\delta - r\left(1 - \frac{2x}{k}\right)\right)^{-1} \tag{6}$$

Now suppose the stock is harvested as a CPR by n users. Following Maler *et al.* [12] and Akpalu [11], the optimization programme for each community is

$$V(x, h_i) = \max_{h_i} \int_0^\infty (p - c(x)) h_i e^{-\delta t} dt, \quad i = 1, 2, ..., n \tag{7}$$

$$\dot{x} = rx\left(1 - \frac{x}{k}\right) - \sum_i^n h_i, \quad H = \sum_i^n h_i \tag{8}$$

The corresponding first order condition from the maximum principle is

$$\frac{\partial Z(.)}{\partial h_i} = p - c(x) - \mu_i = 0, \quad i = 1, 2, ..., n \tag{9}$$

The shadow value assigned to the resource by each symmetric community is $\mu_i = \frac{\mu}{n}$. The symmetric open-loop Nash equilibrium solution is

$$p - c(x) = \frac{\mu}{n} = -c_x \left(rx\left(1 - \frac{x}{k}\right)\right)\left(n\left(\delta - r\left(1 - \frac{2x}{k}\right)\right)\right)^{-1} \tag{10}$$

Equation (10) could be solved for the equilibrium stock level $^{(i.e.,\ x^{**}\ =\ x(k,n))}$. Suppose the resource is harvested as a CPR and let a tax be imposed on cost of harvest (*i.e.*, c(x)(1+τ)) to generate the first best solution. The equilibrium stock equation with the tax is

$$p-c(x)(1+\tau) = -c_x\left(1+\tau\right)\left[rx\left(1-\frac{x}{k}\right)\right]\left\langle n\left(\delta-r\left(1-\frac{2x}{k}\right)\right)\right\rangle^{-1}$$

(11)

or

$$n\left(p-c(x)(1+\tau)\right)(1+\tau)^{-1} = -c_x\left(rx\left(1-\frac{x}{k}\right)\right)\left(\delta-r\left(1-\frac{2x}{k}\right)\right)^{-1}$$

From Equations (6) and (11): $\frac{n(p-c(x^{*})(1+\tau))}{(1+\tau)}=p-c(x^{*})$, which implies

$$tax(\tau) = \frac{p-c\left(x^{*}\right)}{\frac{p}{(n-1)}+c\left(x^{*}\right)}$$

(12)

If n>1, aggregate catch will exceed the socially desirable level and a policy intervention will be required to regulate catch. Using $c(x^{*})=\frac{c}{qx^{*}}$ (where c and q are cost per unit effort and catchability coefficient, respectively), the steady state stock x^{*} is

$$x^{*}=\frac{k}{4}\left[\left(\frac{c}{pqk}+1-\frac{\delta}{r}\right)+\sqrt{\left(\left(\frac{c}{pqk}+1-\frac{\delta}{r}\right)^{2}+\frac{8c\delta}{pqkr}\right)}\right]$$

(13)

The specific tax expression is based on these specific functional forms. The tax depends on the values of the socio-economic parameters (*i.e.*, p, c and δ), which are readily available, and biological parameters (*i.e.*, r, q and k) which are not.

Climate Variability and Optimal Tax Rate

If climate variability impacts the carrying capacity, then the tax rate must reflect potential variability in the climate. As indicated in the introduction, there is overwhelming evidence that climate variability may impact carrying capacity of the stock. Atmospheric forcing may result in a change in atmospheric temperature (see Figure 1). The change in temperature impacts water temperature and subsequently influences seasonal upwelling (or downwelling). This influences primary production, species distribution, fish yield, and increased variability of catches (4).

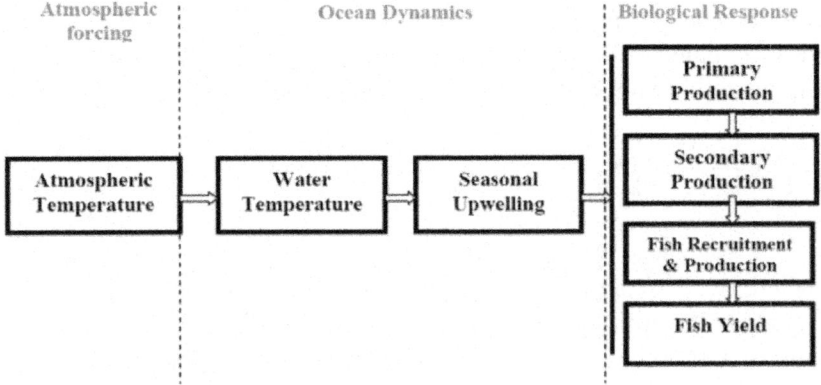

Figure 1: The flow chart of the impact of atmospheric temperature on fish yield. Source: own compilation.

To account for the impact of atmospheric temperature on fish production, we surmise that the carrying capacity is defined as

$$k = \frac{k_0}{1 + \varepsilon \Delta T_t + \eta \sigma_t}$$

(14)

where T and σ are the state of the climate variable and its variances, respectively; Δ is notation for first difference (We used the change in temperature because the levels of the series are within a limited range, given the period considered for the empirical analysis. As expected, the levels were not significant in the empirical analysis (presented in the later section of the paper) but the first differences were.); and k_0, ε, and η are constants. The corresponding optimum path of the stock is

$$x^*(t) = \frac{k_0}{4(1 + \varepsilon \Delta T_t + \eta \sigma_t)} \left[\left(\frac{c(1 + \varepsilon \Delta T_t + \eta \sigma_t)}{pqk_0} + 1 - \frac{\delta}{r} \right) + \sqrt{ \left(\frac{c(1 + \varepsilon \Delta T_t + \eta \sigma_t)}{pqk_0} + 1 - \frac{\delta}{r} \right)^2 + \frac{8c\delta(1 + \varepsilon \Delta T_t + \eta \sigma_t)}{pqk_0 r} } \right]$$

(15)

OBTAINING THE BIOPHYSICAL PARAMETERS

To establish the link between climate variability and fish production, a biological model is employed. In order to estimate the biological parameters, a number of authors have employed models by Schaefer and Fox [13]. These models assume equilibrium or steady state conditions in order to obtain an equation that is used to estimate next period's catch per unit effort without specifying future anticipated effort [13]. However, Schnute [14] has shown that these models may be invalid for non-equilibrium conditions and the assumption that catch per unit effort could be predicted without specifying future anticipated effort contradicts almost all theory on fisheries biology. As a result, the author suggested a modified version, which is Equation (16)

$$\ln\left(\frac{U_{t+1}}{U_t}\right) = r - \frac{r}{qk_0}\left(\frac{U_t + U_{t+1}}{2}\right) - q\left(\frac{E_t + E_{t+1}}{2}\right)$$

(16)

where $U_t = \left(\frac{h_t}{E_t}\right)$ signifies catch per unit effort, and Et is fishing effort. As indicated in the introduction, there is overwhelming evidence that climate variability may impact carrying capacity of fish stock. Using Equation (14) in Equation (16), gives

$$\ln\left(\frac{U_{t+1}}{U_t}\right) = r - \frac{r}{qk_0}\left(\frac{U_t + U_{t+1}}{2}\right) + \frac{r\varepsilon}{qk_0}\left(\frac{U_t + U_{t+1}}{2}\right)\Delta T_t + \frac{r\eta}{qk_0}\left(\frac{U_t + U_{t+1}}{2}\right)\sigma_t - q\left(\frac{E_t + E_{t+1}}{2}\right)$$

(17)

THE EMPIRICAL MODEL AND ESTIMATION METHODS

In this section, the extended Schnute model (*i.e.*, Equation (17)) and an estimation method known as a GME estimator is presented. For the purpose of estimation, Equation (17) is specified as

$$\ln\left(\frac{U_{t+1}}{U_t}\right) = a_0 + a_1\bar{U}_t + a_2\bar{U}_t\Delta T_t + a_3\bar{U}_t\sigma_t + a_4\bar{E}_t + \mu_t$$

(18)

where $\bar{U}_t = \left(\frac{U_t + U_{t+1}}{2}\right)$, $\bar{E}_t = \left(\frac{E_t + E_{t+1}}{2}\right)$, $a_0 = r > 0$, $a_1 = -\frac{r}{\sigma k} < 0$, $a_4 = -q < 0$, and μ_t is an error term. Time series data on catch, fishing effort, and temperature is required to estimate Equation (18).

Empirical Estimations: Generalized Maximum Entropy

The obvious problem with applying ordinary least squares estimation procedure to Equation (18) is endogeneity since the dependent variable interacts with other variables on the right hand side. It is also very likely that some of the variables are highly correlated. As a result, the coefficients are estimated using GME, which are explained in the subsequent sections. The GME method could generate reliable estimates of the parameters of our model. The GME is a semiparametric estimator and belongs to a class of estimators used in engineering and physics. To present the GME estimator, let

$$a_k = \sum_s z_{ks} p_{ks}$$

(19)

where $p_{ks} \geq 0$ are unknown probabilities and $\sum_s p_{ks} = 1$; z_{ks} constitutes a predetermined discrete support space (s) for the parameters; and a_k is as defined in Equation (18). Furthermore, define the error term in Equation (18) as

$$u_i = \sum_g V_{ig} w_{ig}$$

(20)

where $w_{ig} \geq 0$ are unknown probabilities and $\sum_g w_{ig} = 1$; V_{ig} constitutes an a priori discrete support space (g) for the errors; and u_i is as defined in Equation (18). The GME estimator is specified as

$$\max H(p_{ks}, w_{ig}) = -\sum_s p_{ks} \ln(p_{ks}) - \sum_g w_{ig} \ln(w_{ig})$$

(21)

subject to Equation (18), but with the coefficients and the error term substituted by Equations (19) and (20). The limitation of this method is that the values of the parameters are sensitive to arbitrarily chosen support values making policy recommendations sensitive to such values. The estimations are implemented in general algebraic modeling system (GAMS).

Data Types and Sources

Data on catch and effort were collected from the Directorate of Fisheries of the Ministry of Food and Agriculture (MOFA) in Ghana. The Directorate is mandated to carry out research for the assessment for fisheries resources. As noted in the introduction, the artisanal fishery sector is one of the most important sectors within the economy. However, recent landing statistics for the artisanal fleet indicate landings peaked in 1992, and then declined due to overexploitation [15]. The data on temperature was collected from Ghana Meteorological Agency. The agency has 17 weather stations each reporting monthly average temperatures. The figure for the standard deviation of temperature is computed from the 12-month averages for each year. The summary statistics of the data is presented in Table 1.

From Table 1, both catch and effort levels have increased over the last 18 years. In addition, as depicted in Figure 2, catch per unit effort oscilates over time. The periods of decline in the CPUE were 1992 to 1995, 1997 to 2000, and 2002 through 2007. On the other hand, the CPUE increased from 1994 to 1997, 2000 to 2002, and 2007 through 2008. On the average, however, the variable shows a weak pattern of decline from 1992 through 2008.

Table 1: Descriptive Statistics of catch, fishing trips and coastal temperature in Ghana

Variable	1972–2008		1990–2008	
	Mean	**Standard dev.**	**Mean**	**Standard dev.**
Catch (in kg)	195,354	53,378	229,212	34,192
Effort (Number of trips)	1,319,614	1,549,476	1,653,440	1,988,563
Temperature (in °C)	27.0	0.419	27.3	0.291
Std. Dev. of temperature	1.193	0.170	1.193	0.124

Source: Catch and effort data is collected from Directorate of Fisheries, Ghana; and temperature data is collected from Ghana Meteorological Agency.

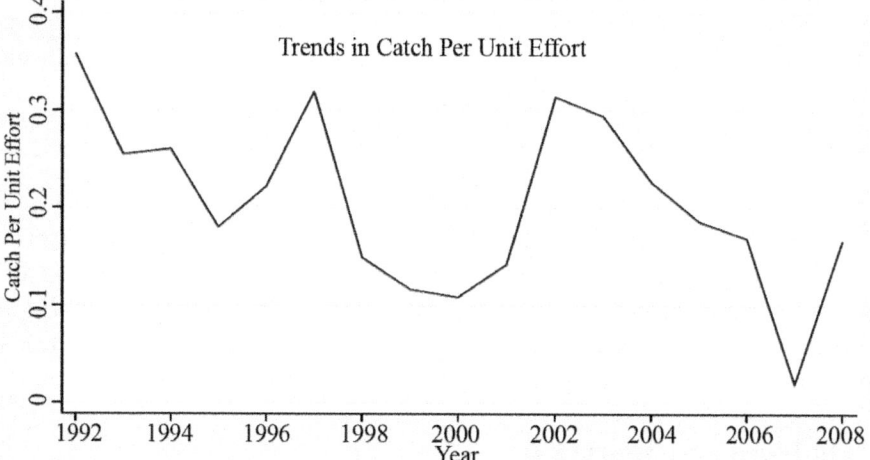

Figure 2: Trends in catch per unit effort of artisanal stocks (1992–2008); Source: own illustration.

Furthermore, from Table 1, the mean temperature within the entire period (1972–2007) is lower than that of the last 18 years (1990–2007). The time trend of the atmospheric coastal temperature has revealed a rising trend over time (see Figure 3). Moreover, although the annual variance of the coastal temperature (shown in Figure 4a) (The variance is calculated as the sum of the squared deviation of monthly temperature from the yearly average, divided by 11 (*i.e.*, degrees of freedom).) looks like a stationary process, a careful examination of a segment of the data (from 1990 through 2008) reveals an upward trend with a gentle slope (shown in Figure 4b). The period of 1990 through 2008 witnessed a decline in the catch per unit effort.

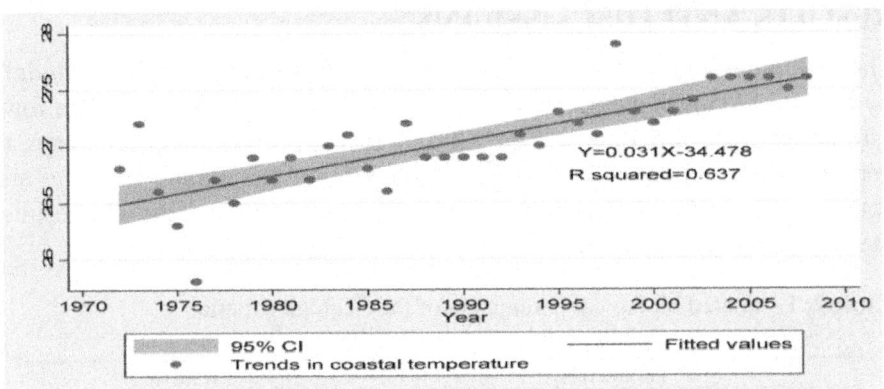

Figure 3: Trends in coastal temperature (1972–2008); Source: own illustration.

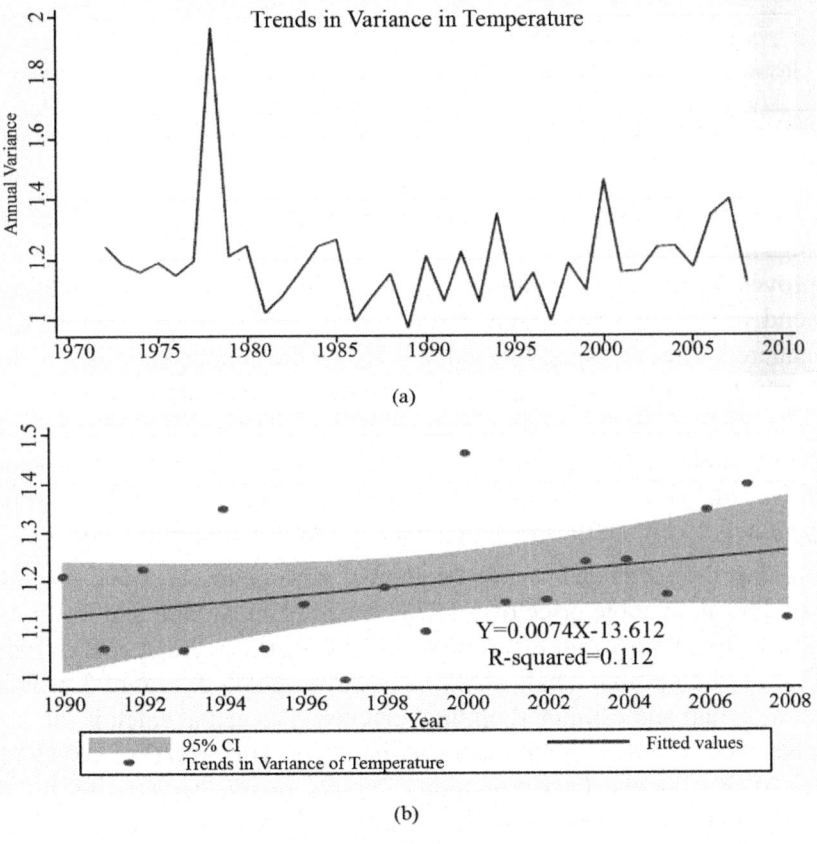

Figure 4: (a) Trends in the variance of coastal temperature (1972–2008); (b) Trends in the variance of coastal temperature (1990–2008); Source: own illustration.

RESULTS AND DISCUSSIONS

The biological parameters were estimated using the General Algebraic Modeling System (GAMS). Since the catch intensified beginning 1990, the data for the estimation spans a period of 1990–2007. For the purpose of comparison, two versions of the model were estimated: one without the climate variables and a complete version with the climate impact on carrying capacity. The results of the estimation are reported in Table 2.

Table 2: Estimated biological parameters of the Schnute equation

Parameters		Estimates	
		GME	
Description	Notation	1	2
Intrinsic growth rate	r	1.91369	1.960
Catchability coefficient *	q	0.627×10^{-6}	0.636×10^{-6}
Carrying capacity (in kg)	k_0	530,066	449,683
Impact of temperature on k	ε		0.166244
Impact of temperature variation on k	η		0.115561
Pseudo R-squared		0.70	0.73

Source: own computations. * Using surplus production functions, Clarke *et al.* [13] estimated the catchability coefficient to be within the range of 0.376×10^{-6} and 0.913×10^{-6}.

The pseudo R-squared indicates that including climate variables (*i.e.*, the change in temperature and the annual variance of the temperature) in the model improves the fit of the estimation. Approximately 73% of the variability in the dependent variables is explained by the regressors if the climate variables are considered. The corresponding value is 70% if the climate variables are ignored. The environmental carrying capacity (k0) indicates that, without accounting for climate impact, the maximum stock the environment could accommodate is approximately 450 tons. Furthermore, the values of the parameters ε and η are positive implying change in temperature and annual variance of temperature impact negatively on the carrying capacity (as per Equation (14)).

Using the estimates for the biological parameters, a social discount rate (δ) of 3%, an average price of US$264 taken from Akpalu and Vondolia [16], and price to cost per unit effort ratio of 0.11 (or the average cost of harvest of US$232), the optimal catch series has been calculated. Figure 5 provides the plots of actual and estimated optimal catches. The actual catch is the observed catch data while the optimal catch is based on $H^*(t)=rx(t)^*(1-x(t)^*k^{-1})$. Note that $x(t)^*$ is obtained from Equation (15). As clearly depicted by the graphs, the actual catches are much higher than the optimal values (since the climate variable is accounted for) indicating a policy instrument is necessary to regulate catch. Thus, ignoring the climate impact may result in overestimation of the stock level as depicted in Figure 6. The estimated stock level that ignores the climate variable is Equation (13), while the same that account for the impact is Equation (15).

The optimal tax path based on Equation (12) has also been calculated. Based on the values for price and cost per unit effort used, the values range from 8.5% to 21%, with the mean tax being 14.2%. The implication is that for harvest levels to mimic the desired or optimal trajectory in Figure 5, the tax rate on cost of harvest must follow the series depicted in Figure 7. Note that the tax evolves over time. Currently premix fuel, which constitutes a significant input in production, is subsidized at an approximate rate of 18%. Withdrawing a portion of the subsidy corresponding to the tax is necessary to lower catches to sustainable levels. There is a large amount of literature advocating for the withdrawal of input subsidies to save fisheries in both developed and developming countries (see e.g., [17,18,19]). Furthermore, the figure shows a direct relationship between the tax rate and the change in temperature. This makes sense because the carrying capacity decreases as the change in temperature increases leading to lower fish production. As a result, the tax rate must increase to regulate harvest.

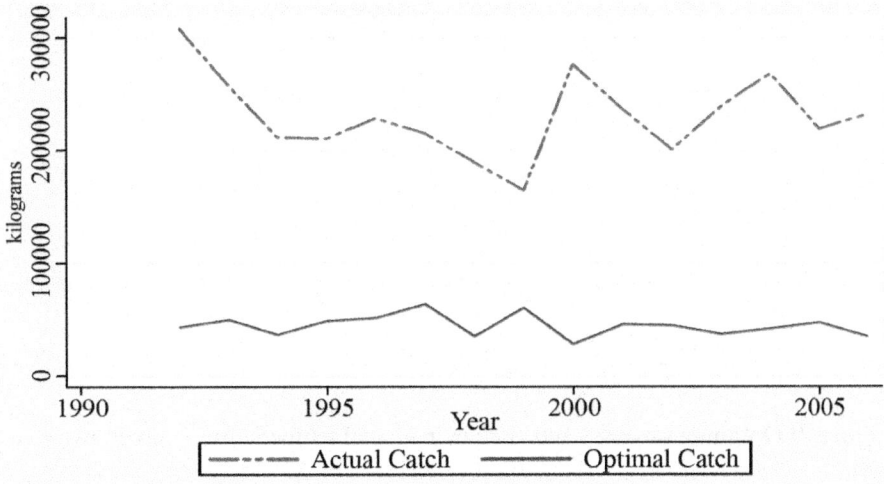

Figure 5: Actual and optimal catches of artisanal stocks in Ghana; Source: own illustration.

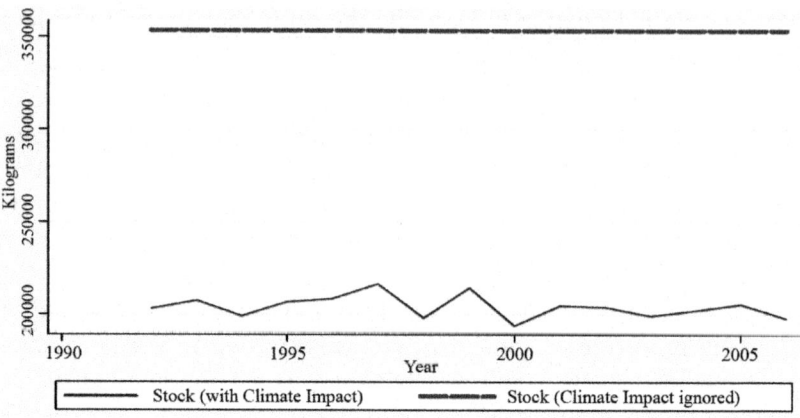

Figure 6: Misperceived stock due to ignorance of climate impact in Ghana. Source: own illustration.

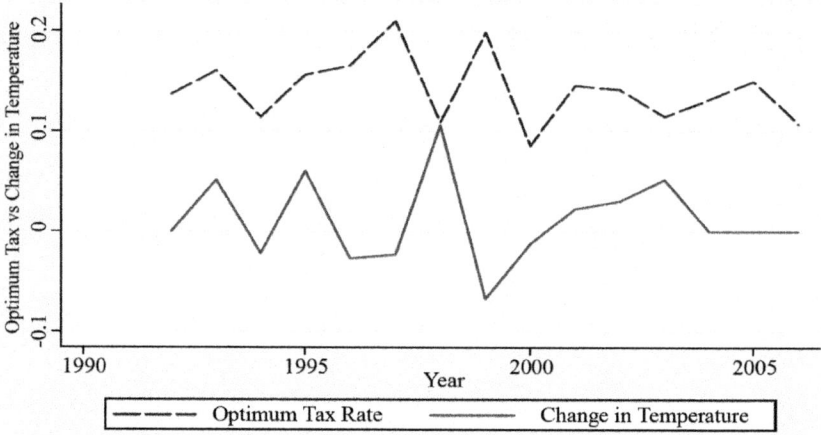

Figure 7: Optimum tax rates and change in annual temperature. Source: own illustration.

Predicting the Coastal Temperature

In the preceding section, it has been shown that if the coastal temperature increases or its annual variance increases, the environmental carrying capacity will decrease causing fish production to decline. As a result, we proceeded to investigate whether or not the local coastal temperature and the variance will rise or fall in the near future based on the historical trends of the series. To forecast the future values, the time series properties of the data were investigated. Table 3 contains the results of the augmented Dickey-Fuller

(ADF) tests. The statistical software STATA 12 was used for the analysis. The results indicate that if trends and constants are included in the tests, the temperature series is stationary, but its annual variance is non-stationary at a 1% and 5% significance level. The first difference of the variance is, however, stationary implying the temperature and variance are integrated of order zero and one respectively.

Table 3: Unit root analysis of annual temperature and variance of annual temperature

Series	ADF (with Drift Term and Trend)			
	Z-Score	Critical Values		
		1%	5%	10%
Temperature (Temp)	−5.455	−4.297	−3.564	−3.218
Variance of temperature (Vtemp)	−3.478	−4.297	−3.564	−3.218
First difference of Vtemp (DVtemp)	−5.455	−4.306	−3.568	−3.221

Source: own computations.

Following the Box-Jenkings approach to univariate time series econometric modeling, the plots of the autocorrelation and partial autocorrelation functions depict that the temperature series follow an autoregressive moving average (ARMA) process. A further analysis reveals that the variable could be modeled as ARMA (1, 10) process. The estimated results are presented in Table 4. The Wald Chi-square test indicates that the line is a good fit at a 1% significance level. The coefficients of the first lag of the series, and the first and tenth lags of the error term are all significant at a 1% level. In addition the drift term, denoting the average temperature, is 27.15 °C and it is also significant at a 1% level.

Table 4: Fitting temperature with autoregressive moving average (ARMA)

Variables	Coefficient
$Temp_{t-1}$	0.90
	(0.048) ***
e_{t-1}	−0.56
	(0.18) ***
e_{t-10}	0.55
	(0.21) ***
Constant	27.15
	(0.28) ***
Wald chi2(2)	345.53 (Prob > 0.00)

Note: Standard errors are in parentheses; *** significant at 1%; Source: own computations.

Based on the results of the univariate analysis, the values of the temperature are forecasted and the forecast and actual values are presented in Figure 8.

From the figure, it is evident that the annual temperature will continue to rise in the near future. This also implies that the artisanal stock is likely to decline; hence higher taxes on cost of harvest may be necessary to protect the stock.

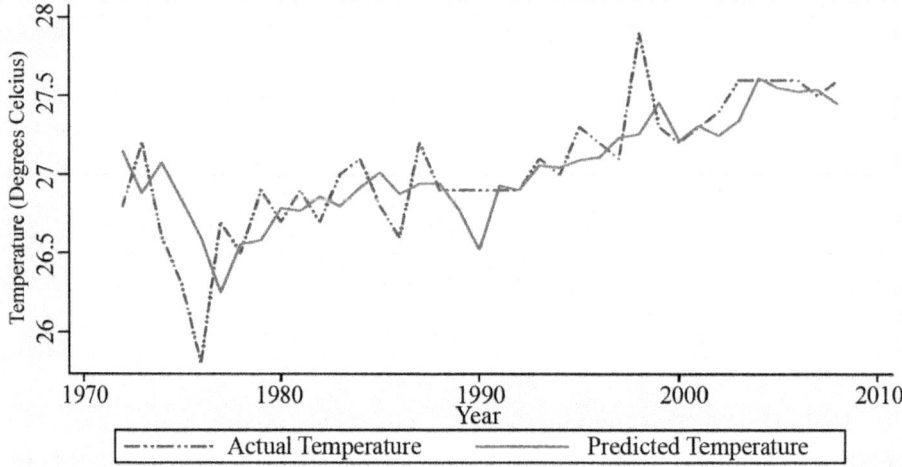

Figure 8: Actual and predicted values of atmospheric temperature. Source: own illustration.

Finally, the time path of the variance of the annual temperature is modelled. The corellogram of the first difference of the variance indicates it is an autoregressive (AR) (1) process without a drift term (see Table 5). The Wald Chi-square test indicates that the line is a good fit at a 99% confidence level. The coefficient of the AR (1) term is negative indicating the first difference of the temperature is declining over time, with a marginal effect of −0.055. This also implies that the variance of annual temperature rises but at a decreasing rate. The plot of the actual and predicted values of the series in Figure 9 shows that the estimated model predicts the actual values quite well.

Table 5: Fitting change in variance of annual temperature with ARMA (1, 0)

Variables	Coefficient
$\Delta VTemp_{t-1}$	−0.55
	(0.19) ***
Constant	−0.000098
	(0.0223)
Wald chi2(2)	7.93 (Prob. > 0.00)

Note: Standard errors are in parentheses; *** significant at 1%; Source: own computations.

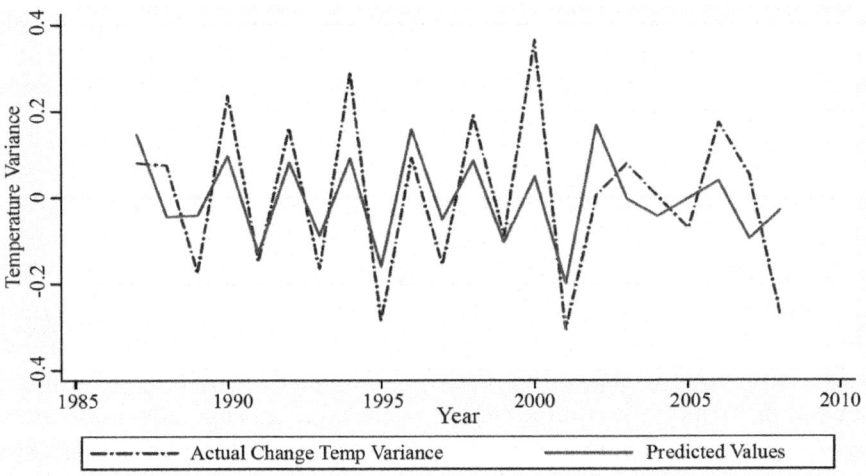

Figure 9: Actual and predicted values of change in variance of annual temperature. Source: own illustration.

CONCLUSIONS

Catch per unit effort of most artisanal fish stocks have declined over the past two decades due to overcapitalization of such stocks. The state of the artisanal fishery in Ghana typifies such occurrence. With the increasing poverty trap, coupled with a high unemployment rate within the coastal regions where off-fishing economic activities hardly exist, fishing is a livelihood of last resort. Indeed, the overfishing problem is expected to worsen.

In addition to human activities, it has been found that the coastal climate is getting warmer with potential consequences for capture fisheries. If the warmer climate increases seasonal upwelling and thereby increases primary food production, it will be good for the fishery. On the other hand, if the warmer climate, for example, bleaches corals and rather reduces the food production capability of the aquatic system, the environmental carrying capacity and fish production will decline. In this study, evidence has been found in support of the later case. A dynamic model of the common pool resources management problem in fisheries has been derived and an optimum tax necessary to internalize the congestion externality as well as account for the changing coastal temperature has been proposed. Using data on artisanal fisheries in Ghana, and selected values for price of fish and cost per unit effort, the tax rate is calculated to be within the range of 8 and 21% on cost per unit harvest. Since premix fuel, which is an important input in catch, is subsidized at 18% of ex-refinery price, withdrawing the subsidy could improve the sustainability of fishery. Moreover, the tax must positively correlate with the rising rate of change in temperature as well as the annual variance of the temperature. It is important to note that these results relate to species of low trophic levels and similar research is required for species of higher trophic levels.

It is noteworthy that our study suffers some data limitations. Sea surface temperature was proxied by atmospheric temperature, balanced data on the relevant variables ended at 2008, and data on catch and fishing effort obtained from the fiesheries directorate are based on some approximations akin to fisheries data elsewhere. Regarding the climate data, a study in Ghana (not yet published) found a strong correlation between atmospheric temperature and sea surface temperature, suggesting that the findings of this study are somewhat robust. Furthermore, it is impossible to determine *a priori* how an increase in the time series data could alter the results. This empirical concern can only be adequately addressed as and when additionl data is available. Finally, the cost associated with collecting fisheries data (on fishing effort and catch) convering the entire population of fishers in Ghana is prohitive and precausions are taken to ensure that samples drawn are representative (This is as per communication with an official of the fisheries directorate).

ACKNOWLEGMENTS

The authors are indebted to Channing Arndt, Rob B. Dellink and Aziz Karimov for their invaluable comments. We would also like to express our profound gratitude to the two anonymous reviewers of the journal for their thoughtful comments.

AUTHOR CONTRIBUTIONS

Wisdom Akpalu formulated the research problem and constructed the theoretical model. The three authors contributed equally to the data analysis and writing up of the report. All authors have read and approved the final manuscript.

REFERENCES

1.   Food and Agriculture Organization (FAO). *The State of World Fisheries and Aquaculture 2006*; FAO: Rome, Italy, 2007.

2.   Organization for Economic Co-operation and Development (OECD). *Why Fish Piracy Persists: The Economics of Illegal, Unreported and Unregulated Fishing*; OECD: Paris, France, 2006.

3.   Arnason, R.; Killeher, K.; Willman, R. *The Sunken Billions: The Economic Justification for Fisheries Reform*; The World Bank: Washington, DC, USA; FAO: Rome, Italy, 2009.

4.   Pauly, D.; Zeller, D. The Global Fisheries Crises as A Rationale for Improving the FAO's Database of Fisheries Statistics. *Fish. Centre Rep.* **2003**, *11*, 1–9.

5.   Chimatiro, S. Post-Compact Interventions through the International Partnership for African Fisheries Governance and Trade (PAF). 2010. unpublished work.

6.   Bannerman, P.O.; Koranteng, K.A.; Yeboah, C.A. Ghana Canoe Frame Survey, 2000. In *Information Report No. 33*; Fisheries Department: Research and Utilization Branch: Tema, Ghana, 2001.

7.   Allison, E.H.; Adger, W.N.; Badjeck, M.C.; Brown, K.; Conway, D.; Dulvy, N.K.; Halls, A.; Perry, A.; Reynolds, J.D.*Effects of Climate Change on the Sustainability of Capture and Enhancement Fisheries Important to the Poor: Analysis of the Vulnerability and Adaptability of Fisher Folk Living in Poverty, Project R4778J. Final Technical Report*; Fisheries Management Science Programme MRAG/DFID: London, UK, 2005.

8.   Bakun, A. Global Climate Change and Intensification of Coastal Ocean Upwelling. *Science* **1990**, *247*, 198–201.

9.   Akpalu, W. Economics of Biodiversity and Sustainable Fisheries Management. *Ecol. Econ.* **2009**, *68*, 2729–2733.

10. Sterner, T. Unobserved Diversity, Depletion and Irreversibility: The Importance of Subpopulations for Management of Cod Stocks. *Ecol. Econ.* **2007**, *16*, 566–574.

11. Akpalu, W. Bioeconomics of Fisheries Management under Common Pool and Territorial Use Rights Regimes. 2013. unpublished work.

12. Maler, K.G.; Xepapadeas, A.; de Zeeuw, A. The Economics of Shallow Lakes. *Environ. Resour. Econ.* **2003**, *26*, 603–624.

13. Clarke, R.P.; Yashimoto, S.S.; Pooley, S.G. A Bioeconomic Analysis of the North-Western Hawaiian Islands Lobster Fishery. *Mar. Resour. Econ.* **1992**, *7*, 115–140.

14. Schnute, J. Improved Estimates from the Schaefer Production Model: Theoretical Considerations. *J. Fish. Res. Board Can.* **1977**, *34*, 583–603.

15. Koranteng, K.A. The Impacts of Environmental Forcing on the Dynamics of Demersal Fishery Resources of Ghana. Ph.D. Thesis, University of Warwick, Warwick, UK, 1998.

16. Akpalu, W.; Vondolia, G. Bioeconomic Model of Spatial Fishery Management Conflicts in Exclusive Economic Zones of Developing Countries. *Environ. Dev. Econ.* **2012**, *17*, 145–161.

17. Beddington, J.R.; Rettig, R.B. *Approaches to the Regulation of Fishing Effort*; Fisheries Technical Paper No. 243; FAO: Rome, Italy, 1983.

18. McGoodwin, J.R. *Crisis in the World's Fisheries*; Stanford University Press: Stanford, CA, USA, 1990.

19. Sumaila, U.; Teh, L.; Watson, R.; Tyedmers, P.; Pauly, D. Fuel subsidies to fisheries globally: Magnitude and impacts on resource sustainability. In *Catching More Bait: A Bottom-Up Re-Estimation of Global Fisheries Subsidies. Fisheries Centre Research Reports*; Sumaila, U.R., Pauly, D., Eds.; Fisheries Centre, The University of British Columbia: Vancouver, BC, Canada, 2006; Volume 14, pp. 35–45.

Chapter 13

EFFECTS OF OCEAN ACIDIFICATION ON TEMPERATE COASTAL MARINE ECOSYSTEMS AND FISHERIES IN THE NORTHEAST PACIFIC

Rowan Haigh[1], Debby Ianson[2], Carrie A. Holt[1], Holly E. Neate[1,3],and Andrew M. Edwards[1,3]

[1]Pacific Biological Station, Fisheries and Oceans Canada, 3190 Hammond Bay Road, Nanaimo, British Columbia, V9T 6N7, Canada

[2]Institute of Ocean Sciences, Fisheries and Oceans Canada, 9860 West Saanich Road, Sidney, British Columbia, V8L 4B2, Canada

[3]Department of Biology, University of Victoria, Station CSC, Victoria, British Columbia, V8W 2Y2, Canada

ABSTRACT

As the oceans absorb anthropogenic CO_2 they become more acidic, a problem termed *ocean acidification* (OA). Since this increase in CO_2 is occurring rapidly, OA may have profound implications for marine ecosystems. In the temperate northeast Pacific, fisheries play key economic and cultural roles and provide significant employment, especially in rural areas. In British Columbia (BC), sport (recreational) fishing generates more income than commercial fishing (including the expanding aquaculture industry). Salmon (fished recreationally and farmed) and Pacific Halibut are responsible for the majority of fishery-related income. This region naturally has relatively acidic (low pH) waters due to ocean circulation, and so may be particularly vulnerable to OA. We have analyzed available data to provide a current description of the marine ecosystem, focusing on vertical distributions of commercially harvested groups in BC in the context of local carbon and pH conditions. We then evaluated the potential impact of OA on this temperate marine system using currently available studies. Our results highlight significant knowledge gaps. Above trophic levels 2–3 (where most local fishery-income is generated), little is known about the direct impact of OA, and more importantly about the combined impact of multi-stressors, like temperature, that are also changing as our climate changes. There is evidence that OA may have indirect negative impacts on finfish through changes at lower trophic levels and in habitats. In particular, OA may lead to increased fish-killing algal blooms that can affect the lucrative salmon aquaculture industry. On the other hand, some species of locally farmed shellfish have been well-studied and exhibit significant negative direct impacts associated with OA, especially at the larval stage. We summarize the direct and indirect impacts of OA on all groups of marine organisms in this region and provide conclusions, ordered by immediacy and certainty.

INTRODUCTION

Fossil fuel burning and changes in land use by humankind have increased atmospheric carbon dioxide (CO_2) at an unprecedented rate, causing our climate to change [1]. A significant portion of this anthropogenic CO_2 (~30%; [2]) has been absorbed by the ocean. When CO_2 enters the ocean it combines with water (H_2O), resulting in an increase in the concentration of hydrogen ions [H^+] and an increase in acidity (decrease in pH [3, 4]. Therefore, as our climate changes, our oceans become more acidic due to anthropogenic contributions, a problem termed Ocean Acidification (OA) [5].

While anthropogenic atmospheric CO_2 dominates contributions to OA on a global scale, other anthropogenic sources may be significant on a local scale [6]. For example, acid rain from vehicle emissions and industry cause an increase in ocean acidity, which is likely relevant, at least near (and downwind of) urbanized regions [7]. Any addition of organic carbon to the ocean, such as sewage, decomposes to dissolved inorganic carbon (DIC), and increases acidity. Agricultural run-off provides nutrients which then fuel (an anthropogenic) increase in production of organic carbon in the ocean [8], again increasing acidity.

Aquatic acidity is most commonly reported as pH. However, pH is difficult to determine accurately in saltwater because of the additional ions present in solution [9]. It is closely linked with carbonate chemistry in the ocean, which is complex. To quantify the *carbon state* (*i.e.* the concentration of each chemical form of DIC present) in seawater, two of four measured parameters—DIC, pH, total alkalinity (TA), and partial pressure of CO_2 (P_{CO2})—must be known, in addition to temperature and salinity. To be more accurate, phosphate and silicic acid concentrations are also required [10]. In the past, pH has most often been determined from DIC and TA (*e.g.* [11]). (TA is the acid neutralizing capacity of the solution, which is not simply related to pH in seawater [10].) Thus, although one can generalize to say that high DIC is usually associated with low pH (or high P_{CO2}), more information, *e.g.* TA, is required to be quantitative.

The carbon state is relevant to biology. Most of the DIC in the ocean occurs in the form of bicarbonate (HCO−3) and carbonate (CO2−3), with less than 1% in the form of CO_2. When pH decreases, the balance between HCO−3 and CO2−3 changes so that there is less CO2−3. This shift has important implications for plants and animals that build calcium carbonate ($CaCO_3$ structures (e.g. shellfish, corals) [12]. Two mineral forms of $CaCO_3$ (aragonite and calcite) are common in biological structures. The aragonitic form is more soluble than calcite given the same environmental conditions [13]; therefore, creatures that use aragonite are more susceptible to OA than those that use calcite [12]. The ease with which these minerals are formed is quantified by the saturation state (Ω), such that as Ω decreases, dissolution increases [14]. The water is *undersaturated* with respect to $CaCO_3$ when the chemical rate of dissolution exceeds the rate of formation [15]. For organisms that precipitate $CaCO_3$, decreasing Ω means that more energy is required to build and maintain their carbonate structures [16, 17].

Marine organisms are also affected by carbon state (defined above) and OA in other ways. All marine animals need to rid themselves of metabolically produced CO_2 through respiration. The effectiveness of this removal is dependent, in part, on the ambient P_{CO2} of the medium (e.g.[18]). Similarly, plants and animals rely on pH to regulate ion transport, and the energy they must expend to maintain intra- and extracellular pH depends on ambient pH (e.g. [19]). Thus, there is no one carbon parameter that best indicates OA impacts on all marine organisms, and so full knowledge of the complete carbon state is desirable (e.g. [20]).

A large and growing number of studies have been undertaken regarding OA. To understand and predict biological impacts, an increasing number of experiments have been completed that attempt to emulate future ocean conditions in the laboratory. Experimental conditions are usually defined by controlling either the P_{CO2} or the pH and recently an internationally accepted guide has been published that describes the techniques used [21]. In most of these experiments, present-day conditions (the control) are set at either atmospheric P_{CO2} (~400 μatm at the time of writing) or the estimated current global average pH of the surface ocean, which is 8.1 [5]. However, marine organisms in the natural environment may experience values that are significantly different depending on location and the depth that they occupy.

In the ocean, DIC (and P_{CO2}) generally increase with depth while pH decreases. In other words, low pH conditions naturally occur at depth. This partitioning of inorganic carbon towards deeper parts of the ocean is due in large part to the 'biological pump' that allows the ocean to hold more carbon [22]. Photosynthesis in the surface draws down DIC (which increases pH) and produces organic forms of carbon. Some of this organic carbon falls to deeper levels, where it decays back to DIC (decreasing pH).

British Columbia—oceanography

British Columbia (BC) makes up 27,000 km (17,000 mi) of the temperate northeast Pacific coastline. Circulation along this coast (Fig. 1) is dynamic so that large changes in carbon parameters occur both in space (*e.g.* [23]) and time (*e.g.* [24, 25]). Coastal upwelling along the west coast of Vancouver Island (WCVI) [26] brings subsurface water high in DIC into the surface mixed layer [27] so that low pH (*e.g.* 7.6) is found at relatively shallow depths, *e.g.* above 125 m (Fig. 2). Furthermore, these subsurface waters are enriched in DIC relative to waters at the same depth in other ocean basins, simply because north Pacific water is relatively 'old' and has had more time to receive organic matter [28, 29]. Upwelled waters are also rich in nutrients that are limiting to phytoplankton growth and so cause high primary production that increases pH at times. In fact, the WCVI enjoys the highest productivity of any zone on the northeast Pacific coast [30]. Consequently, present-day ranges in pH in the surface mixed layer along the outer BC coast span a remarkable range (7.8–8.4; Fig. 2). The low end of this range is significantly lower than the benchmark of present-day average global surface ocean pH (8.1).

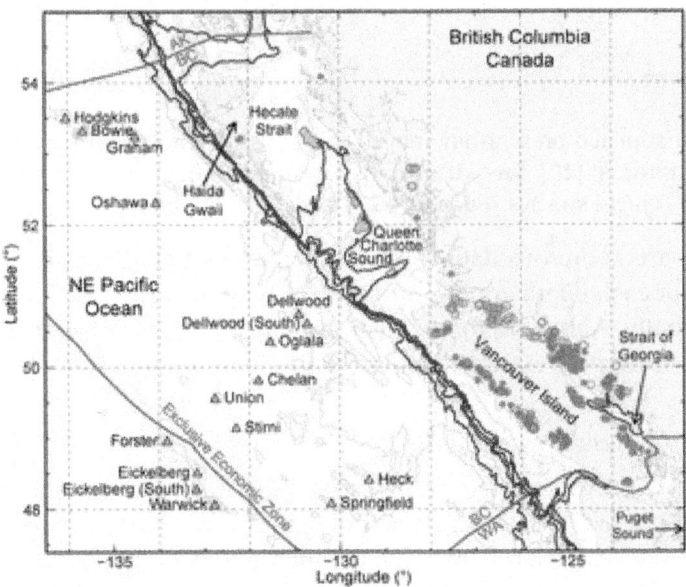

Figure 1: British Columbia (BC) coastline and bathymetry (isobaths in metres: thin grey—100, 200, 300, …, 1000, 1250, 1500, 2000, 2500; thick blue—200, 500, 800, and 1600).

The continental slope along most of BC comprises steep slopes, especially along the west coasts of Haida Gwaii and northern Vancouver Island. Hecate Strait is largely dominated by shallow waters and a flat seafloor. Sponge reef core protected areas in Hecate Strait and Queen Charlotte Sound are shaded pink. The Strait of Georgia forms a large inland sea that is heavily influenced by river runoff and tidal currents. Saltwater finfish farm and hatchery sites are indicated by open red circles, commercial marine shellfish farms are indicated by solid green circles [345]. Select seamounts [346] are marked by blue triangles. Canada's Exclusive Economic Zone (200-nautical miles offshore) is delimited in red. Map was prepared using PBSmapping in R [347].

doi:10.1371/journal.pone.0117533.g001

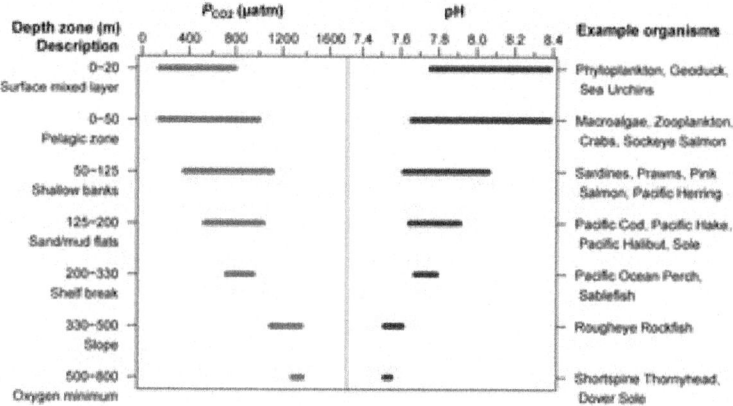

Figure 2: Estimated present-day ranges of P_{CO_2} (red) and pH (blue) during spring [40] and summer [27] for various depth zones along the outer BC continental shelf, with typical species found in each zone (see Methods).

There are numerous data above 50m and few below 125 m. The number of values in each depth zone from top to bottom are: 70, 116, 33, 45, 5, 4 and 2, respectively. Above 50 m, the distributions of values are skewed, such that high P_{CO_2} (low pH) extremes occur less often than the low P_{CO_2} (high pH) extremes.

doi:10.1371/journal.pone.0117533.g002

In protected waters (*e.g.* Strait of Georgia, Fig. 1) less data are available relative to the WCVI. These data show similar (or larger) ranges in surface pH and P_{CO_2} (unpublished data, DI), which are also similar to values found just to the south in the protected waters of Puget Sound, Washington State (WA) [6, 31]. Again, a critical feature in these waterways is low surface pH (high P_{CO_2}) relative to global averages, especially during the winter season [32].

British Columbia—Fishery

Fisheries and aquaculture play an important role in the BC economy, contributing over $650 million (we quote all dollar values in Canadian dollars) to the provincial gross domestic product (GDP) in 2011 [33]. Sport (or recreational) fishing, mainly for salmon and Pacific Halibut, is responsible for approximately 50% of this contribution, while the wild (or capture) fishery makes up ~15% and aquaculture ~10%. Marine ecosystems also play critical cultural roles in BC and their monetary value to tourism is only partially included in these totals (through sport fishing).

Over the past 20 years the wild fishery has declined in terms of both its contribution to the BC GDP and employment, although some individual components are increasing (*e.g.* prawns, Geoduck Clam, Pacific Halibut). Meanwhile aquaculture has nearly tripled its contribution to BC GDP in the same time frame [33]. As a result, published landed values associated with aquaculture are about the same as those from the wild fishery (see Results) and aquaculture now employs slightly more people than does the wild fishery [33].

The wild fishery is for the most part associated with the open coast (outer WCVI and Queen Charlotte Sound, Fig. 1) and is relatively diverse, with no one fishery dominating landed values (see Results). The most important contributors (Pacific Halibut, Geoduck Clams, prawns, crabs, tunas, Sablefish, rockfishes) currently each have landed values in the $20–50 million range [34]. Aquaculture occurs in protected waters: shellfish farming mainly in the northern Strait of Georgia and finfish farms and hatcheries mainly north of that on the northeastern side of Vancouver Island (Fig. 1). In BC, Atlantic Salmon aquaculture clearly dominates all other commercial fisheries (see Results).

Predicting biological impacts due to OA is a highly complex problem that has only become a concern relatively recently (primarily over the past decade). There have been excellent review papers outlining anticipated impacts on a general global scale (*e.g.* [3, 35]) as well as meta-analyses of existing work on the topic (*e.g.* [36]). Cooley and Doney [37] have provided the first estimate of the economic impact of OA, centred on the shellfish fishery, in the United States. However, few studies consider specific ecosystems, particularly in the context of local pH conditions and natural variability, and none focus on the temperate northeast Pacific. Here, we examine the potential impact of OA on temperate coastal ecosystems in the northeast Pacific Ocean, with a focus on BC fisheries. To tackle this issue we:

- describe the current marine ecosystem in BC (especially by depth, Fig. 3);
- define the present-day carbon state with depth in local waters (Fig. 2);

- assess the response by marine organisms in this region to OA by investigating existing biological OA impact studies (on local and non-local species) and comparing anticipated changes in acidity (P_{CO_2}) to those currently experienced along the BC coast.

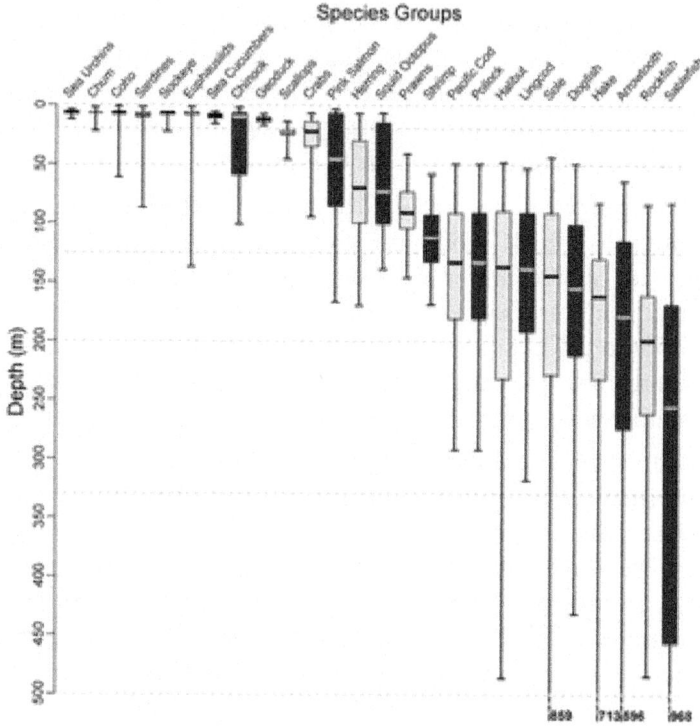

Figure 3: Depth-of-capture, expressed as quantile box plots of depth (m), from fisheries and survey data (where available) for species groups identified in Fig. 4.

For each quantile box, the upper whisker, box top, box delimiter (horizontal line), box bottom and lower whisker correspond to the 0.025, 0.25, 0.5, 0.75, and 0.975 quantiles, respectively. Depth quantiles that lie deeper than the figure limit are indicated along the bottom. Horizontal dashed lines correspond to depth zones in Fig. 2. See Methods for data sources.

doi:10.1371/journal.pone.0117533.g003

We use the best information available at present to address this problem. The quantitative details, including treatments and measured carbon parameters. We provide specific conclusions ordered by immediacy and relevance to the BC fishery.

METHODS

Present State of the BC Marine Ecosystem

Marine organisms were assigned to taxonomic groups and sorted by trophic levels adapted from model-derived output for the BC shelf [38] (Fig. 4). We added several taxonomic groups that are commercially fished [34] (*e.g.* sardine, tuna) and unfished (*e.g.* seagrasses, glass sponges) to this list as necessary. To evaluate species abundance and distribution within these groups, we used published literature (both primary and secondary as cited) where available. When literature was not available we consulted Canadian Department of Fisheries and Oceans (DFO) databases and the expertise of individuals active in the field (see Resultsand *Acknowledgements*). Landed values of fished species were taken from [34] (or [39] for euphausiids).

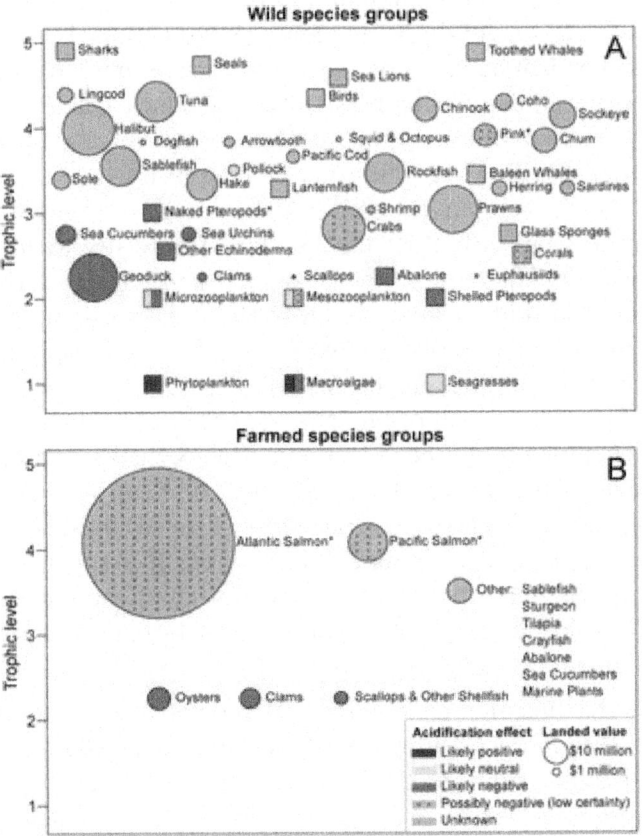

Figure 4: Summary of ocean acidification effects on (A) wild, and (B) farmed species groups in BC waters, including landed value for those that are fished or farmed.

Species groups are arranged vertically by trophic level, adapted from output by Preikshot [38] (courtesy of D. Preikshot, Madrone Environmental Services, Duncan BC). Areas of circles are proportional to the landed values in 2011, based on data in [34] (and [39] for euphausiids). Squares represent species groups that are not commercially harvested. Solid colours represent the likely direct effects of ocean acidification (see Results for explanations). Stippling refers to possible effects. For species marked by an asterisk (*), colours represent indirect effects.

doi:10.1371/journal.pone.0117533.g004

Species depth distributions (Fig. 3) were obtained from DFO databases (Pacific Biological Station, Nanaimo, Canada). Depths associated with commercially-caught groundfish (compiled by RH, May 1, 2014) and shellfish (compiled by Georg Jorgensen, May 6, 2014) are depths-at-capture, most often a mean of the minimum and maximum depths of fishing events (usually trawl or trap). For the commercial species groups (Fig. 4), depths were selected based on fishing methods specific to each group—Sea Urchins (dive), Euphausiids (nets), Sea Cucumbers (dive), Geoduck Clam (dive), Scallops (dive, trawl), Crabs (trap), Squid & Octopus (dive, trap), Prawns (trap), Shrimp (trawl), Pacific Cod (midwater & bottom trawl), Pollock (midwater & bottom trawl), Halibut (bottom trawl), Lingcod (bottom trawl), Sole (bottom trawl), Dogfish(bottom trawl), Hake (midwater trawl), Arrowtooth (bottom trawl), Rockfish (midwater & bottom trawl), Sablefish (bottom trawl). Depths associated with pelagic species (Herring, Sardines, and Salmon—Chinook, Chum, Coho, Sockeye, Pink) come from two sources: the WCVI Sardine Trawl Survey (spanning the WCVI, Fig. 1: −129.14°W to −124.56°W, 48.32°N to 51.14°N), which occurs mid-summer and is conducted during the night (data compiled by Linnea Flostrand, May 8, 2014), and the La Perouse Survey (spanning the BC coast, Fig. 1: −132.89°W to −123.07°W, 43.58°N to 54.64°N), which is a daytime acoustic trawl survey used to verify acoustic targets (data compiled by Jennifer Boldt, May 14, 2014). The two surveys did not capture any SARA-listed species. Mean depths-of-capture are summarised by quantile boxplots where the box represents 50% of the observations, and the region between the whiskers represents 95% (Fig. 3).

Commercial fishing in Canada is regulated by the *Fishery Act*. Specifically, Section 22 (http://laws-lois.justice.gc.ca/eng/regulations/SOR-93-53/page-6.html) identifies all license conditions that DFO uses to manage gear, monitoring, reporting, harvesting, allocation, and catch requirements. DFO's Pacific Region Animal Care Committee requires animal-use protocols, but specifically exempts lethal sampling of fish and invertebrates for stock assessment and sampling from commercial operations where animals are dead or certain to die. Data used here were collected for stock assessments and are therefore exempt from protocols.

Local Inorganic Carbon Distributions

Published inorganic carbon data (DIC, TA) from the outer BC coast in Queen Charlotte Sound (QCS) [40] and along the WCVI [27, 40] (Fig. 1) are used. These data (174 discrete samples) were collected over the continental shelf, slope and offshore, from the surface to 800 m with greater depth resolution in the top 50 m. The carbonate system was defined from TA and DIC (CO2SYS, [41]) and the constants of [42] with conductivity, temperature, depth and nutrient data that were collected concurrently. These data were sorted into depth intervals defined by local bathymetry relevant to local marine organisms (Fig. 2).

Responses of Marine Organisms to OA

We evaluated the potential impact (coded by colour in Fig. 4) of OA on each taxonomic group that occurs in BC, recognizing that uncertainty exists. We also identified the depth distributions that these groups of species occupy, along with associated OA conditions (Fig. 3). Similar to our description of the local marine system, we used published literature where available to assess direct and indirect effects of OA on taxonomic groups. When no publications were available in this rapidly emerging field, we consulted individuals who presented at recent conferences (in particular *2014 Ocean Sciences Meeting*, Honolulu HI and *2014 Salish Sea Ecosystem Conference*, Seattle WA), and we consulted many other experts in their respective fields (cited within Results and *Acknowledgements*).

RESULTS

There are relatively few published carbon data in BC waters. We use these data [27, 40] to estimate present-day ranges in pH and P_{CO2} for depth intervals relevant to local marine organisms (Fig. 2). We then defined three relative P_{CO2} levels, which are based on the present-day ranges in Fig. 2, to group the experimental treatments presented in the literature relative to our local waters (Table 1). For example, Pink Salmon (*Oncorhynchus gorbuscha*) generally occupy depths in both the 0–50 m and the 50–125 m zone (Fig. 3) so for these fish present-day P_{CO2} in our region is ~200–1000 μatm (pH ~7.6–8.4) (Fig. 2) so that a P_{CO2} level of 5000 μatm would be the upper limit of an 'elevated' (Table 1) treatment.

Table 1: Terminology used in the text to quantify levels of P_{CO2} **used in manipulation experiments.**

Terminology	P_{CO2}
present-day	depends on depth range (Fig. 2)
reduced	0.5× present-day
elevated	2–5× present-day
very elevated	5–10× present-day

doi:10.1371/journal.pone.0117533.t001

doi:10.1371/journal.pone.0117533.t001

Vertical distributions of marine organisms on the BC coast are presented with associated impacts of OA, ordered by trophic level (Fig. 4) in the following sections. Depending on trophic level and group, the amount of information available was variable. For many commercially harvested groups (represented by circles in Fig. 4) excellent data were available (*e.g.* finfish,Fig. 3). On the other hand, abundance and species composition of unfished groups are not well characterised, particularly at lower trophic levels (squares in Fig. 4, *e.g.* microzooplankton, corals). For many organisms important in the region, no published OA related studies exist (grey circles and squares in Fig. 4). Where necessary, we have adopted results from OA studies on species elsewhere that are similar to the ones found locally. These caveats are detailed in each section.

Phytoplankton

In the coastal northeast Pacific the predominant class of phytoplankton is diatoms, which are associated with high trophic transfer [43]. Many species (including the dominants:*Skeletonema costatum, Thalassiosira* spp., and *Chaetoceros* spp.) occur along the entire coast of BC [44–50]. Large blooms associated with coastal upwelling are often monospecific (*e.g.* [51]), but in our region they appear to be more diverse and occasionally include large numbers of photosynthetic dinoflagellates [46, 52]. Coccolithophorid blooms have been directly observed in more protected regions [50] and by satellite along the entire BC coast during summer [53]; however, coccolithophores (which calcify) are generally assumed to contribute minimally to overall productivity in the coastal zone (roughly landward of the 800 m isobath,Fig. 1) despite their importance further offshore [54]. Primary production by phytoplankton is exceptionally high in the region [27, 47, 55] and ultimately responsible for the high fish yields along our coast [30].

Phytoplankton species that are harmful to higher trophic levels are also common in the region. Large blooms of diatoms from the genus *Pseudo-nitzschia* occur on the outer coast (*e.g.* [56,57]) while the dinoflagellate *Alexandrium* is more prolific in protected locations [58]. Both*Pseudo-nitzschia* and *Alexandrium* produce neurotoxins that bioaccumulate in higher trophic levels. These toxins can interfere with the reproductive success of fish, seabirds, and mammals and cause mass mortalities [59, 60]. They are also responsible for numerous seasonal shellfish closures in BC (http://www.pac.dfo-mpo.gc.ca/fm-gp/contamination/biotox/index-eng.html). Additionally, significant blooms of *Heterosigma akashiwo* occur in protected waterways [61, 62]. *Heterosigma* releases peroxide free radicals into the water [63], which damage fish gill tissue [64, 65] and cause significant mortality and monetary losses (millions of dollars per year) to salmon aquaculture in BC [66]. Thus, harmful algae already pose a threat to health and food safety along the BC coast [58].

Direct Effects

There have been numerous studies on phytoplankton related to OA and a variety of responses have been observed depending on the species and the experimental treatment (*e.g.* [67–70]). Although natural conditions in most coastal environments, including the BC coast (Figs. 1 & 2), cover an exceptionally large range in carbon states and consequently pH (*e.g.* [6, 27, 71]), experiments in the field are challenging to complete. Thus, most studies have been conducted in the laboratory, often using a single strain of cultured phytoplankton. Also, because coccolithophores calcify (and at least some are easy to culture), they have been studied disproportionately. We sample a relatively small subset of this body of literature to summarise results of most relevance to the mixed, often diatom-dominated, community in the region and briefly describe the current understanding of the mechanisms involved.

Species specific responses by primary producers, including phytoplankton, to increases in ambient CO_2 are highly dependent on their carbon-uptake mechanism. Carbon assimilation relies on the enzyme ribulose biphosphate carboxylase-oxygenase (RuBisCO) to fix CO_2 [72], but this enzyme has a poor affinity for CO_2 [72, 73]. Over geological times scales (i.e. the last 3.5 billions years), as newer phytoplankton species have evolved, their use of RuBisCO has become more effective [72]. Some have carbon-concentrating mechanisms (CCMs), *e.g.* diatoms [74], to help transport and accumulate CO_2 to the active RuBisCO site [75]. The most important CCM for phytoplankton involves carbonic anhydrase to convert abundant HCO_3^- to the limiting CO_2 [76]. Despite CCMs, many photosynthetic phytoplankters, including some diatoms, appear to be carbon-limited under present-day conditions (*e.g.* [72]).

Because of these limitations in carbon uptake, it is anticipated that OA will increase overall production, which may provide more food to higher trophic levels. However, this increase does not appear to be large. Numerous mesocosm experiments, which use natural assemblages, suggest that regardless of species composition, there may be at most a 10–30% increase in primary production due to OA (*e.g.* [77–80]). In addition, a side-effect of elevated P_{CO2} (Table 1) is increased carbon to nitrogen (C:N) ratios in phytoplankton, effectively decreasing its nutritional quality [80].

While it is generally agreed that OA is likely to cause shifts in phytoplankton species composition, it remains unclear what these shifts will be [69]. It is reasonable to expect that species that do not have effective CCMs will do better than species that are already efficient with carbon uptake (diatoms in general). For example, the fish-killing raphidophyte *Heterosigma akashiwo* relies on passive diffusion to obtain CO_2. As a result it responds strongly (increased rates of growth and primary productivity) to an increase in dissolved CO_2 [81, 82] regardless of temperature [82]. In contrast, growth rates for some phytoplankton species reach a maximum value at the low end of present-day P_{CO2} in the upper mixed layer on the outer BC coast (Fig. 2) assuming a salinity of 31–32 [81]. For other species (including several diatoms) these rates remain invariant under elevated P_{CO2} [83].

Competition may be more subtle. For instance, some experiments have shown an increase in the proportion of diatoms relative to smaller phytoplankton with increased P_{CO2} (*e.g.* [84]) while others show the opposite effect (*e.g.* [85]). In addition, Tortell *et al.* [86] found that the prymnesiophyte *Phaeocystis* could outcompete diatoms at reduced P_{CO2} even though both groups have efficient CCMs. Finally, it has been suggested that at least one motile species (*H. akashiwo*) will swim faster under OA and deepen its vertical distribution [87], which may give it (and any species that can take advantage of its absence nearer to the surface) an additional competitive advantage.

Factors associated with climate change, including OA, are expected to increase the frequency and severity of harmful algal blooms [88]. In addition, the production of potent neurotoxins—domoic acid by common and sometimes prolific diatom species of *Pseudo-nitzschia*, and saxitoxin by dinoflagellate species of *Alexandrium*—has been shown to increase markedly under OA conditions [89–91]. In fact, domoic acid production in (at least some) *Pseudo-nitzschia* spp. increases dramatically (5–50× per cell) as P_{CO2} increases [92, 93].

Coccolithophores (prymnesiophytes) are the major calcifiers in the phytoplankton community [94, 95]. The most commonly studied species is *Emiliania huxleyi*, and although it appears to be less prevalent locally in the coastal zone (Fig. 1 in [94]), it plays an important role in the Alaskan Gyre [54]. Numerous experiments (most *in vitro*, some *in situ*) on *Emiliania* have been conducted to determine the effects of carbonate chemistry on calcification. Most (but not all, *e.g.* [96, 97]) suggest decreasing calcification at lower pH values (*e.g.* [70, 98]). Although much remains unknown (*e.g.* [20, 99]), the consensus is that OA will decrease calcification [69]. This observation is reinforced by mesocosm experiments that manipulate coccolithophore populations [67, 100] and by paleolithic records [101].

Phytoplankton Synopsis

We conclude that the overall impact on ecosystems and fisheries due to changes in the phytoplankton community in our region will be negative. While a modest increase in primary production is anticipated (so a direct positive benefit to phytoplankton, Fig. 4A), this increase is not likely to benefit higher trophic levels due to expected shifts in species composition (away from diatoms) and decreased nutritional value of the plankton. More importantly, the fish-killing alga *Heterosigma akashiwo* may gain a competitive advantage, which would seriously threaten salmon aquaculture. In addition, increasing P_{CO2} has been shown to alter the mix of neurotoxins produced by genera such as *Pseudo-nitzschia* and *Alexandrium* to favour the more potent forms, posing a significant threat to higher trophic levels and the shellfish industry as well as overall food safety.

Macroalgae

Three groups of macroalgae are delineated by their pigmentation: green, brown, and red algae, all of which are common in BC. In particular, brown algae constitute the majority of the biomass in intertidal and upper subtidal zones, and are dominated by kelps and rockweeds [102]. Brown algae have soft fleshy morphologies, and both green and red algal groups contain species with hard, calcified structures. Calcified red algae have two morphologies, crust-forming on substrate, and erect and branched. Both red and green algae are found in the intertidal and upper subtidal zones, but red algae extend down to the lower photic zone [103]. The large-blade (brown) macroalgae (*e.g. Laminaria, Macrocystis*) that form dense kelp forests along temperate coasts, common in BC, are the basis of some of the most productive ecosystems on Earth [103, 104]. These forests provide extensive shelter from predation, desiccation and wave action, as well as food, for hundreds of species with representatives from most taxonomic groups [105]. Calcified red algae provide similar protective structures, that are especially important for invertebrate species (*e.g.* urchins and anemones) [106].

Direct Effects

As with phytoplankton, many macroalgal species use carbon concentrating mechanisms (CCMs) to help transport and accumulate the CO_2 required for carbon assimilation [107]. Those relatively rare species without CCMs (most of which are red algae) rely on passive diffusion of CO_2 [108, 109] and so may experience enhanced photosynthesis and growth under OA, whereas those that have CCMs are likely to show no, or only small, positive effects due to reduced energy expenditure [107, 110]. Responses to elevated P_{CO_2}(Table 1) may be more significant at depths where light levels are reduced because energy constrains photosynthesis and CCMs are energetically expensive, though these effects are likely to be species-specific [110]. In addition, UVB (Ultraviolet B, 280–315 nm) exposure near the water surface tends to be harmful to some macroalgae, reducing the positive response to elevated P_{CO_2} [111]. The ultimate effects of OA on photosynthesis and growth of macroalgae will likely depend on interactions with light exposure, UV radiation, and other stressors. There has been less research concerning reproduction and life stages; however, it has been suggested that OA will result in reduced gametophyte growth of giant kelp [112].

For calcifying macroalgae, elevated P_{CO_2} affects the ability to build and maintain the calcified component of their tissues [108]. For example, Hofmann *et al.* [113] observed reduced calcification and growth for a cosmopolitan species of red algae when exposed to elevated P_{CO_2} over a 4-week period. Calcifying red algae are particularly sensitive to OA because unlike most calcifying green algae and invertebrates, red algae deposit a high-magnesium form of calcite into their cell walls, that is more soluble in acidified water than other forms of calcite [28]. However, Kroeker *et al.* [36] found no consistent change in calcification at elevated P_{CO_2} levels for a suite of calcifying macroalgae, perhaps because many species are able to generate microenvironments suitable for calcification despite increases in ambient P_{CO_2} [114–117]. Indeed, the observed reductions in growth with elevated P_{CO_2} (*e.g.* [113]) may result from the increased dissolution of carbonate skeletons rather than reduced production [117]. These effects are likely to interact with other stressors, such as UV radiation and temperature [118]. For example, Gao and Zheng [119] suggest that the carbonate skeleton of the same red algal species serves as a protective layer against UV; thus, CO_2 induced shell dissolution may increase vulnerability to detrimental effects of UV radiation [119].

Indirect Effects

Changes in macroalgal community composition are anticipated given the diversity of responses to OA among species. In general, non-calcifying macroalgae (especially those that rely on diffusion of CO_2 instead of CCMs) are expected to experience increased competitive success compared with calcifying macroalgae [110], resulting in an overall shift of community composition toward non-calcifying species [36]. Furthermore, studies on CO_2-enriched waters surrounding seafloor vents elsewhere support this hypothesis [120]. Most research has focused on losses of crust-forming calcified red algae in particular and replacement with non-calcifying turf-forming algal communities (*i.e.* species that reach heights of <15cm [121]) [36, 115, 122]. In BC, crust-forming red algae release chemical cues that play an important role in the settlement of some invertebrate larvae (*e.g.* abalone [123, 124]), and they bond substrata to provide stable habitats for other benthic species [106], but the resulting ecosystem effects under OA remain highly uncertain. Likewise, the ecological effects of possible declines in erect calcified red macroalgae and replacement by fleshy macroalgal species have received little attention (but see [113, 125]).

In addition to competition, herbivory is another key factor structuring macroalgal communities [126]. Rates of herbivory on macroalgae depend on palatability and the presence of hard carbonate structures for algal defence [127]. OA may reduce structural protection thereby increasing grazing on calcified species [115]. For non-calcified species, OA may increase C:N ratios possibly reducing palatability and hence grazing pressure [115]. However, OA will likely be detrimental to many herbivores, especially calcified species such as echinoderms and molluscs (see below), with resulting beneficial effects on some macroalgal species (*e.g.* [128]).

Given these potential impacts, Harley *et al.* [115] suggest that in the California Current ecosystem, which includes the WCVI, OA may result in a shift from diverse nearshore communities consisting of kelp canopies, understory turf assemblages, crust-forming calcifying algae, and calcifying invertebrates (*e.g.* urchins), to communities dominated by kelp and macroalgal turfs. Where kelp canopies have been lost due to other natural or anthropogenic disturbances (*e.g.* indirect effects of commercial harvest of fish species as found for large regions of the northeast Pacific, [129]), OA may prevent kelp recovery by facilitating expansion of algal turfs which inhibit kelp recruitment [130], as found along the Australian coast [131]. Kelp is the dominant primary producer among macroalgal species in BC, providing food and habitat for commercially important fish species, such as Pacific salmon [132, 133]. However, because responses of benthic communities to OA are highly species-dependent, the results of these studies cannot be extrapolated to other regions with high confidence [115].

In addition to community-level effects from altered competition and herbivory, OA may slow decay rates of some kelp species including those commonly found in BC (*e.g.* bull kelp,*Nereocystis leutkeana*), which could indirectly affect detritivore consumption and nutrient cycling [111]. This delay may result in the accumulation of phytodetritus, possibly reducing food availability for consumers in nearshore waters.

Macroalgae Synopsis

The direct effect of OA is hypothesised to be positive on non-calcifying species due to enhanced availability of CO_2 for carbon assimilation, but negative for calcifying species due to reduced growth and dissolution of protective shells (Fig. 4A). Community composition may shift from calcifying macroalgae species toward non-calcifying species, with an inhibition in the recovery of depleted kelp populations. However, community-level responses will depend on the extent of grazing on fleshy, non-calcifying species, possible changes in grazing due to OA-impacts on invertebrate herbivores, and the expansion of algal turfs. Responses of benthic communities to OA are highly species-dependent, limiting confidence in generalisations and extrapolations among regions and studies.

Seagrasses

Seagrasses belong to a small group of marine angiosperms comprising 60 species worldwide [134]. In BC, there are only two species of eelgrass—the native *Zostera marina* and the introduced species *Z. japonica*—and three species of surfgrass all belonging to the genus*Phyllospadix* [135]. Seagrass beds are well-known as nurseries for juvenile fish and invertebrates [136]. Another advantage conferred by seagrass beds is their ability to modify the seawater carbonate system, increasing aragonite saturation states within their confines [137], which might offer calcifying organisms refugia from the effects of OA.

In contrast to most macroalgae, seagrass cannot take advantage of the abundant HCO−3[138] and so increase their photosynthetic rate when DIC becomes more abundant [139]. With more DIC, seagrass are better able to compensate for light attenuation [139]. As a result, increased P_{CO2} may foster the growth of seagrass beds, despite worldwide losses of seagrass ecosystems due to anthropogenic disturbances along coastal environments [134]. However, OA-related reductions in phenolic compounds [140], which protect seagrasses against herbivory, may result in increased grazing pressure under increased P_{CO2}. The evidence for decreasing phenolics in seagrass under OA is limited and contrary to the trend of increasing phenolics in terrestrial angiosperms under increased atmospheric CO_2 [140].

Seagrass Synopsis

Seagrasses will likely benefit from increased P_{CO2} because higher DIC helps them compensate for light limitation; however, a decrease in protective phenolic compounds may offset any benefit due to increased grazing. The net effect of increased OA will likely be neutral for seagrasses.

Microzooplankton

Microzooplankton (20–200 μm) include heterotrophic protists such as ciliates and non-photosynthetic dinoflagellates. Typical ciliate genera along the BC coast include *Strombidium,Tintinnopsis* and *Strobilidium* [141] while the heterotrophic dinoflagellate species belong chiefly to *Protoperidinium*, which feeds almost exclusively on diatoms [142], and *Gyrodinium*. In nearshore waters, microzooplankton can be very abundant, depending on the time of year and food source (*e.g.* [44]). More importantly, fluctuations in microzooplankton populations, tightly coupled to phytoplankton, can have a large effect on pelagic ecosystems [143] and can influence the success or failure of fish recruitment [144].

Direct Effects

There are no studies that test the direct effects of OA on individual microzooplankton species. That said, foraminifera are amoeboid protists that form $CaCO_3$ shells and, like coccolithophores, are probably at risk from OA (*e.g.* [145]). There is also speculation that microzooplankton motility might be affected by OA [146], with the closest evidence coming from the study of the photosynthetic flagellate *Heterosigma* that demonstrated an increase in swimming speed and an increase in downward migration [87]. Large-scale mesocosm manipulations and on-board experiments that compare present-day and elevated P_{CO2} (Table 1) have found conflicting results—(i) no shifts in composition or abundance [147–149], (ii) almost identical succession patterns [150], and (iii) significant increases in heterotrophic dinoflagellate abundance [151, 152], although in the former (i.e. [151]) an increase in the prey species of diatoms was likely responsible.

Microzooplankton Synopsis

Based on the limited studies for microzooplankton, we expect that most species will be unaffected by OA, except through changes to their prey (phytoplankton). Direct OA effects will likely have a negative effect on foraminifera through reductions in $CaCO_3$ shells.

Mesozooplankton

In our region, the zooplankton community is strongly dominated by calanoid copepods [153,154]. Important species include *Neocalanus plumchrus*, *Acartia longiremis* and *Pseudocalanus*spp. [153, 154]. In protected regions like the Strait of Georgia (Fig. 1) *Calanus pacificus* is also important [154], while on the outer shelf *Calanus marshalae* is significant [153]. Some species spend part of their life cycles (that includes egg production) in relatively deep waters, >300–500 m (*e.g. Neocalanus plumchrus* and *Calanus pacificus*) while others, like *Acartia longiremis*, are always found above ~50 m. Zooplankton productivity is variable and appears to be changing over time [153], with species composition dependent on temperature [154]. Mesozooplankton provide the main trophic link connecting phytoplankton and microzooplankton with larger oceanic predators [155]. They are critical for several commercially-valuable fish species that prey on them directly, such as Pacific Herring, Pacific Hake, Pacific Sardine, various salmon species, and Spiny Dogfish (*Squalus acanthias*) [155].

Direct Effects

Only *Calanus pacificus* has been studied locally so we include experiments on copepods found elsewhere from the common genera *Acartia* and *Calanus*. Although responses to acidic conditions can be species-specific, even within genera (*e.g.* [156]), our summary provides a general indication of possible effects on the mesozooplankton community in our region.

Most OA related mesozooplankton research involves eggs and/or survival rates within individual stages. Egg production rates of adult females appear unaffected by increased P_{CO2}(even under very elevated conditions, Table 1) [156–160], although P_{CO2}-induced increases or decreases were observed depending on temperature [161]. On the other hand, egg hatching rates may decrease with OA [156–160], although increases have also been observed [161]. However, it is possible that hatching is simply *delayed* and so not observed in short-term experiments [160]. Effects of OA on overall egg hatching success are uncertain. In Puget Sound, WA (Fig. 1), egg hatching in *Calanus pacificus* is reduced under elevated P_{CO2} (Anna McLaskey, pers. comm., University of Alaska, Fairbanks AK), whereas egg hatching success in *Calanus helgolandicus* (found in the North Atlantic) appears unaffected [162]. For copepod embryos, survival rates appear unaffected by OA, while developmental rates may decline [163]. In adult copepods, survival rates are not significantly affected even under very elevated experimental conditions (except for one species) [156, 157, 159].

Although impacts on individual life stages may not be significantly different from a control scenario, the cumulative impacts may be significant. In addition, the studies thus far have been relatively short-term, and do not consider the possibility for copepods to respond to environmental changes through adaptive evolution [161]. The lack of detailed information on potential effects on zooplankton physiology "currently restricts our ability to reliably predict future impacts" [162].

Mesozooplankton Synopsis

For copepod species from the genera *Acartia* and *Calanus*, adult survival rates and egg productions rates appear unaffected by OA, even when P_{CO2} is 'very elevated' (Table 1), whereas egg hatching rates are negatively affected and egg hatching success remains uncertain. Cumulative impacts across life stages are unknown. Thus, the effects of OA on mesozooplankton will likely be neutral and possibly negative (Fig. 4A).

Pteropods

In BC waters only three species of pelagic snail, or pteropod, have been regularly observed [164]. *Limacina helicina* (shelled) is by far the most common of these three, occurring throughout most of the year, generally in the upper 100 m [164] and occasionally forming strong blooms (> 1000 m^{-3}) which can dominate the plankton (M. Galbraith, pers. comm., Institute of Ocean Sciences, Sidney BC). *Clione* spp. (naked) is also often present, although at significantly lower numbers. These two species are common in the Strait of Georgia and less so in Hecate Strait (Fig. 1); they are also found on the outer BC shelf and in the Alaskan Gyre (M. Galbraith, pers. comm.). *Clio pyramidata* (shelled), a subtropical species, is present only episodically along the WCVI [165]. Pteropods are an important food source for fish (especially juvenile salmon [166]), birds and marine mammals [167, 168]. Most pteropods produce aragonitic shells [167] and those that don't (naked pteropods) feed almost exclusively on the shelled species, making all pteropods susceptible to OA [164].

Direct Effects

It is difficult to keep pteropods in laboratory conditions [164] due to their delicate feeding structure [167]. Thus, few controlled experiments on live animals have been made until recently, and sample size remains limited. Most of these experiments have been conducted on (variants of) *L. helicina* harvested from Arctic and Antarctic waters.

Shells of dead pteropods dissolve in waters undersaturated with respect to aragonite, (e.g.[169, 170]) as expected. Live individuals, which may form protective biological coatings on the exterior of their shell [171] and/or actively counteract dissolution [170] also show evidence of dissolution when harvested from waters under, or near, saturation with respect to aragonite [172–174]. Similarly, live individuals incubated for short periods under the high end of present-day P_{CO2} (0–100 m, Fig. 2) and elevated P_{CO2} (Table 1) show reduced calcification (e.g. [170, 175]). In one experiment the larval state failed to calcify at all [176].

Despite the negative impacts on shell quality and maintenance, many (and in some cases all, e.g. [175]) animals studied survived their respective treatments (e.g. [170, 177]). However, the reduction of shell formation will impact the pteropods' ability to control buoyancy and withstand predation [167]. In addition, as P_{CO2} rises, increased energetic costs associated with maintaining their shells are likely, particularly as temperature increases [170]. The ability to supply energy to perform these (and other) tasks may be suppressed, [178] although some pteropods are likely to be more resilient than others (e.g. [179]).

Pteropod Synopsis

In summary, there is a clear cause for concern about the future of pteropods and the animals that depend on them. Although in the last several decades pteropods make up, on average, only about 5% of the average annual zooplankton biomass in BC waters (M. Galbraith, pers. comm.), they are an important food source for juvenile Pink Salmon [166] and are related to Pink Salmon survival [180] (see Fish—Indirect effects). Already in our region, where aragonite saturation horizons are frequently shallower than 100 m [11, 31, 32], numbers of the most common pteropod have declined significantly [164].

Molluscs

Molluscs comprise a diverse group of organisms that includes a variety of shellfish as well as predators such as squid and octopus (and pteropods, above). In the northeast Pacific, mussels dominate rocky intertidal zones (e.g. *Mytilus californianus* [181]) while oysters (mainly the Pacific Oyster, *Crassostrea gigas*), clams (family Veneridae) and cockles (family Cardiidae) are commonly found on beaches [182]. Geoduck Clams and scallops live significantly deeper (~10–20 m and 15–45 m, respectively) as do squids and octopuses (~15–140 m, Fig. 3). Shellfish consume plankton through filter-feeding and are able to significantly reduce plankton concentrations on a local scale (e.g. [183]), making them strong indicators of water quality [184, 185]. In turn, shellfish are preyed upon by many animals including sea otters, octopuses, seabirds and sea stars [186, 187].

The annual landed value of molluscs harvested from wild and farmed fisheries in BC is $63 million (Fig. 4), of which 66% is Geoduck Clam (*Panopea abrupta*). Other major harvested clams are Manila Clam (*Venerupis philippinarum*), Native Littleneck Clam (*Leukoma staminea*), Butter Clam (*Saxidomus gigantea*) and Varnish (Savoury) Clam (*Nutallia obscurata*) [188]. The Pacific Oyster was introduced into BC waters in the early 1900s and is used in aquaculture, while the native Olympia Oyster (*Ostrea conchaphila*) is no longer harvested [189, 190] and is listed as *Special Concern* under the Canadian Species at Risk Act (SARA). There are small fisheries for Pink Scallop (*Chlamys rubida*) and Spiny Scallop (*Chlamys hastata*) [191]; a commercially-developed hybrid called "Pacific Scallop" (*Patinopecten caurinus x yessoensis*) is used in aquaculture. There is a small but growing mussel industry, no harvest for Northern Abalone (*Haliotis kamtschatkana*) as it is listed by SARA as *Endangered*, and minor harvests for squid and octopus.

Direct Effects

Shelled molluscs calcify internally and actively increase pH at that site to do so, making them directly vulnerable to OA [16, 17]. Larval shells are particularly vulnerable since they are mostly composed of aragonite [192, 193] and for at least a few species the initial deposit is amorphous $CaCO_3$ (the least stable form of $CaCO_3$) [192]. By adulthood, shells are composed of aragonite and/ or calcite, depending on the species [192, 193]; *e.g.*, oyster shells are mainly calcite [194]. To deal with vulnerability at the larval stage (*e.g.* [195]), mollusc aquaculture in the northeast Pacific relies on hatcheries (often with controlled conditions) to rear larvae that are then distributed to growers.

Experiments to quantify OA effects on shellfish have yielded a range of conclusions [36, 196]; however, with the advancement of the field, results are beginning to converge. Kroeker et al. [36] found that OA significantly reduced calcification (by 40%), growth (by 17%) and development (by 25%) in molluscs. Another recent review [197] found that 37 of 41 studies on calcification by molluscs reported significant negative effects following exposure to increased CO_2 levels. Here we summarise experiments performed on species that are found in the northeast Pacific and elsewhere (*e.g.* scallops). There have been no studies on Geoduck Clams (despite their commercial importance), or on BC scallop species.

Experiments on fertilisation in Pacific Oyster have produced mixed results. Both sperm swimming speed and egg fertilisation success can be unaffected [198] or decline [199, 200] under elevated P_{CO_2} (Table 1). Within two days of fertilisation, Pacific Oyster larvae precipitate >90% of their body weight as $CaCO_3$, using limited energy reserves in eggs [17]. Early development (up to 8 h) remains unaffected at elevated P_{CO2} [201]; however, the number of embryos reaching the planktonic 'D-veliger' larval stage declines [199–201]. Elevated P_{CO2} increases the number of larvae with shells one day after fertilisation (due to an enhanced metabolic rate), yet decreases it three days after [202]. Larval survival of Pacific Oysters is unaffected by P_{CO2} after three and 16 days [202, 203]. Species that do exhibit a decline in larval survival are Northern Abalone [204] and Bay Scallop (*Argopecten irradians*) [205].

Metamorphosis from larvae to juveniles is affected differently for different species under elevated P_{CO2}. For Olympia Oyster, the proportion of metamorphosing larvae declines [206,207] and size at metamorphosis decreases [206]. Similar results, plus a delay in metamorphosis and reduction in survival, are usually seen for Bay Scallop [205, 208–210]. However, for Northern Abalone from the WCVI the proportion of metamorphosing larvae is unaffected [204]. Increased abnormalities in larvae have been observed under elevated P_{CO2} in Pacific Oyster [199–201] and Northern Abalone [204]. In the latter species, shell abnormalities increased substantially, occurring in 99% of larvae at P_{CO2} 1800 μatm [204]. These abnormalities did not appear to affect survival rates in the laboratory, but in the field the abnormal larvae would be more susceptible to predation [204].

The size of D-veliger larvae of Pacific Oyster decreases [199–202, 211] and shell growth of later larval stages generally declines [199, 201] under elevated P_{CO2}, though not always [199,203]. Decreases in larvae shell growth also occur in Olympia Oyster [207, 212], Northern Abalone [204] and Bay Scallop [208–210]. Molecular analyses show that expression of proteins related to calcification and cytoskeleton production can be severely suppressed under high P_{CO2} [211]. For Northern Abalone larvae, settlement (attachment to the experimental container) is unaffected by P_{CO2} [204]. Additional effects on other larvae include decreased O_2 consumption and feeding rates [203], and reduced lipid content [209, 210].

Shell growth and calcification of juvenile and adult molluscs under OA remains uncertain due to limited studies with contrasting results. Pacific Oyster juveniles exhibit increased expansion of shell area (but not thickness) under reduced pH, despite declines in O_2 consumption and feeding rates of larvae [203]. In juvenile Bay Scallops, elevated P_{CO2} (Table 1) does not affect shell and tissue growth but does reduce survival [209]. Declines in calcification rates have been observed for Pacific Oyster juveniles and adults under elevated P_{CO2} [213] and for adult Zhikong Scallops (*Chlamys farreri*) under reduced pH [214].

The byssal threads that mussels use to attach themselves to rocks or vertical lines in aquaculture must be robust so that they do not drop off or get ripped off. The threads of the common mussel (*Mytilus trossulus*) have been shown to weaken under elevated P_{CO2} [215], although they may be more sensitive to temperature during during short-term fluctuations typical of local inlets (L. Newcomb, University of Washington, Seattle pers. comm.).

Metabolic rates of juveniles and adults appear to be generally unaffected by OA alone [216–218]. Also unaffected, at least in juvenile King Scallops, are clearance rates, growth rates, the ratio RNA:DNA (suggesting no effect on growth potential) [217], and various measures related to 'clapping' (rapid closing used for locomotion) by adults—frequency, recovery time between claps and clapping fatigue [218]. The latter study, however, did find a reduction in the force exerted by the clapping under elevated P_{CO2}, which could reduce the scallops' ability to escape predators.

As above, the larval stage is vulnerable to OA. South of BC, at a hatchery for Pacific Oyster in Oregon (USA), carbonate levels experience large fluctuations due to strong coastal upwelling [195]. Negative correlations were found between the aragonite saturation state (Ω_{arag} of water in which larvae were spawned and reared, and the resulting larval production and mid-stage growth [195]. In the laboratory, the shell growth rate of juvenile Olympia Oysters depends on pH exposure at the larval stage but not at the juvenile stage [212]. To test such carry-over effects in a natural system, Olympia Oyster larvae were reared under different P_{CO2} levels, then transferred to field sites after metamorphosis [206]. Juvenile survival was not significantly different between the two larval treatments, but the elevated-P_{CO2} larvae yielded smaller juveniles, suggesting that they suffer irreversible damage (*e.g.* energy deficit, abnormality, inability for compensatory growth) [206].

Indirect Effects

Changes in species composition can be expected under OA. Few studies explore these changes for molluscs, however it has been shown that Eastern Oyster larvae (*Crassostrea virginica*) have higher survival rates than Bay Scallops under elevated P_{CO2}, which is the opposite of the present-day P_{CO2} result (and in the absence of brown tides—in this study caused by a temperate phytoplankton species not found in the northeast Pacific) [210]. Thus, scallops may be affected by OA more than oysters. Scallops are also sensitive to other anthropogenic stressors, such as eutrophication [219], while the impact of these conditions on oysters and other shellfish was not investigated.

OA may increase the vulnerability of shelled molluscs to predation by thinning their protective shells and may also cause food web shifts. For example, Boring Sponges (*Cliona celata*) can bore twice the number of holes in Bay Scallop shells, and remove twice the weight of shell, at pH 7.8 compared to pH 8.1, despite taking longer to attach themselves to the shells [220]. Negative impacts on molluscs could also have large unintended consequences for other species [221]. Shell production and aggregation provide refuge for other organisms such as sponges and crabs, and introduce complexity and heterogeneity into benthic environments, with heterogeneity being important for maintaining species richness [221]. Thus, the direct effects of OA on molluscs may have detrimental effects at the ecosystem level.

Squid and Octopus

In BC, there are at least 30 species of squid and eight species of octopus [222], none of which have been studied for OA effects. Common species in BC waters are Opal Squid (*Loligo opalescens*) and Northern Giant Pacific Octopus (*Enteroctopus dofleini*). Similar to the otoliths of fish (see below), squids have internal calcified structures called statoliths used for sensing gravity and movement [223]. Under elevated P_{CO2} statoliths in embryos of the European Squid, *Loligo vulgaris*, are significantly larger than those formed under present-day P_{CO2} [224]. At higher P_{CO2} (still in the elevated range—Table 1), Kaplan *et al.* [225] observed reduced surface area, malformation, and abnormal crystalline structure in statoliths of Atlantic Longfin Squid, *Doryteuthis pealeii*. Aside from calcification, elevated P_{CO2} also leads to increased heavy metal retention in the protective eggshells and changes to the bioaccumulation of silver, mercury and cobalt in larval tissue [224]. Additionally, elevated P_{CO2} depresses metabolic rates in pelagic squids (*e.g.* [226]). The ultimate effect on fitness is not known.

Mollusc Synopsis

We conclude that the effects of OA on shelled molluscs will be negative based on available studies on oysters, scallops, abalone and mussels (Fig. 4). These negative effects occur at various life-history stages, and go beyond direct effects on calcification of larvae, *e.g.* reduced oxygen consumption and feeding rates of larvae and delayed behavioural responses of adults. It is generally anticipated that effects on larval survival rate and reproduction rate will directly influence population size, population distribution and community structure [227]. No experiments were found on local clam species (including geoducks) but given the results on other molluscs [36] we anticipate that they will also be negatively affected by OA, while effects on squid and octopus remain uncertain (Fig. 4).

Sponges and Coldwater Corals

Sponge reefs are globally unique to the northeast Pacific coast [228–230] and all four groups of cold-water corals: octocorals, stylasterids, stony and black corals, are present in the region. They occur where productivity and water flow are high (*e.g.* they are especially dense on seamounts and the heads of canyons, Fig. 1) and from the surface to depths >2000 m [231]. However, due in part to the depth range, very few benthic habitat mapping data exist along the BC coast (*e.g.* [232], Kim Conway, pers. comm., Pacific Geoscience Centre, Sidney, BC) and so we have used these data and the expertise of others to provide our own general description (below). Sponges and cold-water corals form important habitat for many marine organisms including species of fish that are commercially important (*e.g.* the rockfish Pacific Ocean Perch) in our region [233–236].

The coral and sponge contribution to the benthic fauna in BC appears to be patchy but diverse, based on: DFO trawl survey and observer records [237], comparison with neighbouring regions (*e.g.* [238, 239]), isolated studies (*e.g.* [229, 235]), anecdotal evidence (Lynne Yamanaka, pers. comm., Pacific Biological Station, Nanaimo, BC), and modelling work (*e.g.* [240]). This collection is likely dominated by siliceous sponges, and isolated stands of flexible corals with partly organic skeletons (octocorals), more specifically members of the diverse group Alcyonacea (*e.g.* large tree form coral) and pennatulaceans (sea pens and whips). Alcyonacea and solitary glass sponges occur on bedrock, mainly deeper than ~200 m, while pennatulaceans and glass sponge reefs grow on flat sediment, generally shallower than ~200 m [241].

Stylasterids (*e.g.* [242]) and stony corals (Scleractinia) also occur [237, 243], but primarily in small, solitary patches. The reef-forming scleractinian *Lophelia pertusa* has been found [244], but is rare, possibly influenced by the already low aragonite saturation states in this region [245]. Black corals, which do not calcify and are made of organic proteins, are also present below 500 m [237].

Direct Effects

OA studies have focused on stony corals, primarily *Lophelia pertusa*, which is entirely aragonitic. They show an increased energetic cost for calcification in *L. pertusa* with decreasing pH (and Ω_{arag} [246, 247]; however, *L. pertusa* may adapt to moderate decreases in pH given sufficient time [248]. The holdfasts and some parts of the structure of many octocorals are also made of aragonite [249]. Similarly, some stylasterids precipitate aragonite as well as calcite [250].

However, neither octocorals nor stylasterids have been studied with respect to OA to date. Likewise, there are no OA studies specific to glass sponges.

Sponge and Coral Synopsis

The OA response of the cold-water corals most common in our region (octocorals) has not yet been studied. While the skeletons of these corals are partly organic, they also calcify and so may be affected by OA at some level (Fig. 4A). There are no OA studies on glass sponges to date. Loss of coral and sponge habitat would have a negative impact on many fish species, particularly juvenile rockfish [233–235].\

Echinoderms

Echinoderms form a marine set of invertebrate animals with ~7000 known species worldwide [251] and 217 species recorded in BC [252], half of which occur exclusively at depths > 200 m [253]. The echinoderms comprise five classes: (i) echinoids (sea urchins and sand dollars), (ii) asteroids (sea stars), (iii) holothuroids (sea cucumbers), (iv) crinoids (sea lilies and feather stars), and (v) ophiuroids (brittle stars). A few are considered to be "keystone" species, such as the Purple Sea Star (*Pisaster ochraceus*) [254, 255], which is common along the BC coast. Echinoderms modify ecosystems (*e.g.* by mixing and transforming sediments, grazing kelp forests, preying on mussel beds) and provide food for carnivorous fish, shellfish, and marine mammals (*e.g.* sea otters prey heavily on sea urchins and sea cucumbers). In addition, sea stars and sea urchins act as important grazers in the sub-littoral zone [256].

Direct Effects

Green and Red Sea Urchins (*Strongylocentrotus droebachiensis* and *S. franciscanus*, respectively) harvested in BC generate significant income (Fig. 4A). Clark *et al.*[257] found that larval growth and skeletal calcification were reduced at lower pH levels for select species ranging from the tropics to the poles; no changes in skeletal morphology occurred. Studies on shell thickness are confounded by effects of diet and experiment length [125, 258], but urchins have higher growth rates when fed on calcifying algae and may derive some portion of essential elements (*e.g.* calcium, magnesium) from the algae [258]. Therefore, sea urchins may suffer as the proportion of calcifying macroalgae in their diet declines due to direct OA effects on these algae (see Macroalgae section above). In long-term studies, sea urchins have shown an ability to adapt to elevated P_{CO_2} (Table 1); however, in the transition to new OA conditions, species may suffer from life-cycle carry-over effects. For instance, Dupont *et al.* [259] demonstrated that under elevated P_{CO_2} females acclimated for four months experienced a 4.5 decrease in fecundity and produced offspring that suffered 95% juvenile mortality; however, these effects disappeared after acclimitisation for 16 months. OA may also influence reproduction in echinoderms. For example, as P_{CO_2} increases under OA, higher sperm concentrations are necessary to achieve high fertilisation success in the sea urchin *S. franciscanus*, and the egg's mechanism for blocking fertilisation by multiple sperm cells becomes slower [260].

A number of studies have used genetic markers to infer the possible physiological effects of OA in sea urchins. O'Donnell *et al.* [261] measured the change in expression of a molecular helper-protein in *S. franciscanus* and suggested that the ability to handle temperature stress would be reduced under OA. Todgham and Hofmann [262] measured changes in ~1000 genes of the sea star *S. purpuratus* and found reduced expression under elevated P_{CO_2} in four categories—biomineralisation, cellular stress response, metabolism, and apoptosis (cell death). Also for this species, elevated P_{CO_2} triggered changes in 40 functional classes of proteins, affecting biomineralisation, lipid metabolism, and ion homeostasis [263].

Giant Red Sea Cucumber (*Parastichopus californicus*) harvest also provides significant income in BC (sea cucumbers, Fig. 4A) but there are no studies on OA effects for this species. Elsewhere, a single study found that sperm motility of a reef-dwelling sea cucumber species (*Holothuria* sp.) was impaired at pH values <7.7 [264]. Elevated P_{CO_2} and temperatures have been shown to have positive and additive effects on the relative growth of the keystone sea star *Pisaster ochraceus* [265]. Under increased P_{CO_2}, calcification is reduced [265]; however, growth rate remains unchanged as the endoskeleton is primarily composed of soft tissue with relatively small calcareous elements for rigidity and protection. Brittle stars (ophiuroids) are commonly found in the region, but the effects of OA have only been studied in species found elsewhere. In the eastern Atlantic Ocean, keystone brittle star *Ophiothrix fragilis* was found to be especially sensitive to small changes in pH [266], with 100% mortality of larvae at pH 7.9 vs. 30% mortality in the control (pH = 8.1). Finally, while Dupont *et al.* [251] found that echinoderms studied to date are relatively robust to OA effects, they conclude that the overall impact of OA on this group will be negative and suggest that associated ecosystem impacts may be more severe.

Indirect Effects

Declines in some echinoderms may affect the predators that depend on them, but ecosystem effects remain unknown. For example, on our coast, various nearshore rockfish and numerous flatfish prey on ophiuroids [267], although they only form an important component of the diet for China Rockfish (*Sebastes nebulosus*), Flathead Sole (*Hippoglossoides elassodon*), and Southern Rock Sole (*Lepidopsetta bilineatus*) [267]. Additionally, the deep-water rockfish Longspine Thornyhead (*Sebastolobus altivelis*) relies on brittle stars for a large proportion of its food [268]. In the eastern Atlantic, the inevitable decline in pH may lead to the disappearance of the keystone brittle star *O. fragilis*; the impact on the ecosystem is not really known [266].

Echinoderm Synopsis

Although many echinoderms have not been studied, the existing evidence indicates significant negative effects due to OA, especially at early life stages. Thus, we suggest that this group will be affected negatively (Fig. 4A). Of more concern are the anticipated negative impacts on ecosystems, *e.g.* declines in the population of a keystone species like the Purple Sea Star would have wide-ranging effects on the food web.

Crustaceans

Marine crustaceans are represented in BC by copepods [269], krill (euphausiids) [39], barnacles [270], shrimps, prawns and crabs [271]. Copepods (see Mesozooplankton) and krill form a substantial biomass in the oceans and provide an important source of food for upper trophic levels in temperate marine foodwebs and act as important grazers (e.g., [272]). Crabs are found in the upper 50 m, while adult prawns (*Pandalus platyceros*) and adult shrimp (mainly Smooth Pink—*Pandalus jordani* and Sidestripe—*Pandalopsis dispar*) are deeper (~100 m and 120 m, respectively: Fig. 3). Krill, primarily *Euphausia pacifica*, perform strong diel vertical migration from the surface to depths exceeding 100 m. Krill is harvested on a limited basis in the Strait of Georgia and various inlets [39]. Prawns and shrimps, which are farmed extensively in other parts of the world, are only harvested from the wild in BC; the prawn fishery is substantial (~$40 million, Fig. 4) [34]. The crab fishery in BC is also valuable (~$33 million) [34, 273], with Dungeness Crab (*Cancer magister*) being the most important commercial species.

Direct Effects

Crustacean exoskeletons, composed of chitin and $CaCO_3$ [274], are generally considered to be unaffected by OA. In fact, evidence suggests that this protective covering actually serves as a buffer to the corrosive nature of OA, and some crustaceans can use the increased DIC in seawater to fortify their shells through calcification [275]. This enhancement of the shell contrasts with shell dissolution in molluscs (see Molluscs Section), and is likely due to some crustaceans (crabs, lobsters) having an efficient proton-regulating mechanism [275]. Despite the advantage of localised pH-regulation, the calcification response appears to depend on a variety of additional factors: external organic coatings, skeletal mineralisation composition (*e.g.* magnesium content in calcite), and the degree to which amorphous $CaCO_3$(precursor to calcite/aragonite shells) is utilised [275–277].

Crustacean species' ability to deal with increasing OA also depends on life-history strategies and habitat [278]. Active species or those in highly fluctuating environments (*e.g.* intertidal or estuarine) tend to utilise the oxygen-transporting protein haemocyanin, which also confers additional buffering capacity against high H^+ concentrations. Sedentary species or those in stable environments (*e.g.* deep-sea or polar) tend to have less haemocyanin and consequently less buffering capacity. The latter group relies more on HCO−3 buffering and is probably more sensitive to OA.[278]

Recent studies on Alaskan King Crab (AKC, *Paralithodes camtschaticus*) and Tanner Crab (TC, *Chionoecetes bairdi*) in Alaskan waters highlight the vulnerability of the early life stages to OA [279, 280]. For AKC embryos and larvae, OA produces larger embryos (but not larger mass), smaller egg yolks, higher developmental rates, and higher calcium content [280]. In juveniles of both species, increased mortality occurs with elevated P_{CO2} (Table 1), with 100% mortality in their most extreme treatment [279]. Differences between the two Alaskan crabs (decreased condition index in AKC but not TC and decreased calcium content in TC but not AKC) suggest that AKC puts more energy into osmoregulation and calcification than does TC [279]. Additionally, there is some preliminary evidence that adult AKC females fail to moult [280].

Initial studies are underway on the dominant local species of krill, *Euphausia pacifica*. A recent study in Puget Sound, WA (Fig. 1), found that elevated P_{CO2} slowed the development of hatched nauplii to the first feeding stage (Anna McLaskey, pers. comm., University of Alaska, Fairbanks AK). Also, under higher P_{CO2} the Antarctic krill species, *Euphausia superba*, experiences ingestion rates 3.5 times higher than those under present-day conditions, and consistently higher metabolic rates [281].

For the cold-water barnacle, *Semibalanus balanoides* (common in BC), experimental treatments at elevated CO_2 reduced adult survival and slowed embryonic development, which delayed the time of hatching by 19 days [282]. The cold-water shrimp,*Pandalus borealis* (common and commercially important in BC), also exhibited delayed juvenile development at reduced pH [283]. Other studies find no such delays [284–286], though significant effects have been observed when temperature and P_{CO2} interact [285]. The ability to tolerate OA also depends in part on prior exposure to habitats that experience highly fluctuating P_{CO2} [287].

Indirect Effects

Slow embryonic development [282] could potentially cause a timing mismatch between larval release and prey availability related to the spring phytoplankton bloom [288]. Potentially slower growth and lower fitness in juveniles and young adults may reduce egg production by females over their lifetime [279]. Despite the stability of adult exoskeletons, the post-moult calcification stage in crustaceans may be delayed significantly under elevated P_{CO2} [278], which may increase mortality due to predation on this defenseless life stage (*e.g.*[289]). Additionally, Kunkel *et al.* [290] hypothesise that OA may degrade the thin outer layer of calcite, which helps protect decapods from microbial attack. Finally, stock assessment models that incorporate reduced recruitment survival as a function of OA suggest that there can be a substantial socio-economic cost that is currently not recognised by decision makers [291].

Crustacean Synopsis

Generally, the crustaceans are expected to be sensitive to OA effects at early life cycle stages, while available studies suggest mixed results for adults. However, many local species, such as prawns, have not been studied (Fig. 4A). There is evidence that developmental anomalies in embryos and larvae occur at reduced pH, which may affect the fitness of juveniles and adults; however, the effects are species-specific and phenotypic adaptation is not known. Additionally, changes in growth rate and calcification may increase the susceptibility to predation, and delays in development may decouple life cycle timing between larval release and optimal foraging conditions.

Fish

In BC coastal waters, there are over 300 species of marine fish [292, 293]. The taxonomic groups represented in BC include jawless fish (*e.g.* hagfish (270–1010 m)), cartilaginous fish (*e.g.* ratfish (50–380 m), dogfish (50–430 m), sharks (90–1020 m), skates (50–860 m)), and bony fish. The latter group includes important contributors to BC fisheries—Pacific Herring (*Clupea pallasi*, 5–170 m), salmon (five species of *Oncorhynchus*, mostly in the surface 50 m but some species deeper than 100 m), Pacific Hake (*Merluccius productus*, 80–700 m), Pacific Cod (*Gadus macrocephalus*, 50–300 m), Walleye Pollock (*Theragra chalcogramma*, 50–300 m), rockfish (at least 36 species of *Sebastes* (70–470

m) and two species of *Sebastolobus*(160–1010 m)), Sablefish (*Anoplopoma fimbria*, 70–970 m), Lingcod (*Ophiodon elongatus*, 50–310 m), Arrowtooth Flounder (*Atheresthes stomias*, 60–600 m), soles and flounders (~18 species, 50–860 m), and Pacific Halibut (*Hippoglossus stenolepis*, 50–490 m). Depth distributions for valuable BC fisheries (Fig. 4) appear in Fig. 3. Marine fish species are economically important (GDP of capture fisheries, aquaculture, and sport fishing in BC was over $340 million in 2011 [29]) and ecologically valuable because of their roles providing food sources to higher trophic levels (*e.g.* birds and mammals) and cycling nutrients to other ecosystems (*e.g.* salmon providing nutrients to coastal terrestrial ecosystems [294]).

Direct Effects

In general, we expect that adult fish will be tolerant of OA because they can control ion concentrations through evolved regulatory mechanisms [295, 296]. In particular, active fish exhibit transient elevated metabolic rates and highly variable extracellular CO_2 and proton concentrations. Acid-base imbalances are regulated by specialised gill epithelia, which compensate for pH disturbances caused by exposure to increased environmental P_{CO2} [296]. Although some studies suggest that aerobic performance of tropical fishes may decline under elevated P_{CO2} [297] (Table 1), detrimental effects were not found in a temperate species, Atlantic Cod, under elevated P_{CO2} (*e.g.* [295]).

The effects of lake acidification on diadromous fish (those migrating between marine and fresh water) are well known, but using these observations to suggest OA effects is potentially misleading due to (i) large physiochemical differences between fresh and acidified marine waters and (ii) high physiological variability between diadromous and marine species [298,299]. Also, fluctuations in in [H⁺] seen in lake acidification are orders of magnitude greater than those in the ocean [298].

As with the invertebrates, OA effects in fish are expected to occur during the vulnerable developmental stage, and these effects appear to be species specific. The acid-base regulatory mechanisms of the larval stage remain rudimentary until gills have formed and respiration switches from cutaneous to branchial [300]. Developmental responses are thought to be more the result of CO_2 toxicity rather than through pH acting alone [301, 302].

There are limited OA studies on fish species that occur in our region. Hurst *et al.* [303] showed that the effects of OA on the growth of Walleye Pollock larvae were minor and varied greatly within treatments (Fig. 4). Slightly higher growth rates in elevated P_{CO2} conditions (Table 1) proved non-significant. Other studies on Atlantic temperate fish species (cod and herring), closely related to those in BC waters, found no significant effects on sperm motility, embryogenesis, egg survival, or the development of skeletal, heart, and lung tissue [300, 304,305]. Despite these benign effects, researchers have found some developmental anomalies. Franke and Clemmesen [305] showed an inverse relationship for Atlantic Herring between P_{CO2} and the ratio RNA/DNA at hatching, potentially reducing protein biosynthesis and growth. Frommel *et al.* [300] found significant tissue damage in liver, pancreas, kidney, eye, and gut of Atlantic Cod larvae under elevated P_{CO2} Baumann *et al.* [306] demonstrated that increasing P_{CO2} caused a 74% reduction in survival and an 18% reduction in length of embryos of a ubiquitous estuarine fish called Inland Silverside (*Menidia beryllina*). Any significant developmental effect could alter the abundance and diversity of marine fish populations.

Otoliths (ear bones) are aragonite-based structures that fish use to sense acceleration and orientation. In some species, otoliths grow larger when larval fish are exposed to elevated P_{CO2} (*e.g.* White Sea Bass, a species found in BC waters [307]; Atlantic Cod [308] and tropical clownfish [309]). Under elevated P_{CO2} pH is regulated in the endolymph sac surrounding the otolith resulting in increased $CaCO_3$ precipitation and enhanced otolith growth for those species [309]. An increase in otolith size may enhance hearing range [310], which might help or harm fish depending on sensitivity to important auditory cues or disruptive background noise [310].

Behavioural responses have recently been documented at elevated P_{CO2} for larvae of tropical reef fish. In particular, behaviour to olfactory, auditory, and visual cues changes when larvae are selecting habitats and responding to predators [311–315]. Additionally, elevated P_{CO2} reduces learning abilities related to predator avoidance [316] and changes the propensity of larval reef fish to turn left or right (lateralisation) [317]. These behavioural changes can expose larval fish to increased mortality risk, which has important fitness consequences [313, 318]. Given possible behavioural effects on predators as well as prey under elevated P_{CO2} community-level responses are difficult to predict [318, 319].

Relatively few studies have investigated behavioural changes to OA in temperate species (three exceptions being [320–322]), and none have examined commercially important species in BC waters. The larvae of Threespine Stickleback (*Gasterosteus aculeatus*), a species found in marine and fresh water on the BC coast, exhibit behavioural disturbances (*e.g.* reduction in boldness and curiosity), compromised learning abilities, and declines in lateralisation when reared in elevated P_{CO2} [320]. These responses are surprising given the physiological plasticity of this species, which is expected to confer enhanced acclimatisation abilities to environmental challenges. These results suggest that sensitivity to OA is not limited to species occupying narrow ecological niches, such as tropical reef fish [320].

Elevated P_{CO2} can disrupt the functioning of GABA$_A$ (γ-Aminobutyric acid) receptors, the main inhibitory neurotransmitter receptors in the fish brain [323]. Normally, the opening of these receptors results in an inflow of Cl⁻ and HCO−3 ions over the neuronal membrane, leading to inhibition of the neuron. When concentrations of intracellular Cl⁻ and HCO−3are altered (e.g, when fish with strong acid-base regulatory systems are exposed to higher environmental P_{CO2}, the flow of ions can be reversed, resulting in neuronal excitation instead of inhibition. Such changes have been associated with dramatic shifts in behaviour and sensory preferences in larval tropical reef fish [323], but the effects on temperate species are unknown. Although these receptors are shared by many, if not most fish, the resulting behavioural responses will likely vary due to species-specific differences in acid-base regulatory systems [323].

Indirect Effects

Fish will likely be affected indirectly by OA through food-web interactions. Off the southern WCVI, the pelagic system is dominated by Pacific Hake, Pacific Herring, Spiny Dogfish, and Chinook Salmon (*Oncorhynchus tshawytscha*), all largely dependent on krill production in the region [324]. This area has also been described as a "toxic hot spot" due to consistently high levels of *Pseudonitzschia* species and the presence of domoic acid [325]. These neurotoxins are transferred to higher trophic levels [59], and as P_{CO2} increases under OA the toxicity of these blooms may also increase [93].

Many fish species of the north Pacific Ocean prey on shelled pteropods (*e.g.* cod, pollock, mackerel) and a decline in pteropod abundances may lead to a shift in diet toward greater predation on juvenile fish such as salmon [326]. Pteropods (see Pteropods—Indirect effects) are also an important food source for Pink Salmon in the first year of marine life [166]. Because pteropods often exhibit swarming behaviour, foraging costs are relatively low for Pink Salmon feeding on patches [166, 180], possibly enhancing growth in early marine life and increasing adult biomass [327]. Reductions in pteropod densities may therefore have significant impacts on Pink Salmon biomass (Fig. 4A).

Trophodynamic modelling can suggest possible impacts of OA on fish populations. One study [328] explored various scenarios under OA, one of which assumes a significant mortality on benthic shelled invertebrates (*e.g.* bivalves, corals, sea urchins, sea stars) that leads to a biomass reduction for fish that feed on these species. While both English Sole (*Parophrys vetulus*) and small demersal sharks (*e.g.* Spiny Dogfish) rely on these invertebrates for only 10% of their diet in the model, English Sole experiences a much bigger decline due to a lack of alternative prey items. Another OA scenario in [328] adds an additional mortality on large zooplankton and small phytoplankton, which leads to a large increase in microzooplankton, detritus, and bacteria. In this scenario, the model predicts various higher-order interactions: a reduction of Lingcod due to a decline in macrozooplanktonic prey; an increase in Canary Rockfish (*Sebastes pinniger*) due to an increase in sea urchins and shrimps; and the increase of nearshore rockfish due to a decline in one of its predators, Lingcod. While there are many possible outcomes using such modelling tools, they do highlight how effects from OA on any single biological component can affect the entire trophic web.

Fish Synopsis

In general, we expect that adult fish will be tolerant of OA because of their ability to control internal ion concentrations. However, OA may affect fish during vulnerable developmental stages, though evidence for these effects is weak for species in BC. Perhaps more importantly, behavioural responses to OA have been widely documented in tropical reef fish, resulting in reduced survival. Similar effects may occur in temperate species, though studies in this area are limited. OA-induced reductions in availability of some prey species may reduce fish growth and survival, though these effects may be tempered by prey-switching. Possible increases in HABs would have a negative impact on farmed fish and shellfish; wild fish might increasingly suffer the effects of biotoxin accumulation.

Marine mammals

British Columbia is host to a large and diverse group of marine mammals (~30 species [329]), many of which have experienced dramatic population increases over the last century when hunting and culling practices were discontinued (*e.g.* on Grey Whales (*Eschrichtius robustus*) and Harbour Seals (*Phoca vitulina*), respectively) [330]. In addition to their role as top predator in the marine food web and their contribution to ecotourism, these mammals are iconic symbols of the region. Thus, they are valuable, but their value is difficult to assess (*e.g.*[331]).

In general, marine mammals cover an appreciable geographic range and many are able to dive to remarkable depths [332]. Their physiology is adapted to high pressures and they have an exceptional capacity for O_2 [332]. Because they breathe at the surface, they are not susceptible to acidosis in the way that many other complex marine organisms will be as carbon levels increase (*e.g.* [302]). Therefore, direct impacts of OA on marine mammals are not expected, and have not been investigated (Fig. 4A). Indirect food web impacts are anticipated, *e.g.* for cetaceans that rely heavily on cephalopods or zooplankton such as pteropods [333]. In addition, underwater sound absorption at low frequencies (relevant for marine mammals) will decrease with OA [334]. However, this decrease is projected to be small (less than 0.2 dB) over the next few centuries and negligible in the context of the current noise associated with shipping [335].

Marine Mammal Synopsis

Marine mammals will likely be affected by OA indirectly through food web changes, however direct impacts are not anticipated. While noise levels will increase with OA, this increase will not be large enough over the next few centuries to affect animals that rely on underwater sound.

DISCUSSION

We have described the marine ecosystem in the temperate coastal northeast Pacific region at present, and then its response to OA. However, the available information is limited. For some organisms, no OA studies exist (*e.g.* Geoduck Clam, rockfish). In general there are more studies, with respect to distributions and OA impacts, on species that are easier to observe, are of commercial value (*e.g.* oysters) or that threaten human health (*e.g.* harmful algae). The results of studies like these are often adopted when similar research on native organisms is not available (as we have done), limiting the ability to predict responses with confidence. Furthermore, OA is only one aspect of climate change and predicting shifts in marine ecosystems, and the degree to which they are caused by natural or anthropogenic forcing, is a highly complex problem. In the following, we discuss these and other issues that influence our evaluation.

Caveats

The number of studies related to OA is growing rapidly. While experiments in these studies are highly valuable, translating their results into changes in the real world is challenging. For example, wild populations of marine organisms will adapt (both physiologically in a single lifespan and genetically over multiple generations) to their changing environment, which is difficult or impossible to capture *in vitro*. However, using temperature-dependent adaptation as a guide,

Kelly and Hofmann [336] caution that the ability to adapt to changing pH may be limited.

In addition, food-web interactions and responses to OA are extremely difficult to predict, but will influence marine populations and could tip the balance from an overall negative impact to a positive one for a given species if a key predator is removed. Ecosystem effects resulting from OA have previously been identified as a key knowledge gap [337]. Furthermore, different life stages, particularly the juvenile stage (*e.g.* echinoderms), often display increased susceptibility to OA, but the impact of exposure of one life stage to low pH conditions on the subsequent life stages has only rarely been studied (but see [206, 212]). Similarly, even in organisms that have been comparatively well studied, not all life stages have been considered and certainly not within the context of the variability in natural conditions (Fig. 2).

Manipulated experiments generally consider present-day atmospheric conditions ($\sim$360–400μatm) to be the control P_{CO2} level and all treatments above that to be 'elevated'. MeanwhileP_{CO2} varies significantly with depth, and is naturally high in the north Pacific [28]. We quantify 'elevated' based on the local P_{CO2} levels at the depths of the organisms in question (Table 1). The combined effect of coastal upwelling, and local remineralisation of high production [27], results in exceptionally high (and variable) subsurface P_{CO2} on the outer BC shelf (Fig. 2). In local and connected inshore waters subsurface P_{CO2} is also high (unpublished data, DI; [6,31]). Thus, many marine organisms in our region are currently experiencing conditions that are viewed as 'elevated' in the literature (Fig. 2). In addition, laboratory treatments often specify environmental conditions (*e.g.* temperature, P_{CO2}) that do not occur in nature and are unlikely to occur, at least locally (*e.g.* [31]). Exposure time may also limit the interpretation of results, as there are distinct differences between treatments that are 'shocked' and those that are allowed to acclimate (*e.g.* [87, 248]).

Finally, defining the carbon state in seawater is not trivial [10] and requires that at least two of the four carbon parameters (DIC, TA, P_{CO2}, pH) be measured. The quality of the measurements and manipulation in the laboratory work cited here is variable. While the high degree of accuracy and precision required by chemical oceanographers [10] is in general not necessary to obtain insight from biological manipulation experiments, the equations that define the carbon system lead to compounding errors when calculating one of the unknowns. Thus, a moderate uncertainty in P_{CO2} may translate to an estimated pH that has little, or no meaning.

Climate Change—the Whole Picture

The ocean has absorbed a significant portion of the anthropogenically produced carbon [2] and that has caused on average a 30% change in surface ocean acidity [5]. However the annual variability in surface P_{CO2} and pH in dynamic regions like the BC [24] and WA [31] coasts is generally more than two orders of magnitude greater than the annual atmospheric increase in CO_2. In other words, we expect the OA trend to be present, but overlaid is a signal with large amplitude.

Climate change may alter this dynamic natural cycle so that negative impacts associated with high acidity are experienced earlier in the coastal northeast Pacific than elsewhere, regardless of OA. There are critical times during the year when carbon conditions (particularly in the upper mixed layer; 20–30 m on the outer coast; ~10 m or less in protected waterways) change dramatically. For example, the spring bloom in the Strait of Georgia causes a large and rapid increase in surface pH (Ben Moore Maley pers. comm., University of British Columbia, Vancouver BC) and the timing of this event varies significantly from year to year [338]. On the outer shelf, the onset of summer upwelling brings lower pH water over the continental shelf and decreases pH (on average) throughout the entire water column. Climate change may alter the strength, timing [339–341], or even the variability in the timing, of such events. Thus, the influence of climate change on weather may play a critical role, that will only be exacerbated as OA progresses.

In addition to changing weather, sea surface temperatures are expected to increase and subsurface O_2 is expected to decrease (leading to increased occurrence of hypoxia) with climate change, concurrent with OA. Temperature has a large effect on marine organisms because metabolism increases as the ocean warms, consequently increasing energetic costs. As a result, changes in present-day distributions of marine organisms have already been linked to changes in temperature [342]. Thus, a 'multi-stressor' approach is required to understand the net effect of climate change on marine organisms. The net effect of all three stressors (warming, hypoxia and OA) may be synergistic and has been generally described as a narrowing of the thermal ranges in which organisms can perform well, and a decrease in maximal performance [343]. Lastly, changes in human behaviour (*e.g.* fishing) as climate change and OA progress may also play an important, and possibly additive, role in shaping future marine ecosystems (*e.g.* [344]).

CONCLUSIONS

There remain significant knowledge gaps with respect to the biological impacts of OA on marine ecosystems globally, and locally. The most critical impacts will likely be indirect as a result of food web changes, and so are highly complex and difficult to predict even with extensive study. Furthermore, OA related changes will occur in concert with other climate change impacts that may be even more severe (see above). In particular, increasing temperature and decreasing dissolved oxygen are likely to produce synergistic effects.

The northeast Pacific region naturally has waters low in pH (undersaturated with respect to aragonite) near the surface. Thus, it is potentially more vulnerable to OA than other regions. We summarise the most relevant risks and identify key knowledge gaps, given present-day knowledge, to Pacific Canadian fisheries and marine ecosystems in the order of immediacy and certainty.

- Shellfish aquaculture is highly susceptible to OA due to the direct impact of OA on shell formation and the dependence of the industry on hatchery production. These impacts are already experienced in BC (and WA). Wild shellfish experience similar difficulties but have the opportunity to adapt (*e.g.* [197]) and so will likely not be affected as rapidly and severely.

- There are no studies on Geoduck Clams, which are responsible for a lucrative wild fishery and a growing aquaculture industry in BC (although the latter is still in its infancy).

- The commercial BC fishery is dominated monetarily by salmon aquaculture. While uncertainty remains low, it is anticipated that the fish-killing alga *Heterosigma akashiwo*will gain a competitive advantage under OA, making blooms more frequent. Such blooms are already a significant issue for this industry in BC.

- Neurotoxins produced by other harmful algae are expected to become more potent under OA. Such blooms already cause shellfish closures in BC. If this increase in toxicity occurs, the shellfish industry will be affected. In addition, these toxins may cause decreased reproductive success, and even mass mortality, at higher trophic levels including fish, seabirds and marine mammals.

- Food web changes due to OA (*e.g.* in BC changes in the species composition of phytoplankton and decline of pteropods) are anticipated but remain unknown, as are the impacts of these lower level changes on higher trophic levels.

- Finfish are likely to experience OA impacts through foodweb changes. In BC examples include: the decline of pteropods, that are directly preyed upon by some fish (particularly Pink Salmon), and the anticipated decline of some echinoderms, that are eaten by various species of rockfish and flatfish.

- Habitat changes may also have a critical negative impact, in particular for juvenile fish. While these impacts remain highly uncertain, there may be a shift from upright macroalgae to algal turf. Also, local coral species (in BC primarily octocorals) that provide vertical structure may decline. Direct impacts of OA on finfish may also occur, but only at relatively high levels of CO_2.

- There are few direct OA studies on local finfish species and none on Pacific Halibut and salmon, which drive the sport fishing industry. Similarly there are no studies on the adaptation of these local species to OA and multiple stressors, like temperature and O_2, that will be changing at the same time. Because sport fishing dominates fishery related income in BC, this knowledge gap is significant.

- Behavioural changes at various trophic levels have been observed (*e.g.* increased downward swimming in phytoflagellates, decreased detection and avoidance of predators in larval fish) and postulated (*e.g.* increased movement to OA refugia such as eelgrass meadows). Such behavioural changes might alter the structure of marine communities in BC, and present another knowledge gap.

- Crabs may experience negative impacts under OA while other crustaceans significant to the harvest fishery in BC, like prawns, have not been well studied but appear to be more strongly sensitive to temperature than OA. In general, the juvenile stages of crustaceans are most vulnerable to OA, growing more slowly because they need to expend more energy under OA.

ACKNOWLEDGMENTS

We thank the following people for their helpful conversations and correspondence: Richard Beamish, Jim Boutillier, Stephen Cairns, Kim Conway, Carol Cooper, Paul Covert, Lyanne Curtis, Nancy Davis, Jason Dunham, Jessica Finney, John Ford, Ian Forster, Moira Galbraith, Nicky Haigh, Christopher Harley, John Holmes, Catriona Hurd, Hyewon Kim, Joanne Lessard, Sally Leys, Dave Mackas, Shayne MacLellan, Patrick Mahoux, Erin McClelland, Kristi Miller, James Murray, Linda Nichol, Miriam O, Chris Pearce, Angelica Peña, Ian Perry, Dave Preikshot, Peter Ross, George Somero, Bob Stone, Karyn Suchy, Terri Sutherland, Curtis Suttle, Ron Tanasichuk, Phil Tortell, Verena Tunnicliffe, and Lynne Yamanaka. Additionally, we thank the following people for their help with data: Leslie Barton, Jennifer Boldt, Linnea Flostrand, Moira Galbraith, Georg Jorgensen, Lisa Lacko, Norm Olsen, and

Dave Preikshot. The manuscript benefited significantly from the helpful comments of two anonymous reveiwers. This work results from the project 'Ocean Acidification Impacts on Marine Ecosystems', funded by Fisheries and Oceans Canada's International Governance Strategy program.

AUTHOR CONTRIBUTIONS

Conceived and designed the experiments: DI RH CH AE. Performed the experiments: RH CH DI AE HN. Analyzed the data: DI RH CH AE. Wrote the paper: DI RH CH AE.

REFERENCES

1.   Intergovernmental Panel on Climate Change (2013) Climate Change 2013: The Physical Science Basis. Contribution of Working Group I to the Fifth Assessment Report of the Intergovernmental Panel on Climate Change. Cambridge, UK and New York, USA: Cambridge University Press, 1535 pp.

2.   Canadell JG, Le Quere C, Raupach MR, Field CB, Buitenhuis ET, et al. (2007) Contributions to accelerating atmospheric CO2 growth from economic activity, carbon intensity, and efficiency of natural sinks. Proc Natl Acad Sci USA 104: 18866–18870. doi: 10.1073/pnas.0702737104. pmid:17962418

3.   Doney SC, Feely VJFRA, Kleypas JA (2009) Ocean acidification: the other CO2 problem. Ann Rev Mar Sci 1: 169–192. doi: 10.1146/annurev. marine.010908.163834. pmid:21141034

4.   Intergovernmental Panel on Climate Change (2011) Workshop Report of the Intergovernmental Panel on Climate Change Workshop on Impacts of Ocean Acidification on Marine Biology and Ecosystems. Stanford, USA: Working Group II Technical Support Unit, Carnegie Institution, 164 pp.

5.   Raven J, Caldeira K, Elderfield H, Hoegh-Guldberg O, Liss P, et al. (2005) Ocean acidification due to increasing carbon dioxide. The Royal Society 12/05: viii + 60p.

6.   Feely RA, Alin SR, Sabine CL, Warner M, Devol A, et al. (2010) The combined effects of ocean acidification, mixing and respiration on pH and carbonate saturation in an urbanized estuary. Estuar Coast Shelf Sci 88: 442–449. doi: 10.1016/j.ecss.2010.05.004.

7.   Doney SC, Mahowald N, Lima I, Feely RA, Mackenzie FT, et al. (2007) Impact of anthropogenic atmospheric nitrogen and sulfur deposition on ocean acidification and the inorganic carbon system. Proc Natl Acad Sci

USA 104: 14580–14585. doi: 10.1073/pnas.0702218104. pmid:17804807

8.   Howarth RW, Sharpley A, Walker D (2002) Sources of nutrient pollution to the coastal waters in the United States: Implications for achieving coastal water quality goals. Estuaries 25: 656–676. doi: 10.1007/BF02804898.

9.   Brewer PG (2013) A short history of ocean acidification science in the 20th century: a chemist's view. Biogeosciences 10: 7411–7422. doi: 10.5194/bg-10-7411-2013.

10. Dickson AG, Sabine CL, Christian JR (2007) Determination of dissolved organic carbon and total dissolved nitrogen in sea water. In: Guide to best practices of ocean CO2 measurements, PICES Special Publication 3.

11. Feely RA, Sabine CL, Hernandez-Ayon JM, Ianson D, Hales B (2008) Evidence for upwelling of corrosive 'acidified' water onto the continental shelf. Science 320: 1490–1492. doi: 10.1126/science.1155676. pmid:18497259

12. Kleypas J, Buddemeier RW, Archer D, Gattuso JP, Langdon C, et al. (1999) Geochemical consequences of increased atmospheric carbon dioxide on coral reefs. Science 284: 118–120. doi: 10.1126/science.284.5411.118. pmid:10102806

13. Mucci A (1983) The solubility of calcite and aragonite in seawater at various salinities, temperatures and 1 atmosphere total pressure. Am J Sci 238: 780–799. doi: 10.2475/ajs.283.7.780.

14. Sarmiento JL, Gruber N (2006) Ocean Biogeochemical Dynamics. Princeton University Press, 503 pp.

15. Feely RA, Byrne RH, Acker JG, Beltzer PR, Chen CTA, et al. (1998) Winter-summer variations of calcite and aragonite saturation in the northeast Pacific. Mar Chem 25: 227–241. doi: 10.1016/0304-4203(88)90052-7.

16. Weiner S, Dove PM (2003) An overview of biomineralization processes and the problem of the vital effect, Mineralog. Soc. Am. Geochem. Soc., volume 54. pp. 1–29.

17. Waldbusser GG, Brunner EL, Haley BA, Hales B, Langdon CJ, et al. (2013) A developmental and energetic basis linking larval oyster shell formation to acidification sensitivity. Geophys Res Lett 40: 2171–2176. doi: 10.1002/grl.50449.

18. Hofmann AF, Peltzer ET, Brewer PG (2013) Kinetic bottlenecks to chemical exchange rates for deep-sea animals—Part 2: Carbon dioxide. Biogeosciences 10: 2409–2425. doi: 10.5194/bg-10-5049-2013.

19. Claiborne JB, Edwards SL, Morrison-Shetlar AI (2002) Acid-base regulation in fishes: cellular and molecular mechanisms. J Exp Zool 293: 302–319. doi: 10.1002/jez.10125. pmid:12115903

20. Bach LT, Riebesell U, Shulz KG (2011) Distinguishing between the effects of ocean acidification and ocean carbonation in the coccolithophore Emiliania huxleyi. Limnol Oceanogr 56: 2040–2050. doi: 10.4319/lo.2011.56.6.2040.

21. Riebesell U, Fabry VJ, L H, Gattuso JP (2011) Guide to best practices for ocean acidification research and data reporting. Publications Office of the European Union, Luxembourg: EPOCA—European Project on OCean Acidification, 260 pp.

22. Volk T, Hoffert MI (1985) Ocean carbon pumps: analysis of relative strengths and efficiencies in ocean-driven atmospheric pCO2 changes. In: Sundquist ET, Broecker WS, editors, The carbon cycle and atmospheric CO2, natural variations archean to present. Washington, DC, volume AGU Monograph 32, pp. 99–110.

23. Nemcek N, Ianson D, Tortell PD (2008) A high-resolution survey of DMS, CO2, and O2/Ar distributions in productive coastal waters. Global Biogeochem Cycles 22: GB2009. doi: 10.1029/2006GB002879.

24. Ianson D, Allen SE (2002) A two-dimensional nitrogen and carbon flux model in a coastal upwelling region. Glob Biogeochem Cycles 16: 10.1029/2001GB001451. . doi: 10.1029/2001GB001451.

25. Tortell PD, Merzouk A, Ianson D, Pawlowicz R, Yelland DR (2012) Influence of regional climate forcing on surface water pCO2, δO2/Ar and dimethylsulfide (DMS) along the southern British Columbia coast. Cont Shelf Res 47: 119–132. doi: 10.1016/j.csr.2012.07.007.

26. Thomson RE (1981) Oceanography of the British Columbia Coast. 56. Ottawa: Canadian Special Publication of Fisheries and Aquatic Sciences, 291 pp.

27. Ianson D, Harris S, Allen SE, Orians K, Varela D, et al. (2003) The inorganic carbon system in the coastal upwelling region west of Vancouver Island, Canada. Deep Sea Res I 50: 1023–1042. doi: 10.1016/S0967-0637(03)00114-6.

28. Feely RA, Sabine CL, Lee K, Berelson W, Kleypas J, et al. (2004) Impact of anthropogenic CO2 on the CaCO3 system in the oceans. Science 305: 362. doi: 10.1126/science.1097329. pmid:15256664

29. Feely RA, Sabine CL, Byrne RH, Millero FJ, Dickson AG, et al. (2012) Decadal changes in the aragonite and calcite saturation state

of the Pacific ocean. Global Biogeochem Cycles 26: GB3001. doi: 10.1029/2011GB004157.

30. Ware DM, Thomson RE (2005) Bottom-up ecosystem trophic dynamics determine fish production in the Northeast Pacific. Science 308: 1280–1284. doi: 10.1126/science.1109049. pmid:15845876

31. Reum JCP, Alin SR, Feely RA, Newton J, Warner M, et al. (2014) Seasonal carbonate chemistry covariation with temperature, oxygen, and salinity in a fjord estuary: Implications for the design of ocean acidification experiments. PLoS ONE 9: e89619. doi: 10.1371/journal. pone.0089619. pmid:24586915

32. Ianson D (2013) The increase in carbon along the Canadian Pacific coast. In: Christian JR, Foreman MGG, editors, Climate Trends and Projections for the Pacific Large Area Basin. volume 3032 of Can. Tech. Rep. Fish. Aquat. Sci., pp. xi + 113p.

33. Stroomer C, Wilson M (2013) British Columbia's fisheries and aquaculture sector, 2012 Edition. British Columbia Statistics, Canada, 104 pp.

34. BC Ministry of Agriculture (2012) British Columbia Seafood Industry 2011 Year in Review. British Columbia Ministry of Agriculture, ii + 14 pp. URL http://www.env.gov.bc.ca/omfd/reports/Seafood-YIR-2011.pdf.

35. Fabry VJ, Seibel BA, Feely RA, Orr JC (2008) Impacts of ocean acidification on marine fauna and ecosystem processes. ICES J Mar Sci 65: 414–432. doi: 10.1093/icesjms/fsn048.

36. Kroeker KJ, Kordas RL, Crim R, Hendriks IE, Ramajo L, et al. (2013) Impacts of ocean acidification on marine organisms: quantifying sensitivities and interaction with warming. Glob Change Biol 19: 1884–1896. doi: 10.1111/gcb.12179.

37. Cooley SR, Doney SC (2009) Anticipating ocean acidification's economic consequences for commercial fisheries. Environ Res Lett 4: 024007. doi: 10.1088/1748-9326/4/2/024007.

38. Preikshot DB (2007) The Influence of Geographic Scale, Climate and Trophic Dynamics upon North Pacific Oceanic Ecosystem Models. Ph.D. thesis, University of British Columbia, Canada.

39. Fisheries and Oceans Canada (2013) Pacific Region Integrated Fisheries Management Plan—Euphausiids—January 1, 2013 to December 31, 2017. http://www.pac.dfo-mpo.gc.ca/fm-gp/mplans/2013/krill-sm-2013-17-eng.pdf, 52 pp.

40. Ianson D, Feely RA, Sabine CL, Juranek L (2009) Features of coastal upwelling regions that determine net air-sea CO2 flux. J Oceanogr 65: 677–687. doi: 10.1007/s10872-009-0059-z.

41. Lewis ER, Wallace DWR (1998) Program Developed for CO2 System Calculations. DOE (U.S. Department of Energy). ORNL/CDIAC-105.

42. Lueker TJ, Dickson AG, Keeling CD (2000) Ocean pCO2 calculated from dissolved inorganic carbon, alkalinity, and equations for K1 and K2: validation based on laboratory measurements of CO2 in gas and seawater at equilibrium. Marine Chemistry 70: 105–119. doi: 10.1016/S0304-4203(00)00022-0.

43. Ryther JH (1969) Photosynthesis and fish production in the sea. Science 166: 72–76. doi: 10.1126/science.166.3901.72. pmid:5817762

44. Haigh R, Taylor FJR (1991) Mosaicism of microplankton communities in the northern Strait of Georgia, British Columbia. Mar Biol 110: 301–314. doi: 10.1007/BF01313717.

45. Haigh R, Taylor FJR, Sutherland TF (1992) Phytoplankton ecology of Sechelt Inlet, a fjord system on the British Columbia coast. I. General features of the nano- and microplankton. Mar Ecol Prog Ser 89: 117–134. doi: 10.3354/meps089117.

46. Harris SJ, Varela DE, Whitney FW, Harrison PJ (2009) Nutrient and phytoplankton dynamics off the west coast of Vancouver Island during the 1997/98 ENSO event. Deep Sea Res II 56: 2487–2502. doi: 10.1016/j.dsr2.2009.02.009.

47. Harrison PJ, Fulton JD, Taylor FJR, Parsons TR (1983) Review of the biological oceanography of the Strait of Georgia: pelagic environment. Can J Fish Aquat Sci 40: 1064–1094. doi: 10.1139/f83-129.

48. Ianson D, Pond S, Parsons TR (2001) The spring phytoplankton bloom in the coastal temperate ocean: growth criteria and seeding from shallow embayments. J Oceanogr 57: 723–734. doi: 10.1023/A:1021288510407.

49. Perry RI, Dilke BR, Parsons TR (1983) Tidal mixing and summer plankton distributions in Hecate Strait, British Columbia. Can J Fish Aquat Sci 40: 871–887. doi: 10.1139/f83-114.

50. Peterson TD, Toews HNJ, Robinson CLK, Harrison PJ (2007) Nutrient and phytoplankton dynamics in the Queen Charlotte Islands (Canada) during the summer upwelling seasons of 2001–2002. J Plankton Res 29: 219–239. doi: 10.1093/plankt/fbm010.

51. Lassiter AM, Wilkerson FP, Dugdale RC, Hogue VE (2006) Phytoplankton assemblages in the CoOP-WEST coastal upwelling area. Deep Sea Res II 53: 3063–3077. doi: 10.1016/j.dsr2.2006.07.013.

52. Olson MB, Lessard EJ, Wong CHJ, Bernhardt MJ (2006) Copepod feeding selectivity on microzooplankton, including the toxigenic diatoms

Pseudo-nitzschia spp., in the coastal Pacific Northwest. Mar Ecol Prog Ser 326: 207–220. doi: 10.3354/meps326207.

53. Gower J (2013) Phytoplankton blooms on the BC coast. In: Irvine JR, Crawford WR, editors, State of physical, biological, and selected fishery resources of Pacific Canadian marine ecosystems in 2012, Canadian Science Advisory Secretariat, Research Document 2013/032. pp. 60–61.

54. Lipsen MS, Crawford DW, Gower J, Harrison PJ (2007) Spatial and temporal variability in coccolithophore abundance and production of PIC and POC in the NE subarctic Pacific during El Niño (1998), La Niña (1999) and 2000 during wind-driven coastal upwelling. Prog Oceanogr 75: 304–325. doi: 10.1016/j.pocean.2007.08.004.

55. Hickey BM, Banas NS (2008) Why is the northern end of the California current system so productive? Oceanography 21: 90–107. doi: 10.5670/oceanog.2008.07.

56. Hickey BM, Trainer VL, Kosro PM, Adams NG, Connolly TP, et al. (2013) A springtime source of toxic Pseudo-nitzschia cells on razor clam beaches in the Pacific Northwest. Harmful Algae 25: 1–14. doi: 10.1016/j.hal.2013.01.006.

57. Trainer VL, Wells ML, Cochlan WP, Trick CG, Bill BD, et al. (2009) An ecological study of a massive bloom of toxigenic Pseudo-nitzschia cuspidata off the Washington State coast. Limnol Oceanogr 54: 1461–1474. doi: 10.4319/lo.2009.54.5.1461.

58. Taylor FJR, Harrison PJ (2002) Harmful algal blooms in western Canadian coastal waters. In: Taylor FJR, Trainer VL, editors, Harmful Algal Blooms in the PICES Region of the North Pacific, North Pacific Marine Science Organization, PICES Scientific Report 23. pp. 77–88.

59. Bejarano AC, VanDola FM, Gulland FM, Rowles TK, Schwacke LH (2008) Production and toxicity of the marine biotoxin domoic acid and its effects on wildlife: a review. Hum Ecol Risk Assess 14: 544–567. doi: 10.1080/10807030802074220.

60. Deeds JR, Landsberg JH, Etheridge SM, Pitcher GC, Longan SW (2008) Non-traditional vectors for paralytic shellfish poisoning. Mar Drugs 6: 308–348. doi: 10.3390/md20080015. pmid:18728730

61. Taylor FJR, Haigh R (1993) The ecology of fish-killing blooms of the chloromonad flagellate Heterosigma in the Strait of Georgia and adjacent waters. In: Smayda TJ, Shimuzu Y, editors, Toxic Phytoplankton Blooms in the Sea, Elsevier Science. pp. 705–710.

62. Rensel JEJ (2007) Fish kills from the harmful alga Heterosigma akashiwo in Puget Sound: recent blooms and review. Technical report,

National Oceanic and Atmospheric Administration (NOAA), Center for Sponsored Coastal Ocean Research (CSCOR), v+58 p.

63. Twiner MJ, Trick CG (2000) Possible physiological mechanisms for production of hydrogen peroxide by the ichthyotoxic flagellate Heterosigma akashiwo. J Plankton Res 22: 1961–1975. doi: 10.1093/plankt/22.10.1961.

64. Yang CZ, Albright LJ, Yousif AN (1995) Oxygen-radical-mediated effects of the toxic phytoplankter Heterosigma carterae on juvenile rainbow trout Oncorhynchus mykiss. Dis Aquat Org 23: 101–108. doi: 10.3354/dao023101.

65. Kim D, Nakamura A, Okamoto T, Komatsu N, Oda T, et al. (1999) Toxic potential of the raphidophyte Olisthodiscus luteus: mediation by reactive oxygen species. J Plankton Res 21: 1017–1027. doi: 10.1093/plankt/21.6.1017.

66. Haigh N, Esenkulova S (2013) Economic losses to the British Columbia salmon aquaculture industry due to harmful algal blooms 2009–2012. In: Workshop 6. Economic impacts of harmful algal blooms on fisheries and aquaculture. North Pacific Marine Science Organization (PICES), Nanaimo, BC, October 11–20, 2013.

67. Engel A, Zondervan I, Aerts K, Beaufort L, Benthien A, et al. (2005) Testing the direct effect of CO2 concentration on a bloom of the coccolithophorid Emiliania huxleyi in mesocosm experiments. Limnol Oceanogr 50: 493–507. doi: 10.4319/lo.2005.50.2.0493.

68. Nielsen LT, Jakobsen HH, Hansen PJ (2010) High resilience of two coastal plankton communities to twenty-first century seawater acidification: evidence from microcosm studies. Mar Biol Res 6: 542–555. doi: 10.1080/17451000903476941.

69. Riebesell UM, Tortell PD (2011) Effects of ocean acidification on pelagic organisms and ecosystems. In: Gattuso JP, Hansson L, editors, Ocean Acidification, Oxford University Press. pp. 99–121.

70. Riebesell U, Zondervan I, Rost B, Tortell PD, Zeebe RE, et al. (2000) Reduced calcification of marine plankton in response to increased atmospheric CO2. Nature 407: 364–367. doi: 10.1038/35030078. pmid:11014189

71. Hinga KR (2002) Effects of pH on coastal marine phytoplankton. Mar Ecol Prog Ser 238: 281–300. doi: 10.3354/meps238281.

72. Beardall J, Raven JA (2004) The potential effects of global climate change on microalgal photosynthesis, growth and ecology. Phycologia 43: 26–40. doi: 10.2216/i0031-8884-43-1-26.1.

73. Moroney JV, Somanchi A (1999) How do algae concentrate CO2 to increase the efficiency of photosynthetic carbon fixation. Plant Phys 119: 9–16. doi: 10.1104/pp.119.1.9.

74. Tortell PD, Rau GH, Morel FMM (2000) Inorganic carbon acquisition in coastal Pacific phytoplankton communities. Limnol Oceanogr 45: 1485–1500. doi: 10.4319/lo.2000.45.7.1485.

75. Raven JA, Cockell CS, de la Rocha CL (2008) The evolution of inorganic carbon concentrating mechanisms in photosynthesis. Phil Trans R Soc Lond B, Biol Sci 363: 2641–2650. doi: 10.1098/rstb.2008.0020.

76. Nimer NA, Iglesias-Rodriguez MD, Merrett MJ (1997) Bicarbonate utilization by marine phytoplankton species. J Phycol 33: 625–631. doi: 10.1111/j.0022-3646.1997.00625.x.

77. Bellerby RGJ, Schulz KG, Riebesell U, Neill C, Nondal G, et al. (2008) Marine ecosystem community carbon and nutrient uptake stoichiometry under varying ocean acidification during the PeECE III experiment. Biogeosciences 5: 1517–1527. doi: 10.5194/bg-5-1517-2008.

78. Egge JK, Thingstad TF, Larsen A, Engel A, Wohlers J, et al. (2009) Primary production during nutrient-induced blooms at elevated CO2 concentrations. Biogeosciences 6: 877–885. doi: 10.5194/bg-6-877-2009.

79. Hein M, Sand-Jensen K (1997) CO2 increases oceanic primary production. Nature 388: 526–527. doi: 10.1038/41457.

80.

81. Riebesell U, Schulz KG, Bellerby RGJ, Botros M, Fritsche P, et al. (2007) Enhanced biological carbon consumption in a high CO2 ocean. Nature 450: 545–548. doi: 10.1038/nature06267. pmid:17994008

82. Clark DR, Flynn KJ (2000) The relationship between the dissolved inorganic carbon concentration and growth rate in marine phytoplankton. Proc R Soc Lond, B, Biol Sci 267: 953–959. doi: 10.1098/rspb.2000.1096.

83. Fu FX, Zhang Y, Warner ME, Feng Y, Sun J, et al. (2008) A comparison of future increased CO2 and temperature effects on sympatric Heterosigma akashiwo and Prorocentrum minimum. Harmful Algae 7: 76–90. doi: 10.1016/j.hal.2007.05.006.

84. Burkhardt S, Riebesell U, Zondervan I (1999) Effects of growth rate, CO2 concentration, and cell size on the stable carbon isotope fractionation in marine phytoplankton. Geochim Cosmochim Acta 63: 3729–3741. doi: 10.1016/S0016-7037(99)00217-3.

85. Tortell PD, Payne CD, Li Y, Trimborn S, Rost B, et al. (2008) CO2

sensitivity of Southern Ocean phytoplankton. Geophys Res Lett 35: L04605. doi: 10.1029/2007GL032583.

86. Hare CE, Leblanc K, DiTullio GR, Kudela RM, Zhang Y, et al. (2007) Consequences of increased temperature and CO2 for phytoplankton community structure in the Bering Sea. Mar Ecol Prog Ser 352: 9–16. doi: 10.3354/meps07182.

87. Tortell PD, DiTullio GR, Sigman DM, Morel FMM (2002) CO2 effects on taxonomic composition and nutrient utilization in an equatorial Pacific phytoplankton assemblage. Mar Ecol Prog Ser 236: 37–43. doi: 10.3354/meps236037.

88. Kim H, Spivack AJ, Menden-Deuer S (2013) pH alters the swimming behaviors of the raphidophyte Heterosigma akashiwo: implications for bloom formation in an acidified ocean. Harmful Algae 26: 1–11. doi: 10.1016/j.hal.2013.03.004.

89. Hallegraeff GM (2010) Ocean climate change, phytoplankton community responses, and harmful algal blooms: a formidable predictive challenge. J Phycol 46: 220–235. doi: 10.1111/j.1529-8817.2010.00815.x.

90. Hwang DF, Lu YH (2000) Influence of environmental and nutritional factors on growth, toxicity, and toxin profile of dinoflagellate Alexandrium minutum. Toxicon 38: 1491–1503. doi: 10.1016/S0041-0101(00)00080-5. pmid:10775750

91. Fu FX, Place AR, Garcia NS, Hutchins DA (2010) CO2 and phosphate availability control the toxicity of the harmful bloom dinoflagellate Karlodinium veneficum. Aquat Microb Ecol 59: 55–65. doi: 10.3354/ame01396.

92. Tatters AO, Flewelling LJ, Fu F, Granholm AA, Hutchins DA (2013) High CO2 promotes the production of paralytic shellfish poisoning toxins by Alexandrium catenella from Southern California waters. Harmful Algae 30: 37–43. doi: 10.1016/j.hal.2013.08.007.

93. Sun J, Hutchins DA, Feng Y, Seubert EL, Caron DA, et al. (2011) Effects of changing pCO2 and phosphate availability on domoic acid production and physiology of the marine harmful bloom diatom Pseudo-nitzschia multiseries. Limnol Oceanogr 56: 829–840. doi: 10.4319/lo.2011.56.3.0829.

94. Tatters AO, Fu FX, Hutchins DA (2012) High CO2 and silicate limitation synergistically increase the toxicity of Pseudo-nitzschia fraudulenta. PLoS ONE 7: e32116. doi: 10.1371/journal.pone.0032116. pmid:22363805

95. Brown CW, Yoder JA (1994) Coccolithophorid blooms in the global ocean. J Geophys Res 99: 7467–7482. doi: 10.1029/93JC02156.

96. Westbroek P, Young JR, Linschooten K (1989) Coccolith production (biomineralization) in the marine alga Emiliania huxleyi. J Protozool 36: 368–373. doi: 10.1111/j.1550-7408.1989.tb05528.x.

97. Iglesias-Rodriguez MD, Halloran PR, Rickaby REM, Hall IR, Colmenero-Hidalgo E, et al. (2008) Phytoplankton calcification in a high-CO2 world. Science 320: 336–340. doi: 10.1126/science.1154122. pmid:18420926

98. Shi D, Xu Y, Morel FMM (2009) Effects of the pH/pCO2 control method on medium chemistry and phytoplankton growth. Biogeosciences 6: 1199–1207. doi: 10.5194/bg-6-1199-2009.

99. Zondervan I, Zeebe RE, Rost B, Riebesell U (2001) Decreasing marine biogenic calcification: a negative feedback on rising atmospheric pCO2. Glob Biogeochem Cycles 5: 507–516. doi: 10.1029/2000GB001321.

100. Ridgwell A, Schmidt DN, Turley C, Brownlee C, Maldonado MT, et al. (2009) From laboratory manipulations to Earth system models: scaling calcification impacts of ocean acidification. Biogeosciences 6: 2611–2623. doi: 10.5194/bg-6-2611-2009.

101. Delille B, Harlay J, Zondervan I, Jacquet S, Chou L, et al. (2005) Response of primary production and calcification to changes of pCO2 during experimental blooms of the coccolithophorid Emiliania huxleyi. Glob Biogeochem Cycles 19: GB2023, 1–14. doi: 10.1029/2004GB002318.

102. Beaufort L, Probert I, de Garidel-Thoron T, Bendif EM, Ruiz-Pino D, et al. (2011) Sensitivity of coccolithophores to carbonate chemistry and ocean acidification. Nature 476: 80–83. doi: 10.1038/nature10295. pmid:21814280

103. Bates C (2004) E-flora atlas of flora of British Columbia. An introduction to the (macro) algae of British Columbia. In: Klinkenberg B, editor, E-Flora BC: Atlas of the Plants of British Columbia [www.eflora.bc.ca]. Lab for Advanced Spatial Analysis, Department of Geography, University of British Columbia, Vancouver. [Date Accessed: 2013-06-06], Royal British Columbia Museum. URL http://www.geog.ubc.ca/biodiversity/eflora/algae.html.

104. Druehl LD (2000) Pacific Seaweeds: A Guide to Common Seaweeds of the West Coast. Harbour Publishing, Madeira Park BC.

105. Mann KH (1973) Seaweeds: their productivity and strategy for growth. Science 182: 975–981. doi: 10.1126/science.182.4116.975. pmid:17833778

106. Foster MS, Schiel DR (1985) The ecology of giant kelp forests in California: a community profile. US Fish Wildl Serv Biol Rep 85 (7.2): xv+152 p.

107. Nelson WA (2009) Calcified macroalgae—critical to coastal ecosystems and vulnerable to change: a review. Mar Freshw Res 60: 787–801. doi: 10.1071/MF08335.

108. Cornwall CE, Hepburn CD, Pritchard D, Currie KI, McGraw CM, et al. (2012) Carbon-use strategies in macroalgae: differential responses to lowered pH and implications for ocean acidification. J Phycol 48: 137–144. doi: 10.1111/j.1529-8817.2011.01085.x.

109. Hurd CL, Hepburn CD, Currie KI, Raven JA, Hunter KA (2009) Testing the effects of ocean acidification on algal metabolism: considerations for experimental designs. J Phycol 45: 1236–1251. doi: 10.1111/j.1529-8817.2009.00768.x.

110. Kubler JE, Johnston AM, Raven JA (1999) The effects of reduced and elevated CO2 and O2 on the seaweed Lomentaria articulata. Plant Cell Environ 22: 1303–1310. doi: 10.1046/j.1365-3040.1999.00492.x.

111. Hepburn CD, Pritchard DW, Cornwall CE, McLeod RJ, Beardall J, et al. (2011) Diversity of carbon use strategies in a kelp forest community: implications for a high CO2 ocean. Glob Change Biol 17: 2488–2497. doi: 10.1111/j.1365-2486.2011.02411.x.

112. Swanson AK, Fox CH (2007) Altered kelp (Laminariales) phlorotannins and growth under elevated carbon dioxide and ultraviolet-B treatments can influence associated intertidal food webs. Glob Change Biol 13: 1696–1709. doi: 10.1111/j.1365-2486.2007.01384.x.

113. Roleda MY, Morris JN, McGraw CM, Hurd CL (2012) Ocean acidification and seaweed reproduction: increased CO2 ameliorates the negative effect of lowered pH on meiospore germination in the giant kelp Macrocystis pyrifera (Laminariales, Phaeophyceae). Glob Change Biol 18: 854–864. doi: 10.1111/j.1365-2486.2011.02594.x.

114. Hofmann LC, Straub S, Bischof K (2012) Competition between calcifying and noncalcifying temperate marine macroalgae under elevated CO2 levels. Mar Ecol Prog Ser 464: 89–105. doi: 10.3354/meps09892.

115. Cornwall CE, Hepburn CD, Pilditch CA, Hurd CL (2013) Concentration boundary layers around complex assemblages of macroalgae: Implications for the effects of ocean acidification on understory coralline algae. Limnol Oceanogr 58: 121–130. doi: 10.4319/lo.2013.58.1.0121.

116. Harley CDG, Anderson KM, Demes KW, Jorve JP, Kordas RL, et al. (2012) Effects of climate change on global seaweed communities. J Phycol 48: 1064–1078. doi: 10.1111/j.1529-8817.2012.01224.x.

117. Hurd CL, Cornwall CE, Currie K, Hepburn CD, McGraw CM, et al. (2011) Metabolically induced pH fluctuations by some coastal calcifiers exceed

projected 22nd century ocean acidification: a mechanism for differential susceptibility? Glob Change Biol 17: 3254–3262. doi: 10.1111/j.1365-2486.2011.02473.x.

118. Roleda MY, Boyd PW, Hurd CL (2012) Before ocean acidification: calcifier chemistry lessons. J Phycol 48: 840–843. doi: 10.1111/j.1529-8817.2012.01195.x.

119. Gao KS, Helbling EW, Hader DP, Hutchins DA (2012) Responses of marine primary producers to interactions between ocean acidification, solar radiation, and warming. Mar Ecol Prog Ser 470: 167–189. doi: 10.3354/meps10043.

120. Gao KS, Zheng YQ (2010) Combined effects of ocean acidification and solar UV radiation on photosynthesis, growth, pigmentation and calcification of the coralline alga Corallina sessilis (Rhodophyta). Glob Change Biol 16: 2388–2398. doi: 10.1111/j.1365-2486.2009.02113.x.

121. Koch M, Bowes G, Ross C, Zhang XH (2013) Climate change and ocean acidification effects on seagrasses and marine macroalgae. Glob Change Biol 19: 103–132. doi: 10.1111/j.1365-2486.2012.02791.x.

122. Lessard J, Campbell A (2007) Describing northern abalone, Haliotis kamtschatkana, habitat: focusing rebuilding efforts in British Columbia, Canada. J Shell Res 26: 677–686. doi: 10.2983/0730-8000(2007)26%5B677:DNAHKH%5D2.0.CO;2.

123. Kuffner IB, Andersson AJ, Jokiel PL, Rodgers KS, Mackenzie FT (2008) Decreased abundance of crustose coralline algae due to ocean acidification. Nat Geosci 1: 114–117. doi: 10.1038/ngeo100.

124. Roberts R (2001) A review of settlement cues for larval abalone (Haliotis spp.). J Shellfish Res 20: 571–586.

125. Morse ANC (1991) How do planktonic larvae know where to settle? Am Sci 79: 154–167.

126. Asnaghi V, Chiantore M, Mangialajo L, Gazeau F, Francour P, et al. (2013) Cascading effects of ocean acidification in a rocky subtidal community. PLoS ONE 8: e61978. doi: 10.1371/journal.pone.0061978. pmid:23613994

127. Lubchenco J, Gaines SD (1981) A unified approach to marine plant-herbivore interactions. I. Populations and communities. Annu Rev Ecol Syst 12: 405–437. doi: 10.1146/annurev.es.12.110181.002201.

128. Duffy JE, Hay ME (1990) Seaweed adaptations to herbivory—chemical, structural, and morphological defenses are often adjusted to spatial or temporal patterns of attack. Bioscience 40: 368–375. doi:

10.2307/1311214.

129. Johnson VR, Russell BD, Fabricius KE, Brownlee C, Hall-Spencer JM (2012) Temperate and tropical brown macroalgae thrive, despite decalcification, along natural CO2 gradients. Glob Change Biol 18: 2792–2803. doi: 10.1111/j.1365-2486.2012.02716.x.

130. Jackson JBC, Kirby MX, Berger WH, Bjorndal KA, Botsford LW, et al. (2001) Historical overfishing and the recent collapse of coastal ecosystems. Science 293: 629–637. doi: 10.1126/science.1059199. pmid:11474098

131. Connell S, Kroeker KJ, Fabricius KE, Kline DI, Russell BD (2013) The other ocean acidification problem: CO2 as a resource among competitors for ecosystem dominance. Phil Trans R Soc Lond B, Biol Sci 368: 20120442. doi: 10.1098/rstb.2012.0442.

132. Connell SD, Russell BD (2010) The direct effects of increasing CO2 and temperature on non-calcifying organisms: increasing the potential for phase shifts in kelp forests. Proc R Soc Lond, B, Biol Sci 277: 1409–1415. doi: 10.1098/rspb.2009.2069.

133. Shaffer J (2004) Preferential use of nearshore kelp habitats by juvenile salmon and forage fish. In: Toni D, Fraser DA, editors, Proceedings of the 2003 Georgia Basin / Puget Sound Research Conference, 31 March–3 April, Vancouver, British Columbia. pp. 1–11. URL http://www.caseinlet. org/uploads/SalmonKelp_Shaffer_1_.pdf.

134. Duggins DO, Simenstad CA, Estes JA (1989) Magnification of secondary production by kelp detritus in coastal marine ecosystems. Science 245: 170–173. doi: 10.1126/science.245.4914.170. pmid:17787876

135. Orth RJ, Carruthers TJB, Dennison WC, Duarte CM, Fourqurean JW, et al. (2006) A global crisis for seagrass ecosystems. Bioscience 56: 987–996. doi: 10.1641/0006-3568(2006)56%5B987:AGCFSE%5D2.0.CO;2.

136. Lucas BG, Johannessen D, Lindstrom S (2007) Appendix E: Marine Plants. In: Lucas BG, Verrin S, Brown R, editors, Ecosystem Overview: Pacific North Coast Integrated Management Area (PNCIMA), Canadian Technical Report of Fisheries and Aquatic Sciences 2667, iv + 23 p.

137. Beck MW, KLH Jr, Able KW, Childers DL, Eggleston DB, et al. (2001) The identification, conservation, and management of estuarine and marine nurseries for fish and invertebrates. BioSci 51: 633–641. doi: 10.1641/0006-3568(2001)051%5B0633:TICAMO%5D2.0.CO;2.

138. Hendriks IE, Olsen YS, Ramajo L, Basso L, Steckbauer A, et al. (2014) Photosynthetic activity buffers ocean acidification in seagrass meadows. Biogeosci 11: 333–346. doi: 10.5194/bg-11-333-2014.

139. Beer S (1989) Photosynthesis and photorespiration of marine angiosperms. Aquat Bot 34: 153–166. doi: 10.1016/0304-3770(89)90054-5.

140. Zimmerman RC, Kohrs DG, Steller DL, Alberte RS (1997) Impacts of CO_2 enrichment on productivity and light requirements of eelgrass. Plant Physiol 115: 599–607. pmid:12223828

141. Arnold T, Mealey C, Leahey H, Miller AW, Hall-Spencer JM, et al. (2012) Ocean acidification and the loss of phenolic substances in marine plants. PLoS ONE 7: e35107. doi: 10.1371/journal.pone.0035107. pmid:22558120

142. Martin AJ, Montagnes DJS (1993) Winter ciliates in a British Columbian fjord: six new species and an analysis of ciliate putative prey. J Eukaryot Microbiol 40: 535–549. doi: 10.1111/j.1550-7408.1993.tb06105.x.

143. Jacobson DM, Anderson DM (1986) Thecate heterotrophic dinoflagellates: feeding behavior and mechanisms. J Phycol 22: 249–258. doi: 10.1111/j.1529-8817.1986.tb00021.x.

144. Irigoien X, Flynn KJ, Harris RP (2005) Phytoplankton blooms: a 'loophole' in microzooplankton grazing impact? J Plankton Res 27: 313–321. doi: 10.1093/plankt/fbi011.

145. Bakun A, Broad K (2003) Environmental 'loopholes' and fish population dynamics: comparative pattern recognition with focus on El Niño effects in the Pacific. Fish Oceanogr 12: 458–473. doi: 10.1046/j.1365-2419.2003.00258.x.

146. Moy AD, Howard WR, Bray SG, Trull TW (2009) Reduced calcification in modern Southern Ocean planktonic foraminifera. Nat Geosci 2: 276–280. doi: 10.1038/ngeo460.

147. Caron DA, Hutchins DA (2013) The effects of changing climate on microzooplankton grazing and community structure: drivers, predictions and knowledge gaps. J Plankton Res 35: 235–252. doi: 10.1093/plankt/fbs091.

148. Suffrian K, Simonelli P, Nejstgaard JC, Putzeys S, Carotenuto Y, et al. (2008) Microzooplankton grazing and phytoplankton growth in marine mesocosms with increased CO_2 levels. Biogeosciences 5: 1145–1156. doi: 10.5194/bg-5-1145-2008.

149. Aberle N, Schulz KG, Stuhr A, Malzahn AM, Ludwig A, et al. (2013) High tolerance of microzooplankton to ocean acidification in an Arctic coastal plankton community. Biogeosciences 10: 1471–1481. doi: 10.5194/bg-10-1471-2013.

150. Rossoll D, Sommer U, Winder M (2013) Community interactions dampen

acidification effects in a coastal plankton system. Mar Ecol Prog Ser 486: 37–46. doi: 10.3354/meps10352.

151. Rose JM, Feng Y, Gobler CJ, Gutierrez R, Hare CE, et al. (2009) Effects of increased pCO2 and temperature on the North Atlantic spring bloom. II. Microzooplankton abundance and grazing. Mar Ecol Prog Ser 388: 27–40. doi: 10.3354/meps08134.

152. Kim JM, Lee K, Yang EJ, Shin K, Noh JH, et al. (2010) Enhanced production of oceanic dimethylsulfide resulting from CO2-induced grazing activity in a high CO2 world. Environ Sci Technol 44: 8140–8143. doi: 10.1021/es102028k. pmid:20883015

153. Gravinese PM, Foy M, Lessard E, Murray JW (2014) The effects of elevated pCO2 on microzooplankton biomass, abundance, and community structure—a mesocosm study in the Salish Sea. In: Ocean Sciences Meeting, Session 033, presented Feb. 25, 2014, Honolulu, HI.

154. Mackas DL, Tsuda A (1999) Mesozooplankton in the eastern and western subarctic Pacific: community structure, seasonal life histories, and interannual variability. Prog Oceanogr 43: 335–363. doi: 10.1016/S0079-6611(99)00012-9.

155. Mackas D, Galbraith M, Faust D, Masson D, Young K, et al. (2013) Zooplankton time series from the Strait of Georgia: results from year-round sampling at deep water locations, 1990–2010. Prog Oceanogr 115: 129–159. doi: 10.1016/j.pocean.2013.05.019.

156. Mackas DL, Thomson RE, Galbraith M (2001) Changes in the zooplankton community of the British Columbia continental margin, 1985–1999, and their covariation with oceanographic conditions. Can J Fish Aquat Sci 58: 685–702. doi: 10.1139/f01-009.

157. Zhang D, Li S, Wang G, Guo D (2011) Impacts of CO2-driven seawater acidification on survival, egg production rate and hatching success of four marine copepods. Acta Oceanol Sin 30: 86–94. doi: 10.1007/s13131-011-0165-9.

158. Kurihara H, Shimode S, Shirayama Y (2004) Effects of raised CO2 concentration on the egg production rate and early development of two marine copepods (Acartia steueri and Acartia erythraea). Mar Pollut Bull 49: 721–727. doi: 10.1016/j.marpolbul.2004.05.005. pmid:15530515

159. Mayor DJ, Matthews C, Cook K, Zuur AF, Hay S (2007) CO2-induced acidification affects hatching success in Calanus finmarchicus. Mar Ecol Prog Ser 350: 91–97. doi: 10.3354/meps07142.

160. Kurihara H, Ishimatsu A (2008) Effects of high CO2 seawater on the copepod (Acartia tsuensis) through all life stages and subsequent

generations. Mar Pollut Bull 56: 1086–1090. doi: 10.1016/j.marpolbul.2008.03.023. pmid:18455195

161. Weydmann A, Søreide JE, Kwasniewski S, Widdicombe S (2012) Influence of CO2-induced acidification on the reproduction of a key Arctic copepod Calanus glacialis. J Exp Mar Biol Ecol 428: 39–42. doi: 10.1016/j.jembe.2012.06.002.

162. Vehmaa A, Brutemark A, Engström-Öst J (2012) Maternal effects may act as an adaptation mechanism for copepods facing pH and temperature changes. PLoS ONE 7: e48538. doi: 10.1371/journal.pone.0048538. pmid:23119052

163. Mayor DJ, Everett NR, Cook KB (2012) End of century ocean warming and acidification effects on reproductive success in a temperate marine copepod. J Plankton Res 34: 258–262. doi: 10.1093/plankt/fbr107.

164. Pedersen SA, Hansen BH, Altin D, Olsen AJ (2013) Medium-term exposure of the North Atlantic copepod Calanus finmarchicus (Gunnerus, 1770) to CO2-acidified seawater: effects on survival and development. Biogeosciences 10: 7481–7491. doi: 10.5194/bg-10-7481-2013.

165. Mackas DL, Galbraith MD (2012) Pteropod time-series from the NE Pacific. ICES J Mar Sci 69: 448–459. doi: 10.1093/icesjms/fsr163.

166. Mackas DL, Tsurumi M, Galbraith MD, Yelland D (2005) Zooplankton distribution and dynamics in a North Pacific eddy of coastal origin. 2. Mechanisms of eddy colonization by and retention of offshore species. Deep Sea Res II 52: 1011–1035. doi: 10.1016/j.dsr2.2005.02.008.

167. Armstrong JL, Boldt JL, Cross AD, Moss JH, David ND, et al. (2005) Distribution, size and interannual, seasonal and diel food habits of northern Gulf of Alaska juvenile pink salmon Oncorhynchus gorbuscha. Deep Sea Res II 52: 247–265. doi: 10.1016/j.dsr2.2004.09.019.

168. Lalli CM, Gilmer RW (1989) Pelagic Snails: The Biology of Holoplanktonic and Gastropod Mollusks. Stanford, CA: Stanford University Press, 259 pp.

169. Hunt BP, Pakhomov EA, Hosie GW, Seigel W, Ward P, et al. (2008) Pteropods in Southern Ocean ecosystems. Prog Oceanogr 78: 193–221. doi: 10.1016/j.pocean.2008.06.001.

170. Byrne RH, Acker JG, Betzer PR, Feely RA, Cates MH (1984) Water column dissolution of aragonite in the Pacific Ocean. Nature 312: 321–326. doi: 10.1038/312321a0.

171. Lischka S, Riebesell U (2012) Synergistic effects of ocean acidification and warming on overwintering pteropods in the Arctic. Glob Change

Biol 18: 3517–3528. doi: 10.1111/gcb.12020.

172. Sato-Okoshi W, Okoshi K, H S, Akiha F (2010) Shell structure characteristics of pelagic and benthic molluscs from Antarctic waters. Polar Sci 4: 257–261. doi: 10.1016/j.polar.2010.05.006.

173. Bednarsek N, Tarling GA, Bakker DCE, Fielding S, Jones EM, et al. (2012) Extensive dissolution of live pteropods in the Southern Ocean. Nat Geosci 5: 881–885. doi: 10.1038/ngeo1635.

174. Roger LM, Richardson AJ, McKinnon AD, Knott B, Matear R, et al. (2012) Comparison of the shell structure of two tropical Thecosomata (Creseis acicula and Diacavolinia longirostris) from 1963 to 2009: potential implications of declining aragonite saturation. ICES J Mar Sci 69: 465–474. doi: 10.1093/icesjms/fsr171.

175. Bednarsek N, Feely RA, Reum JCP, Peterson B, Menkel J, et al. (2014) Limacina helicina shell dissolution as an indicator of declining habitat suitability owing to ocean acidification in the California Current Ecosystem. Proc R Soc Lond, B, Biol Sci 281. doi: 10.1098/rspb.2014.0123

176. Comeau S, Gorsky G, Jeffree R, Teyssié JL, Gattuso JP (2009) Impact of ocean acidification on a key arctic pelagic mollusc (Limacina helicina). Biogeosciences 6: 1877–1882. doi: 10.5194/bg-6-1877-2009.

177. Comeau S, Gorsky G, Alliouane S, Gattuso JP (2010) Larvae of the pteropod Cavolinia inflexa exposed to aragonite undersaturation are viable but shell-less. Mar Biol 78: 2341–2345. doi: 10.1007/s00227-010-1493-6.

178. Comeau S, Gattuso JP, Nisumaa AM, Orr J (2012) Impact of aragonite saturation state changes on migratory pteropods. Proc R Soc Lond, B, Biol Sci 279: 732–738. doi: 10.1098/rspb.2011.0910.

179. Seibel BA, Mass AE, Dierssen HM (2012) Energetic plasticity underlies a variable response to ocean acidification in the pteropod Limacina helicina antarctica. PLoS ONE 7: e30464. doi: 10.1371/journal.pone.0030464. pmid:22536312

180. Maas AE, Wishner KF, Seibel BA (2012) The metabolic response of pteropods to acidification reflects natural CO2-exposure in oxygen minimum zones. Biogeosciences 9: 747–757. doi: 10.5194/bg-9-747-2012.

181. Armstrong JL, Myers KW, Beauchamp DA, Davis ND, Walker RV, et al. (2008) Interannual and spatial feeding patterns of hatchery and wild juvenile pink salmon in the Gulf of Alaska in years of high and low survival. Trans Am Fish Soc 137: 1299–1316. doi: 10.1577/T07-196.1.

182. Helmuth B, Harley C, Halpin PM, O'Donnell M, Hoffmann G (2002) Climate change and latitudinal patterns of intertidal thermal stress. Science 298: 1015–1017. doi: 10.1126/science.1076814. pmid:12411702

183. Harbo R (1996) Shells and Shellfish of the Pacific northwest: a field guide. Madiera Park, BC: Harbour Publishing.

184. Bacher C, Grant J, Hawkins A, Fang J, Zhu M, et al. (2003) Modelling the effect of food depletion on scallop growth in Sungo Bay China. Auat Living Resour 16: 10–24. doi: 10.1016/S0990-7440(03)00003-2.

185. Wilson JG (1994) The role of bioindicators in estuarine management. Estuaries 17: 94–101. doi: 10.2307/1352337.

186. Lindahl O, Hart R, Hernroth B, Kollberg S, Loo LO, et al. (2005) Improving marine water quality by mussel farming: a profitable solution for Swedish society. Ambio 34: 131–138. doi: 10.1579/0044-7447-34.2.131. pmid:15865310

187. Johnson AM (1982) Status of Alaska sea otter populations and developing conflicts with fisheries. In: Sabol K, editor, Transactions of the 47th North American Wildlife and natural resources conference. Washington DC, USA, volume 42, pp. 293–299.

188. Baker P (1995) Review of ecology and fishery of the Olympia oyster Ostrea lurida with annotated bibliography. J Shellfish Res 14: 501–518.

189. Fisheries and Oceans Canada (2013) Pacific Region Integrated Fisheries Management Plan—Intertidal Clams—January 1, 2013 to December 31, 2015. http://www.pac.dfo-mpo.gc.ca/fm-gp/mplans/2013/intertidal_clam-palourde_intercotidale-2013-15-eng.pdf, 100 pp.

190. Gillespie GE (1999) Status of the Olympia Oyster, Ostrea conchaphila, in Canada. Can Stock Assess Sec Res Doc 99/150: 36 p.

191. Gillespie GE, Bower SM, Marcus KL, Kieser D (2012) Biological synopsises for three exotic molluscs, Manila Clam (Venerupis philippinarum), Pacific Oyster (Crassostrea gigas) and Japanese Scallop (Mizuhopecten yessoensis) licensed for Aquaculture in British Columbia. Can Sci Advis Sec Res Doc 2012/013: v + 97 p.

192. Fisheries and Oceans Canada (2013) Pacific Region Exploratory Fishery Guidelines—Pink and Spiny Scallop by Trawl—August 1, 2013 to July 31, 2014. http://www.pac.dfo-mpo.gc.ca/fm-gp/mplans/2013/scallop-petoncle-2013-eng.pdf, 30 pp.

193. Weiss IM, Tuross N, Addadi L, Weiner S (2002) Mollusc larval shell formation: amorphous calcium carbonate is a precursor phase for aragonite. J Exp Zool 293: 478–491. doi: 10.1002/jez.90004. pmid:12486808

194. Wilt FH (2005) Developmental biology meets materials science: morphogenesis of biomineralized structures. Dev Biol 280: 15–25. doi: 10.1016/j.ydbio.2005.01.019. pmid:15766744

195. Stenzel HB (1963) Aragonite and calcite as constituents of adult oyster shells. Science 142: 232–233. doi: 10.1126/science.142.3589.232. pmid:17834841

196. Barton A, Hales B, Waldbusser GG, Langdon C, Feely RA (2012) The Pacific oyster, Crassostrea gigas, shows negative correlation to naturally elevated carbon dioxide levels: Implications for near-term ocean acidification effects. Limnol Oceanogr 57: 698–710. doi: 10.4319/lo.2012.57.3.0698.

197. Kroeker KJ, Kordas RL, Crim RN, Singh GG (2010) Meta-analysis reveals negative yet variable effects of ocean acidification on marine organisms. Ecol Lett 13: 1419–1434. doi: 10.1111/j.1461-0248.2010.01518.x. pmid:20958904

198. Parker LM, Ross PM, O'Connor WA, Pörtner HO, Scanes E, et al. (2013) Predicting the response of molluscs to the impact of ocean acidification. Biology 2: 651–692. doi: 10.3390/biology2020651. pmid:24832802

199. Havenhand JN, Schlegel P (2009) Near-future levels of ocean acidification do not affect sperm motility and fertilization kinetics in the oyster Crassostrea gigas. Biogeosciences 6: 3009–3015. doi: 10.5194/bg-6-3009-2009.

200. Parker LM, Ross PM, O'Connor WA (2010) Comparing the effect of elevated pCO2 and temperature on the fertilization and early development of two species of oysters. Mar Biol 157: 2435–2452. doi: 10.1007/s00227-010-1508-3.

201. Barros P, Sobral P, Range P, Chícharo L, Matias D (2013) Effects of sea-water acidification on fertilization and larval development of the oyster Crassostrea gigas. J Exp Mar Biol Ecol 440: 200–206. doi: 10.1016/j.jembe.2012.12.014.

202. Kurihara H, Kato S, Ishimatsu A (2007) Effects of increased seawater pCO2 on early development of the oyster Crassostrea gigas. Aquat Biol 1: 91–98. doi: 10.3354/ab00009.

203. Timmins-Schiffman E, O'Donnell MJ, Friedman CS, Roberts SB (2013) Elevated pCO2 causes developmental delay in early larval Pacific oysters, Crassostrea gigas. Mar Biol 160: 1973–1982. doi: 10.1007/s00227-012-2055-x.

204. Ginger KWK, Vera CBS, R D, Dennis CKS, Adela LJ, et al. (2013) Larval and post-larval stages of Pacific Oyster (Crassostrea gigas) are

resistant to elevated CO2. PLoS ONE 8: e64147. doi: 10.1371/journal. pone.0064147. pmid:23724027

205. Crim RN, Sunday JM, Harley CDG (2011) Elevated seawater CO2 concentrations impair larval development and reduce larval survival in endangered northern abalone (Haliotis kamtschatkana). J Exp Mar Biol Ecol 400: 272–277. doi: 10.1016/j.jembe.2011.02.002.

206. Gobler CJ, DePasquale EL, Griffith AW, Baumann H (2014) Hypoxia and acidification have additive and synergistic negative effects on the growth, survival, and metamorphosis of early life stage bivalves. PLoS ONE 9: e83648. doi: 10.1371/journal.pone.0083648. pmid:24416169

207. Hettinger A, Sanford E, Hill TM, Lenz EA, Russell AD, et al. (2013) Larval carry-over effects from ocean acidification persist in the natural environment. Glob Change Biol 19: 3317–3326. doi: 10.1111/gcb.12307

208. Hettinger A, Sanford E, Hill TM, Hosfelt JD, Russell AD, et al. (2013) The influence of food supply on the response of Olympia oyster larvae to ocean acidification. Biogeosciences 10: 6629–6638. doi: 10.5194/bg-10-6629-2013.

209. Talmage SC, Gobler CJ (2009) The effects of elevated carbon dioxide concentrations on the metamorphosis, size, and survival of larval hard clams (Mercenaria mercenaria), bay scallops (Argopecten irradians), and Eastern oysters (Crassostrea virginica). Limnol Oceanogr 54: 2072–2080. doi: 10.4319/lo.2009.54.6.2072.

210. Talmage SC, Gobler CJ (2011) Effects of elevated temperature and carbon dioxide on the growth and survival of larvae and juveniles of three species of Northwest Atlantic bivalves. PLoS ONE 6: e26941. doi: 10.1371/journal.pone.0026941. pmid:22066018

211. Talmage SC, Gobler CJ (2012) Effects of CO2 and the harmful alga Aureococcus anophagefferens on growth and survival of oyster and scallop larvae. Mar Ecol Prog Ser 464: 121–134. doi: 10.3354/meps09867.

212. Dineshram R, Wong KKW, Xiao S, Yu Z, Qian PY, et al. (2012) Analysis of Pacific oyster larval proteome and its response to high-CO2. Mar Pollut Bull 64: 2160–2167. doi: 10.1016/j.marpolbul.2012.07.043. pmid:22921897

213. Hettinger A, Sanford E, Hill TM, Russell AD, Sato KNS, et al. (2012) Persistent carry-over effects of planktonic exposure to ocean acidification in the Olympia oyster. Ecology 93: 2758–2768. doi: 10.1890/12-0567.1. pmid:23431605

214. Gazeau F, Quiblier C, Jansen JM, Gattuso JP, Middelburg JJ, et al. (2007) Impact of elevated CO2 on shellfish calcification. Geophys Res Lett 34:

L07603. doi: 10.1029/2006GL028554.

215. Mingliang Z, Jianguang F, Jihong Z, Bin L, Shengmin R, et al. (2011) Effect of marine acidification on calcification and respiration of Chlamys farreri. J Shellfish Res 30: 267–271. doi: 10.2983/035.030.0211.

216. O'Donnell MJ, George MN, Carrington E (2013) Mussel byssus attachment weakened by ocean acidification. Nature Clim Change 3: 1471–1481. doi: 10.1038/nclimate1846

217. Lannig G, Eilers S, Pörtner HO, Sokolova IM, Bock C (2010) Impact of ocean acidification on energy metabolism of oyster, Crassostrea gigas— changes in metabolic pathways and thermal response. Mar Drugs 8: 2318–2339. doi: 10.3390/md8082318. pmid:20948910

218. Sanders MB, Bean TP, Hutchinson TH, Le Quesne WJF (2013) Juvenile king scallop, Pecten maximus, is potentially tolerant to low levels of ocean acidification when food is unrestricted. PLoS ONE 8: e74118. doi: 10.1371/journal.pone.0074118. pmid:24023928

219. Schalkhausser B, Bock C, Stemmer K, Brey T, Pörtner HO, et al. (2013) Impact of ocean acidification on escape performance of the king scallop, Pecten maximus, from Norway. Mar Biol 160: 1995–2006. doi: 10.1007/s00227-012-2057-8.

220. Zhai WD, Zhao HD, Zheng N, Xu Y (2012) Coastal acidification in summer bottom oxygen-depleted waters in northwestern-northern Bohai Sea from June to August in 2011. Chin Sci Bull 57: 1062–1068. doi: 10.1007/s11434-011-4949-2.

221. Duckworth AR, Peterson BJ (2013) Effects of seawater temperature and pH on the boring rates of the sponge Cliona celata in scallop shells. Mar Biol 160: 27–35. doi: 10.1007/s00227-012-2053-z.

222. Gutiérrez JL, Jones CG, Strayer DL, Iribarne OO (2003) Mollusks as ecosystem engineers: the role of shell production in aquatic habitats. Oikos 101: 79–90. doi: 10.1034/j.1600-0706.2003.12322.x.

223. Cosgrove J (2009) Checklist of cephalopods (octopuses and squids) of British Columbia (November, 2009). In: Klinkenberg B, editor, E-Fauna BC: Electronic Atlas of the Fauna of British Columbia [www.efauna.bc.ca]. Lab for Advanced Spatial Analysis, Department of Geography, University of British Columbia, Vancouver. [Date Accessed: 2014-03-06], jacosgrove@telus.net: Royal British Columbia Museum. URL http://www.royalbcmusem.bc.ca/.

224. Arkhipkin AI, Bizikov VA (2000) Role of the statolith in functioning of the acceleration receptor system in squids and sepioids. Journal of Zoology 250: 31–55. doi: 10.1111/j.1469-7998.2000.tb00575.x.

225. Lacoue-Labarthe T, Réveillac E, Oberhänsli F, Teyssié JL, Jeffree R, et al. (2011) Effects of ocean acidification on trace element accumulation in the early-life stages of squid Loligo vulgaris. Aquat Toxicol 105: 166–176. doi: 10.1016/j.aquatox.2011.05.021. pmid:21718660

226. Kaplan MB, Mooney TA, McCorkle DC, Cohen AL (2013) Adverse effects of ocean acidification on early development of squid (Doryteuthis pealeii). PLoS ONE 8: e63714. doi: 10.1371/journal.pone.0063714. pmid:23741298

227. Rosa R, Seibel BA (2008) Synergistic effects of climate-related variables suggest future physiological impairment in a top oceanic predator. Proc Natl Acad Sci USA 105: 20776–20780. doi: 10.1073/pnas.0806886105. pmid:19075232

228. Kurihara H (2008) Effects of CO2-driven ocean acidification on the early developmental stages of invertebrates. Mar Ecol Prog Ser 373: 275–284. doi: 10.3354/meps07802.

229. Conway KW, Barrie JV, Austin WC, Luternauer JL (1991) Holocene sponge bioherms on the western Canadian continental shelf. Cont Shelf Res 11: 771–790. doi: 10.1016/0278-4343(91)90079-L.

230. Leys SP, Wilson K, Holeton C, Reiswig HM, Austin WC, et al. (2004) Patterns of glass sponge (Porifera, Hexactinellida) distribution in coastal waters of British Columbia, Canada. Mar Ecol Prog Ser 283: 133–149. doi: 10.3354/meps283133.

231. Stone RP, Conway DJ, Barrie JV (2013) The boundary reefs: glass sponge reefs on the international border between Canada and the United States. NOAA Tech Mem NMFS-AFSC-264: 31.

232. Roberts JM, Wheeler AJ, Freiwald A (2006) Reefs of the deep: the biology and geology of cold-water coral ecosystems. Science 312: 543–547. doi: 10.1126/science.1119861. pmid:16645087

233. Jamieson GS, Pellegrin N, Jessen S (2006) Taxonomy and zoogeography of cold water corals in explored areas of coastal British Columbia. Can Sci Advis Sec Res Doc 2006/062: ii + 45 p.

234. Stone RP (2006) Coral habitat in the Aleutian Islands of Alaska: depth distribution, fine-scale species associations. Coral Reefs 25: 229–238. doi: 10.1007/s00338-006-0091-z.

235. Rooper CN, Boldt JL, Zimmermann M (2007) An assessment of juvenile Pacific Ocean perch (Sebastes alutus) habitat use in a deepwater nursery. Estuar Coast Shelf Sci 75: 371–380. doi: 10.1016/j.ecss.2007.05.006.

236. Du Preez C, Tunnicliffe V (2011) Shortspine thornyhead and rockfish

(Scorpaenidae) distribution in response to substratum, biogenic structures and trawling. Mar Ecol Prog Ser 425: 217–231. doi: 10.3354/meps09005.

237. Miller RJ, Hocevar J, Sone RP, Fedorov DV (2012) Structure-forming corals and sponges and their use as fish habitat in Bering Sea submarine canyons. PLoS ONE 7: e33885. doi: 10.1371/journal.pone.0033885. pmid:22470486

238. Finney JL, Boutillier P (2010) Distribution of cold-water coral, sponges and sponge reefs in British Columbia with options for identifying significant encounters. Can Sci Advis Sec Res Doc 2010/090: vi + 9 p.

239. Stone RP, Shotwell SK (2007) State of deep coral ecosystems in Alaska region: Gulf of Alaska, Bering Sea and Aleutian Islands. In: Lumsden SE, Hourigan TF, Bruckner AW, Dorr G, editors, The State of Deep Coral Ecosystems of the United States, NOAA Technical Memorandum CRCP-3. Silver Spring MD. pp. 65–108.

240. Cairns SD (2011) A revision of the Primoidae (Octocorallia: Alcyoncea) from the Aleutian Islands and Bering Sea. Smithson Contrib Zool 634: 55. doi: 10.5479/si.00810282.634

241. Finney JL (2010) Overlap of predicted cold-water coral habitat and bottom-contact fisheries in British Columbia. Can Sci Advis Sec Res Doc 2010/067: vi + 26 p.

242. Conway KW, Krautter M, Barrie JV, Whitney F, Thomson RE, et al. (2005) Sponge reefs in the Queen Charlotte Basin, Canada: controls on distribution, growth and development. In: Freiwald A, Roberts JM, editors, Coldwater Corals and Ecosystems, Springer Berlin Heidelberg. pp. 605–621.

243. Cairns SD, Lindner A (2011) A revision of the Stylasteridae (Cnidarian, Hydrozoa, Filifera) from Alaska and adjacent waters. ZooKeys 158: 1–88. doi: 10.3897/zookeys.158.1910. pmid:22303109

244. Cairns SD (1994) Scleractinia of the temperate North Pacific. Smithson Contrib Zool 557: 150. doi: 10.5479/si.00810282.557.i

245. Conway KW, Barrie JV, Hill PR, Austin WC, Pickard K (2007) Mapping sensitive benthic habitats in the Strait of Georgia, coastal British Columbia: deep-water sponge and coral reefs. Geol Surv Can, Curr Res A2: 6p. doi: 10.4095/223389

246. Guinotte JM, Orr J, Cairns SD, Freiwald A, Morgan L, et al. (2006) Will human-induced changes in seawater chemistry alter the distribution of deep-sea scleractinian corals? Front Ecol Environ 4: 141–146. doi: 10.1890/1540-9295(2006)004%5B0141:WHCISC%5D2.0.CO;2.

247. McCulloch M, Trotter J, Montagna P, Falter J, Dunbar R, et al. (2012) Resilience of cold-water scleractinian corals to ocean acidification: Boron isotopic systematics of pH and saturation state up-regulation. Geochim Cosmochim Acta 87: 21–34. doi: 10.1016/j.gca.2012.03.027.

248. Maier C, Hegeman J, Weinbauer MG, Gattuso JP (2009) Calcification of the cold-water Lophelia pertusa under ambient and reduced pH. Biogeosciences 6: 1671–1680. doi: 10.5194/bg-6-1671-2009.

249. Form AU, Riebesell U (2012) Acclimation to ocean acidification during long-term CO2 exposure in the cold-water coral Lophelia pertusa. Glob Change Biol 18: 843–853. doi: 10.1111/j.1365-2486.2011.02583.x.

250. Bayer FM, Macintyre IG (2001) The mineral component of the axis and holdfast of some gorgonacean octocorals (Coelenterata: Anthozoa), with special reference to the family Gorgoniidae. Proc Biol Soc Wash 114: 309–345.

251. Cairns SD, Macintyre IG (1992) Phylogenetic implications of calcium carbonate mineralogy in the Stylasteridae (Cnidaria: Hydrozoa). Palaios 7: 96–107. doi: 10.2307/3514799.

252. Dupont S, Ortega-Martínez O, Thorndyke M (2010) Impact of near-future ocean acidification on echinoderms. Ecotoxicology 19: 449–462. doi: 10.1007/s10646-010-0463-6. pmid:20130988

253. Lambert P (2007) Checklist of the echinoderms of British Columbia (April 2007). In: Klinkenberg, B, editor, E-Fauna BC: Electronic Atlas of the Fauna of British Columbia [www.efauna.bc.ca]. Lab for Advanced Spatial Analysis, Department of Geography, University of British Columbia, Vancouver. [Date Accessed: 2013-12-02], plambert@pacificcoast.net: Royal British Columbia Museum. URL http://www.royalbcmusem.bc.ca/.

254. Lambert P, Boutillier J (2011) Deep-sea echinodermata of British Columbia, Canada. Can Tech Rep Fish Aquat Sci 2929: viii + 143 p.

255. Paine RT (1966) Food web complexity and species diversity. Am Nat 100: 65–75. doi: 10.1086/282400.

256. Paine RT (1969) A note on trophic complexity and community stability. Am Nat 103: 91–93. doi: 10.1086/282586.

257. Lawrence JM (1975) On the relationship between marine plants and sea urchins. Oceanogr Mar Biol Ann Rev 13: 213–286.

258. Clark D, Lamare M, Barker M (2009) Response of sea urchin pluteus larvae (Echinodermata: Echinoidea) to reduced seawater pH: a comparison among a tropical, temperate, and a polar species. Mar Biol

156: 1125–1137. doi: 10.1007/s00227-009-1155-8.

259. Asnaghi V, Mangialajo L, Gattuso JP, Francour P, Privitera D, et al. (2014) Effects of ocean acidification and diet on thickness and carbonate elemental composition of the test of juvenile sea urchins. Mar Environ Res 93: 78–84. doi: 10.1016/j.marenvres.2013.08.005. pmid:24050836

260. Dupont S, Dorey N, Stumpp M, Melzner F, Thorndyke M (2013) Long-term and trans-life-cycle effects of exposure to ocean acidification in the green sea urchin Strongylocentrotus droebachiensis. Mar Biol 160: 1835–1843. doi: 10.1007/s00227-012-1921-x.

261. Reuter KE, Lotterhos KE, Crim RN, Thompson CA, Harley CDG (2011) Elevated pCO2 increases sperm limitation and risk of polyspermy in the red sea urchin Strongylocentrotus franciscanus. Glob Change Biol 17: 163–171. doi: 10.1111/j.1365-2486.2010.02216.x.

262. O'Donnell MJ, Hammond LM, Hofmann GE (2009) Predicted impact of ocean acidification on a marine invertebrate: elevated CO2 alters response to thermal stress in sea urchin larvae. Mar Biol 156: 439–446. doi: 10.1007/s00227-008-1097-6.

263. Todgham AE, Hofmann GE (2009) Transcriptomic response of sea urchin larvae Strongylocentrotus purpuratus to CO2-driven seawater acidification. J Exp Biol 212: 2579–2594. doi: 10.1242/jeb.032540. pmid:19648403

264. Pespeni MH, Sanford E, Gaylord B, Hill TM, Hosfelt JD, et al. (2013) Evolutionary change during experimental ocean acidification. Proc Natl Acad Sci USA 110: 6937–6942. doi: 10.1073/pnas.1220673110. pmid:23569232

265. Morita M, Suwa R, Iguchi A, Nakamura M, Shimada K, et al. (2010) Ocean acidification reduces sperm flagellar motility in broadcast spawning reef invertebrates. Zygote 18: 103–107. doi: 10.1017/S0967199409990177. pmid:20370935

266. Gooding RA, Harley CDG, Tang E (2009) Elevated water temperature and carbon dioxide concentration increase the growth of a keystone echinoderm. Proc Natl Acad Sci USA 106: 9316–9321. doi: 10.1073/pnas.0811143106. pmid:19470464

267. Dupont S, Havenhand J, Thorndyke W, Peck L, Thorndyke M (2008) Near-future level of CO2-driven ocean acidification radically affects larval survival and development in the brittlestar Ophiothrix fragilis. Mar Ecol Prog Ser 373: 285–294. doi: 10.3354/meps07800.

268. Love MS (2011) Certainly More Than You Want to Know About The Fishes of The Pacific Coast: A Postmodern Experience. Really Big Press,

Santa Barbara, California.

269. Smith KL Jr, Brown NO (1983) Oxygen consumption of pelagic juveniles and demersal adults of the deep-sea fish Sebastolobus altivelis, measured at depth. Mar Biol 76: 325–332. doi: 10.1007/BF00393036.

270. Gardner GA, Szabo I (1982) British Columbia pelagic marine copepoda: an identification manual and annotated bibliography. Can Spec Publ Fish Aquat Sci 62: 536 p.

271. Baldwin A (2009) Checklist of the barnacles of British Columbia (updated October 2009). In: Klinkenberg, B, editor, E-Fauna BC: Electronic Atlas of the Fauna of British Columbia [www.efauna.bc.ca]. Lab for Advanced Spatial Analysis, Department of Geography, University of British Columbia, Vancouver. [Date Accessed: 2014-02-05], plambert@pacificcoast.net: Royal British Columbia Museum. URL http://www.royalbcmusem.bc.ca/.

272. Baldwin A (2011) Checklist of the shrimps, crabs, lobsters and crayfish of British Columbia 2011 (order Decapoda). In: Klinkenberg B, editor, E-Fauna BC: Electronic Atlas of the Fauna of British Columbia [www.efauna.bc.ca]. Lab for Advanced Spatial Analysis, Department of Geography, University of British Columbia, Vancouver. [Date Accessed: 2014-02-03], plambert@pacificcoast.net: Royal British Columbia Museum. URL http://www.royalbcmusem.bc.ca/.

273. Nakagawa Y, Endo Y, Taki K (2001) Diet of Euphausia pacifica hansen in Sanriku waters off northeastern Japan. Plankton Biol Ecol 48: 68–77.

274. Fisheries and Oceans Canada (2014) Pacific Region Integrated Fisheries Management Plan—Crab by Trap—January 1, 2014 to December 31, 2014. http://www.pac.dfo-mpo.gc.ca/fm-gp/mplans/2014/crab-crabe-2014-eng.pdf, 227 pp.

275. Muzzarelli RAA (1977) Chitin. Pergamon Press, Oxford.

276. Ries JB, Cohen AL, McCorkle DC (2009) Marine calcifiers exhibit mixed responses to CO2-induced ocean acidification. Geology 37: 1131–1134. doi: 10.1130/G30210A.1.

277. Ries JB (2011) Skeletal mineralogy in a high-CO2 world. J Exp Mar Biol Ecol 403: 54–64. doi: 10.1016/j.jembe.2011.04.006.

278. Findlay HS, Wood HL, Kendall MA, Spicer JI, Twitchett RJ, et al. (2011) Comparing the impact of high co2 on calcium carbonate structures in different marine organisms. Mar Biol Res 7: 565–575. doi: 10.1080/17451000.2010.547200.

279. Whiteley NM (2011) Physiological and ecological responses of

crustaceans to ocean acidification. Mar Ecol Prog Ser 430: 257–271. doi: 10.3354/meps09185.

280. Long WC, Swiney KM, Harris C, Page HN, Foy RJ (2013) Effects of ocean acidification on juvenile Red King Crab (Paralithodes camtschaticus) and Tanner crab (Chionoecetes bairdi) growth, condition calcification and survival. PLOS ONE 8: e60959. doi: 10.1371/journal.pone.0060959. pmid:23593357

281. Long WC, Swiney KM, Foy RJ (2013) Effects of ocean acidification on the embryos and larvae of Red King Crab (Paralithodes camtschaticus). Mar Poll Bull 69: 38–47. doi: 10.1016/j.marpolbul.2013.01.011.

282. Saba GK, Schofield O, Torres JJ, Ombres EH, Steinberg DK (2012) Increased feeding and nutrient excretion of adult Antarctic krill, Euphausia superba, exposed to enhanced carbon dioxide (CO2). PLoS ONE 7: e52224. doi: 10.1371/journal.pone.0052224. pmid:23300621

283. Findlay HS, Kendall MA, Spicer JI, Widdicombe S (2009) Future high CO2 in the intertidal may compromise adult barnacle Semibalanus balanoides survival and embryonic development rate. Mar Ecol Prog Ser 389: 193–202. doi: 10.3354/meps08141.

284. Bechmann RK, Taban IC, Westerlund S, Godal BF, Arnberg M, et al. (2011) Effects of ocean acidification on early life stages of shrimp (Pandalus borealis) and mussel (Mytilus edulis). J Toxicol Environ Health A 74: 424–438. doi: 10.1080/15287394.2011.550460. pmid:21391089

285. McDonald MR, McClintock JB, Amsler CD, Rittschof D, Angus RA, et al. (2009) Effects of ocean acidification over the life history of the barnacle Amphibalanus amphitrite. Mar Ecol Prog Ser 385: 179–187. doi: 10.3354/meps08099.

286. Pansch C, Nasrolahi A, Appelhans YS, Wahl M (2012) Impacts of ocean warming and acidification on the larval development of the barnacle Amphibalanus improvisus. J Exp Mar Biol Ecol 420–421: 48–55. doi: 10.1016/j.jembe.2012.03.023.

287. Pansch C, Schlegel P, Havenhand J (2013) Larval development of the barnacle Amphibalanus improvisus responds variably but robustly to near-future ocean acidification. ICES J Mar Sci 70: 805–811. doi: 10.1093/icesjms/fst092.

288. Pansch C, Schaub I, Havenhand J, Wahl M (2014) Habitat traits and food availability determine the response of marine invertebrates to ocean acidification. Glob Change Biol 20: 765–777. doi: 10.1111/gcb.12478.

289. Cushing DH (1969) The regularity of the spawning season of some fishes. ICES J Mar Sci 33: 81–92. doi: 10.1093/icesjms/33.1.81.

290. Ryer CH, van Montfrans J, Moody KE (1997) Cannibalism, refugia and the molting blue crab. Mar Ecol Prog Ser 147: 77–85. doi: 10.3354/meps147077.

291. Kunkel JG, Nagel W, Jercinovic MJ (2012) Mineral fine structure of the American lobster cuticle. J Shellfish Res 31: 515–526. doi: 10.2983/035.031.0211.

292. Punt AE, Poljak D, Dalton MG, Foy RJ (2014) Evaluating the impact of ocean acidification on fishery yields and profits: The example of the red king crab in Bristol Bay. Ecol Mod 285: 39–53. doi: 10.1016/j.ecolmodel.2014.04.017.

293. Hart JL (1973) Pacific fishes of Canada. Bull Fish Res Bd Can 180: ix + 740 p.

294. Peden A (2002) An introduction to the marine fish of British Columbia. In: Klinkenberg B, editor, E-Fauna BC: Electronic Atlas of the Fauna of British Columbia [www.efauna.bc.ca], Lab for Advanced Spatial Analysis, Department of Geography, University of British Columbia, Vancouver. [Date Accessed: 2013-11-27].

295. Naiman RJ, Bilby RE, Schindler DE, Helfield JM (2002) Pacific salmon, nutrients, and the dynamics of freshwater and riparian ecosystems. Ecosystems 5: 399–417. doi: 10.1007/s10021-001-0083-3.

296. Melzner F, Göbel S, Langenbuch M, Gutowska MA, Pörtner HO, et al. (2009) Swimming performance in Atlantic Cod (Gadus morhua) following long-term (4–12 months) acclimation to elevated seawater PCO2. Aquat Toxicol 92: 30–37. doi: 10.1016/j.aquatox.2008.12.011. pmid:19223084

297. Melzner F, Gutowska MA, Langenbuch M, Dupont S, Lucassen M, et al. (2009) Physiological basis for high CO2 tolerance in marine ectothermic animals: pre-adaptation through lifestyle and ontogeny? Biogeosciences 6: 2313–2331. doi: 10.5194/bg-6-2313-2009.

298. Munday PL, Crawley NE, Nilsson GE (2009) Interacting effects of elevated temperature and ocean acidification on the aerobic performance of coral reef fishes. Mar Ecol Prog Ser 388: 235–242. doi: 10.3354/meps08137.

299. Ishimatsu A, Hayashi M, Kikkawa T (2008) Fishes in high-CO2 and acidified oceans. Mar Ecol Prog Ser 373: 295–302. doi: 10.3354/meps07823.

300. Leduc AOHC, Munday PL, Brown GE, Ferrari MCO (2013) Effects of acidification on olfactory-mediated behaviour in freshwater and marine ecosystems: a synthesis. Phil Trans R Soc Lond B, Biol Sci 368:

20120447. doi: 10.1098/rstb.2012.0447.

301. Frommel AY, Maneja R, Lowe D, Malzahn AM, Geffen AJ, et al. (2012) Severe tissue damage in Atlantic cod larvae under increasing ocean acidification. Nature Clim Change 2: 42–46. doi: 10.1038/nclimate1324.

302. Kikkawa T, Kita J, Ishimatsu A (2004) Comparison of the lethal effect of CO2 and acidification on red sea bream (Pagrus major) during the early developmental stages. Mar Pollut Bull 48: 108–110. doi: 10.1016/S0025-326X(03)00367-9. pmid:14725881

303. Ishimatsu A, Kikkawa T, Hayashi M, Lee KS, Kita J (2004) Effects of CO2 on marine fish: larvae and adults. J Oceanogr 60: 731–741. doi: 10.1007/s10872-004-5765-y.

304. Hurst TP, Fernandez ER, Mathis JT (2013) Effects of ocean acidification on hatch size and larval growth of walleye pollock (Theragra chalcogramma). ICES J Mar Sci 70: 812–822. doi: 10.1093/icesjms/fst053.

305. Frommel AY, Stiebens V, Clemmesen C, Havenhand J (2010) Effect of ocean acidification on marine fish sperm (Baltic cod: Gadus morhua). Biogeosciences 7: 3915–3919. doi: 10.5194/bg-7-3915-2010.

306. Franke A, Clemmesen C (2011) Effect of ocean acidification on early life stages of Atlantic herring (Clupea harengus L.). Biogeosciences 8: 3697–3707. doi: 10.5194/bg-8-3697-2011.

307. Baumann H, Talmage SC, Gobler CJ (2012) Reduced early life growth and survival in a fish in direct response to increased carbon dioxide. Nature Clim Change 2: 38–41. doi: 10.1038/nclimate1291.

308. Checkley DMJ, Dickson AG, Takahashi M, Radich JA, Eisenkolb N, et al. (2009) Elevated CO2 enhances otolith growth in young fish. Science 324: 1683–1683. doi: 10.1126/science.1169806. pmid:19556502

309. Maneja RH, Frommel AY, Geffen AJ, Folkvord A, Piatkowski U, et al. (2013) Effects of ocean acidification on the calcification of otoliths of larval Atlantic cod Gadus morhua. Mar Ecol Prog Ser 477: 251–258. doi: 10.3354/meps10146.

310. Munday PL, Hernaman V, Dixson DL, Thorrold SR (2011) Effect of ocean acidification on otolith development in larvae of a tropical marine fish. Biogeosciences 8: 1631–1641. doi: 10.5194/bg-8-1631-2011.

311. Bignami S, Enochs IC, Manzello DP, Sponaugle S, Cowen RK (2013) Ocean acidification alters the otoliths of a pantropical fish species with implications for sensory function. Proc Natl Acad Sci USA 110: 7366–7370. doi: 10.1073/pnas.1301365110. pmid:23589887

312. Dixson DL, Munday PL, Jones GP (2010) Ocean acidification disrupts the innate ability of fish to detect predator olfactory cues. Ecol Lett 13: 68–75. doi: 10.1111/j.1461-0248.2009.01400.x. pmid:19917053

313. Ferrari MCO, McCormick MI, Munday PL, Meekan MG, Dixson DL, et al. (2012) Effects of ocean acidification on visual risk assessment in coral reef fishes. Funct Ecol 26: 553–558. doi: 10.1111/j.1365-2435.2011.01951.x.

314. Munday PL, Dixson DL, McCormick MI, Meekan M, Ferrari MCO, et al. (2010) Replenishment of fish populations is threatened by ocean acidification. Proc Natl Acad Sci USA 107: 12930–12934. doi: 10.1073/pnas.1004519107. pmid:20615968

315. Munday PL, Dixson DL, Donelson JM, Jones GP, Pratchett MS, et al. (2009) Ocean acidification impairs olfactory discrimination and homing ability of a marine fish. Proc Natl Acad Sci USA 106: 1848–1852. doi: 10.1073/pnas.0809996106. pmid:19188596

316. Simpson SD, Munday PL, Wittenrich ML, Manassa R, Dixson DL, et al. (2011) Ocean acidification erodes crucial auditory behaviour in a marine fish. Biol Lett 7: 917–920. doi: 10.1098/rsbl.2011.0293. pmid:21632617

317. Ferrari MCO, Manassa RP, Dixson DL, Munday PL, McCormick MI, et al. (2012) Effects of ocean acidification on learning in coral reef fishes. PLoS ONE 7: e31478. doi: 10.1371/journal.pone.0031478. pmid:22328936

318. Domenici P, Allan B, McCormick MI, Munday PL (2012) Elevated carbon dioxide affects behavioural lateralization in a coral reef fish. Biol Lett 8: 78–81. doi: 10.1098/rsbl.2011.0591. pmid:21849307

319. Ferrari MCO, McCormick MI, Munday PL, Meekan MG, Dixson DL, et al. (2011) Putting prey and predator into the CO2 equation—qualitative and quantitative effects of ocean acidification on predator-prey interactions. Ecol Lett 14: 1143–1148. doi: 10.1111/j.1461-0248.2011.01683.x. pmid:21936880

320. Cripps IL, Munday PL, McCormick MI (2011) Ocean acidification affects prey detection by a predatory reef fish. PLoS ONE 6: e22736. doi: 10.1371/journal.pone.0022736. pmid:21829497

321. Jutfelt F, de Souza KB, Vuylsteke A, Sturve J (2013) Behavioural disturbances in a temperate fish exposed to sustained high-CO2 levels. PLoS ONE 8: e65825. doi: 10.1371/journal.pone.0065825. pmid:23750274

322. Maneja RH, Frommel AY, Browman HI, Clemmesen C, Geffen AJ, et al. (2013) The swimming kinematics of larval Atlantic cod, Gadus morhua

L., are resilient to elevated seawater pCO2. Mar Biol 160: 1963–1972. doi: 10.1007/s00227-012-2054-y.

323. Dixson DL, Jennings AR, Munday PL (2014) Odor tracking in sharks is reduced under future ocean acidification conditions. Glob Change Biol. doi: 10.1111/gcb.12678

324. Nilsson GE, Dixson DL, Domenici P, McCormick MI, Sorensen C, et al. (2012) Near-future carbon dioxide levels alter fish behaviour by interfering with neurotransmitter function. Nature Clim Change 2: 201–204. doi: 10.1038/nclimate1352.

325. McFarlane GA, Ware DM, Thomson RE, Mackas DL, Robinson CLK (1997) Physical, biological and fisheries oceanography of a large ecosystem (west coast of Vancouver Island) and implications for management. Oceanologica Acta 20: 191–200.

326. Trainer VL, Hickey BM, Lessard EJ, Cochlan WP, Trick CG, et al. (2009) Variability of Pseudo-nitzschia and domoic acid in the Juan de Fuca eddy region and its adjacent shelves. Limnol Oceanogr 54: 289–308. doi: 10.4319/lo.2009.54.1.0289.

327. Willette TM, Cooney RT, Patrick V, Mason DM, Thomas GL, et al. (2001) Ecological processes influencing mortality of juvenile pink salmon (Oncorhynchus gorbuscha) in Prince William Sound, Alaska. Fish Oceanogr 10: 14–41. doi: 10.1046/j.1054-6006.2001.00043.x.

328. Aydin KY, McFarlane GA, King JR, Megrey BA, Myers KW (2005) Linking oceanic food webs to coastal production and growth rates of Pacific salmon (Oncorhynchus spp.), using models on three scales. Deep Sea Research Part II: Topical Studies in Oceanography 52: 757–780. doi: 10.1016/j.dsr2.2004.12.017.

329. Kaplan IC, Levin PS, Burden M, Fulton EA (2010) Fishing catch shares in the face of global change: a framework for integrating cumulative impacts and single species management. Can J Fish Aquat Sci 67: 1968–1982. doi: 10.1139/F10-118.

330. Nagorsen D (2009) Mammals of British Columbia (2009). In: Klinkenberg B, editor, E-Fauna BC: Electronic Atlas of the Fauna of British Columbia [www.efauna.bc.ca]. Lab for Advanced Spatial Analysis, Department of Geography, University of British Columbia, Vancouver. [Date Accessed: 2014-04-14], plambert@pacificcoast.net: Royal British Columbia Museum. URL http://www.royalbcmusem.bc.ca/.

331. Ianson D, Flostrand L (2010) Coastal waters off the west coast of Vancouver Island, British Columbia. In: Ecosystem and Status Trends Report. Can Sci Advis Sec Res Doc, p. 56.

332. Loomis J, Larson D (1994) Total economic values of increasing gray whale populations: results from a contingent valuation survey of visitors and households. Marine Resource Economics 9: 275–286.

333. Ford JK (2014) Marine Mammals of British Columbia. Victoria, Canada: Royal BC Museum, Victoria Canada, 460 pp.

334. Bass CL, Simmonds MP, Isaac SJ (2006) An overview of the potential consequences for cetaceans of oceanic acidification. IWC, Scientific Committee SC/58/E10: 6 p.

335. Hester KC, Peltzer ET, Kirkwood WJ, Brewer PG (2008) Unanticipated consequences of ocean acidification: a noisier ocean at lower pH. Geophys Res Lett 35: L19601. doi: 10.1029/2008GL034913.

336. Joseph JE, Chiu C (2010) A computational assessment of the sensitivity of ambient noise level to ocean acidification. J Acoust Soc Am 128: 3. doi: 10.1121/1.3425738

337. Kelly MW, Hofmann GE (2012) Adaptation and the physiology of ocean acidification. Funct Ecol 27: 980–990. doi: 10.1111/j.1365-2435.2012.02061.x.

338. Godbold JA, Calosi P (2013) Ocean acidification and climate change: advances in ecology and evolution. Phil Trans R Soc Lond B, Biol Sci 368: 20120448. doi: 10.1098/rstb.2012.0448.

339. Allen SE, Wolfe MA (2013) Hindcast of the timing of the spring phytoplankton bloom in the Strait of Georgia, 1968–2010. Prog Oceanogr 26: 81–87. doi: 10.1016/j.pocean.2013.05.026

340. Merryfield WJ, Pal B, Foreman MGG (2009) Projected future changes in surface marine winds off the west coast of Canada. J Geophys Res 114: C06008. doi: 10.1029/2008jc005123

341. Foreman MGG, Pal B, Merryfield WJ (2011) Trends in upwelling and downwelling winds along the British Columbia shelf. J Geophys Res 116: C10023. doi: 10.1029/2011JC006995.

342. Bylhouwer B, Ianson D, Kohfeld K (2013) Changes in the onset and intensity of wind-driven coastal upwelling and downwelling along the North American Pacific coast. J Geophys Res Oceans 118: 1–16. doi: 10.1002/jgrc.20194

343. Beaugrand G (2009) Decadal changes in climate and ecosystems in the North Atlantic Ocean and adjacent seas. Deep Sea Research II 56: 656–673. doi: 10.1016/j.dsr2.2008.12.022.

344. Pörtner HO (2009) Ecosystem effects of ocean acidification in times of ocean warming: a physiologist's view. Mar Ecol Prog Ser 373: 203–217.

doi: 10.3354/meps07768

345. Griffith GP, Fulton EA, Gorton R, Richardson AJ (2012) Predicting interactions among fishing, ocean warming, and ocean acidification in a marine system with whole-ecosystem models. Conserv Biol 26: 1145–1152. doi: 10.1111/j.1523-1739.2012.01937.x. pmid:23009091

346. CRIMS (2013) Saltwater Finfish and Shellfish Tenures (geospatial dataset, accessed 2013–12–19). In: Coastal Resource Information Management System, DataBC, Ministry of Forests, Lands and Natural Resource Operations, Province of British Columbia. https://apps.gov.bc.ca/pub/geometadata/ (finfish: UID = 4025; shellfish: UID = 4031). URL https://apps.gov.bc.ca/pub/geometadata/.

347. International Hydrographic Organization Data Centre (2014) IHO-IOC GEBCO Gazetteer of Undersea Feature Names. In: General Bathymetric Chart of the Oceans, www.gebco.net. URL http://www.gebco.net/data_and_products/undersea_feature_names/.

348. R Core Team (2014) R: A Language and Environment for Statistical Computing. R Foundation for Statistical Computing, Vienna, Austria. URL http://www.R-project.org/.

CITATION

CHAPTER 1

Hans-Joachim Rätz (2012). The Obligation of Sustainable Fisheries Management: Review of Endured Failures and Challenges in Exploitation of the Living Sea, Sustainable Development - Education, Business and Management - Architecture and Building Construction - Agriculture and Food Security, Prof. Chaouki Ghenai (Ed.), ISBN: 978-953-51-0116-1, InTech, DOI: 10.5772/29569.

CHAPTER 2

Javier Sánchez-Hernández, María J. Servia, Rufino Vieira-Lanero and Fernando Cobo (2012). Ontogenetic Dietary Shifts in a Predatory Freshwater Fish Species: The Brown Trout as an Example of a Dynamic Fish Species, New Advances and Contributions to Fish Biology, Prof. Hakan Turker (Ed.), ISBN: 978-953-51-0909-9, InTech, DOI: 10.5772/54133.

CHAPTER 3

Peter C. Sakaris (2013). A Review of the Effects of Hydrologic Alteration on Fisheries and Biodiversity and the Management and Conservation of Natural Resources in Regulated River Systems, Current Perspectives in Contaminant Hydrology and Water Resources Sustainability, Dr. Paul Bradley (Ed.), ISBN: 978-953-51-1046-0, InTech, DOI: 10.5772/55963.

CHAPTER 4

William C. Straka III, Curtis J. Seaman, Kimberly Baugh, Kathleen Cole, Eric Stevens, and Steven D. Miller, Utilization of the Suomi National Polar-Orbiting Partnership (NPP) Visible Infrared Imaging Radiometer Suite (VIIRS) Day/Night Band for Arctic Ship Tracking and Fisheries Management, doi:10.3390/rs70100971.

CHAPTER 5

Myrna Leticia Bravo-Olivas, Rosa María Chávez-Dagostino, Carlos Antonio López-Fletes and Elaine Espino-Barr, Fishprint of Coastal Fisheries in Jalisco, Mexico, doi:10.3390/su6129218.

CHAPTER 6

Jansen T, Campbell A, Kelly C, Hátún H, Payne MR (2012) Migration and Fisheries of North East Atlantic Mackerel (Scomber scombrus) in Autumn and Winter. PLoS ONE 7(12): e51541. doi:10.1371/journal.pone.0051541

CHAPTER 7

Bender MG, Machado GR, Silva PJdA, Floeter SR, Monteiro-Netto C, Luiz OJ, et al. (2014) Local Ecological Knowledge and Scientific Data Reveal Overexploitation by Multigear Artisanal Fisheries in the Southwestern Atlantic. PLoS ONE 9(10): e110332. doi:10.1371/journal.pone.0110332.

CHAPTER 8

Oksanen SM, Ahola MP, Oikarinen J, Kunnasranta M (2015) A Novel Tool to Mitigate By-Catch Mortality of Baltic Seals in Coastal Fyke Net Fishery. PLoS ONE 10(5): e0127510. doi:10.1371/journal.pone.0127510.

CHAPTER 9

Kough AS, Paris CB, Butler MJ IV (2013) Larval Connectivity and the International Management of Fisheries. PLoS ONE 8(6): e64970. doi:10.1371/journal.pone.0064970.

CHAPTER 10

James M. Tolan (2013). Estuarine Fisheries Community-Level Response to Freshwater Inflows, Water Resources Planning, Development and Management, Prof. Ralph Wurbs (Ed.), ISBN: 978-953-51-1092-7, InTech, DOI: 10.5772/52313.

CHAPTER 11

Gerardo Rodríguez-Quiroz, Eugenio Alberto Aragón-Noriega, Miguel A. Cisneros-Mata and Alfredo Ortega Rubio (2012). Fisheries and Biodiversity in the Upper Gulf of California, Mexico, Oceanography, Prof. Marco Marcelli (Ed.), ISBN: 978-953-51-0301-1, InTech, DOI: 10.5772/23064.

CHAPTER 12

Wisdom Akpalu, Isaac Dasmani and Ametefee K. Normanyo, Optimum Fisheries Management under Climate Variability: Evidence from Artisanal Marine Fishing in Ghana, doi:10.3390/su7067942.

CHAPTER 13

Haigh R, Ianson D, Holt CA, Neate HE, Edwards AM (2015) Effects of Ocean Acidification on Temperate Coastal Marine Ecosystems and Fisheries in the Northeast Pacific. PLoS ONE 10(2): e0117533. doi:10.1371/journal.pone.0117533.

INDEX